GAS MIXTURES
Preparation and Control

Gary O. Nelson
President
Miller-Nelson Research, Inc.
Monterey, California

LEWIS PUBLISHERS
Boca Raton   Ann Arbor   London   Tokyo

Library of Congress Cataloging-in-Publication Data

Nelson, Gary O.
  Gas mixtures: preparation and control / by Gary O. Nelson.
      p.   cm.
  Includes bibliographical references and index.
  ISBN 0-87371-298-6
  1. Gases.   I. Title.
QD531.N45   1992
542'.7—dc20   91-44350
                                                                                                    CIP

COPYRIGHT © 1992 by LEWIS PUBLISHERS, INC.
ALL RIGHTS RESERVED

This book represents information obtained from authentic and highly regarded sources. Reprinted material is quoted with permission, and sources are indicated. A wide variety of references are listed. Every reasonable effort has been made to give reliable data and information, but the author and the publisher cannot assume responsibility for the validity of all materials or for the consequences of their use.

Neither this book nor any part may be reproduced or transmitted in any form or by any means, electronic or mechanical, including photocopying, microfilming, and recording, or by any information storage and retrieval system, without permission in writing from the publisher.

LEWIS PUBLISHERS, INC.
121 South Main Street, Chelsea, MI 48118

PRINTED IN THE UNITED STATES OF AMERICA
1 2 3 4 5 6 7 8 9 0
Printed on acid-free paper

PREFACE

The preparation of known gas mixtures has long been a major concern to those involved in analytical and experimental work. Many interesting and useful methods have been developed, but they are scattered throughout the literature. The purpose of this book is to build upon my first effort published in 1971 and bring to light some of the current gas blending techniques.

This book makes practical suggestions about the production and calculation of controlled test atmospheres. It is not only aimed at those working in the laboratory on a routine basis but also those professionals involved in research. This book will serve as a valuable guide not only for those who work in the field of air pollution, analytical chemistry, environmental health, and industrial hygiene but will be useful for anyone who simply wants to make standard multicomponent gas and vapor mixtures.

I am greatly indebted to a number of people who helped make this book possible. Wilma Leon, once again, provided me with her usual top quality illustrations. Also thanks to Dr. Peter Swearengen for his constant encouragement and badgering to complete the project. I would also like to thank my daughters Susan and Linda for their interest and enthusiasm in this effort. A special note of thanks to my wife Diane whose love and encouragement kept me going during the slow tedious periods.

This book is dedicated to my mom who endured far more than any mother should during my upbringing. Although she is not a serious student of the gas laws, she nevertheless keeps a copy of the book on the coffee table in case the mood strikes her.

Gary O. Nelson
Monterey, California

BIOGRAPHY

Gary O. Nelson was born on March 1, 1937 in Caldwell, Idaho. At the age of 5 he moved to Los Angeles and finished John Muir High School in 1955. His interests included science (especially chemistry), swimming, and surfing. From there he attended Stanford University and graduated with a B.S. in biological science in 1959.

After graduation, he attended the U.S. Coast Guard Officer Candidate School and was on active duty for one year. He served in the reserve for ten years and attained the rank of lieutenant commander before his honorable discharge in 1970.

In 1960, he attended the University of California at Berkeley and received a B.S. in chemistry in 1962. He completed his M.S. in chemistry at San Jose State University (night program) in 1967.

In 1962, he began work at Lawrence Livermore National Laboratory in their Industrial Hygiene Group. He later moved to the Special Projects Group. During his 20 years at the laboratory he worked on various projects involving gas mixture generation, carbon evaluation, and instrument development. During this time he authored more than 80 internal reports and journal articles. His first book "Controlled Test Atmospheres" was published in 1971. To date, he has more than 70 presentations at scientific meetings.

In 1976, while still working at the laboratory, he founded Miller-Nelson Research Inc. He left the laboratory in 1981 to operate the business on a full time basis and is currently still president. The chief products of the company are glove, instrument, and gas sorbent evaluation. The company also manufactures small scale humidity control systems. He currently resides in Carmel Valley, California.

Table of Contents

1. Introduction and General Principles 1
   1.1 Fundamental Gas Laws 1
       1.1.1 Ideal Gases 4
             1.1.1.1 Pressure, Temperature, and Volume 4
             1.1.1.2 Density 4
             1.1.1.3 Concentration 5
       1.1.2 Nonideal Gases 8
             1.1.2.1 Atmospheric Pressure 8
             1.1.2.2 High Pressure 8
   1.2 Safety Considerations 9
   1.3 Scientific Notation and Units of Measurement 10

2. Air Purification 11
   2.1 Removal of Water Vapor 11
       2.1.1 Solid Desiccants 11
             2.1.1.1 Anhydrous Magnesium Perchlorate 15
             2.1.1.2 Calcium Sulfate 15
             2.1.1.3 Silica Gel 15
             2.1.1.4 Activated Alumina 15
             2.1.1.5 Molecular Sieve 16
       2.1.2 Liquid Desiccants 16
       2.1.3 Cooling 16
   2.2 Removal of Particulates 16
   2.3 Removal of Organic Vapors 19
   2.4 Removal of Miscellaneous Contaminants 19
   2.5 A Typical Laboratory Compressed-Air System 22
   References 22

3. Flow Rate and Volume Measurements 25
   3.1 Primary Standards 25
       3.1.1 Spirometers 25
       3.1.2 Pitot Tubes 29
       3.1.3 Frictionless Pistons 31

		3.1.4	Syringe Drive Systems	33
		3.1.5	Aspirator Bottles	34
	3.2	Intermediate Standards		34
		3.2.1	Wet Test Meters	34
		3.2.2	Dry Gas Meters	38
	3.3	Secondary Standards		40
		3.3.1	Rotameters	40
		3.3.2	Mass Flow Meters	50
		3.3.3	Orifice Meters	54
		3.3.4	Critical Orifices	56
		3.3.5	Porous Plugs	59
		3.3.6	Controlled Leaks	62
		3.3.7	Miscellaneous Devices	62
	References			64
4.	Static Systems for Producing Gas Mixtures			67
	4.1	Systems at Atmospheric Pressure		67
		4.1.1	Single Rigid Chambers	67
			4.1.1.1 Sample Introduction	68
			4.1.1.2 Gas Blenders	72
			4.1.1.3 Miscellaneous Dispensers	72
			4.1.1.4 Mixing Devices	72
			4.1.1.5 Concentration Calculations	74
			4.1.1.6 Volume Dilution Calculations	78
			4.1.1.7 Discussion	82
		4.1.2	Rigid Chambers in Series	83
		4.1.3	Nonrigid Chambers	85
			4.1.3.1 Sample Introduction	87
			4.1.3.2 Sample Decay	87
	4.2	Pressurized Systems		89
		4.2.1	Gravimetric Methods	92
		4.2.2	Partial Pressure Methods	94
		4.2.3	Volumetric Methods	100
	4.3	Partially Evacuated Systems		101
	Reference			105
5.	Dynamic Systems for Producing Gas Mixtures			109
	5.1	Gas Stream Mixing		110
		5.1.1	Single Dilution	110
		5.1.2	Double Dilution	114
	5.2	Injection Methods		116
		5.2.1	Liquid Reservoirs	116
		5.2.2	Syringe Drive Systems	118
		5.2.3	Liquid Pumps	122
		5.2.4	Miscellaneous Injection Methods	123

		5.2.4.1	Electrolytic Methods	123
		5.2.4.2	Gravity Feed Methods	125
		5.2.4.3	Liquid and Gas Pistons	126
		5.2.4.4	Pulse Diluters	126
	5.2.5	Injection Ports		126
	5.2.6	Vaporization Techniques		128
	5.2.7	The Ultimate System		131
	5.2.8	Calculations		133
5.3	Diffusion Methods			137
5.4	Permeation Methods			146
	5.4.1	Device Construction		148
	5.4.2	Calibration		150
	5.4.3	Calculations		152
	5.4.4	Discussion		157
5.5	Evaporation Methods			158
5.6	Electrolytic Methods			166
5.7	Chemical Methods			171
References				176

6. Specialized Systems 185
    6.1 Ozone 185
        6.1.1 Methods of Generation 185
        6.1.2 Methods of Analysis 188
    6.2 Nitrogen Dioxide 190
        6.2.1 Static Systems 190
        6.2.2 Dynamic Systems 194
    6.3 Mercury 196
    6.4 Hydrogen Cyanide 198
        6.4.1 Static Systems 198
        6.4.2 Dynamic Systems 201
    6.5 Hydrogen Fluoride 201
    6.6 Formaldehyde 205
    6.7 Chlorine Dioxide 207
    6.8 Bromine and Iodine 209
    6.9 Humidity 209
        6.9.1 Static Systems 211
        6.9.2 Dynamic Systems 212
        6.9.3 Calculations 215
        6.9.4 Methods of Measurement 221
    References 223

Appendices 229
    Appendix A: Conversion Factors 229
    Appendix B: Atomic Weights and Numbers 233
    Appendix C: Values of the Molar Gas Constant 235

Appendix D:	Density of Dry Air	237
Appendix E:	Densities of Common Gases and Their Deviation from the Perfect Gas Law	239
Appendix F:	Diffusion Coefficients at 25°C and 760 mmHg in Air	243
Appendix G:	Constants for Vapor Pressure Calculations	247
Appendix H:	Vapor Pressures of Water at Various Temperatures	251
Appendix I:	Mass of Water Vapor in Saturated Air at 1 atm Pressure	253
Appendix J:	Relative Humidity from Wet- and Dry-Bulb Thermometer Readings	255
Appendix K:	Gas Flow Conversion Factors for Mass Flow Controllers	259
Appendix L:	Table of Atomic Volumes Used for Diffusion Coefficient Calculation	263
Appendix M:	Density of Dry Air as a Function of Altitude at 20°C	265
Appendix N:	Equilibrium Relative Humidity of Selected Saturated Salt Solutions	267
Appendix O:	Characteristics of Commercially Available Permeation Sources	269
Appendix P:	Cylinder Size and Capacities	279
Appendix Q:	Viscosity of Gases and Vapors	283
Appendix R:	Miran 1A Operating Instructions	285
Index		287

CHAPTER 1

Introduction and General Principles

Gas and vapor mixtures with air can be produced by a wide variety of methods. Each situation must be dealt with separately depending on the experimental circumstances. Before a scientist devises any system, he must decide what concentrations, volumes, flow rates, accuracies, and equilibrium times are required. In addition, he must determine if his material resources and budget are sufficient to fabricate a system to meet his needs. The summary presented in Table 1.1 can be used as a guide to acquaint the reader with the basic systems discussed in this book. Once the proper system requirements are determined, the reader can better identify which chapter and section will best fit the immediate experimental needs.

1.1 FUNDAMENTAL GAS LAWS

The laboratory production of standard gas mixtures requires a working knowledge of the behavior of gases in relation to temperature and pressure. One must also be able to convert concentration, flow rate, volume, and density systems from one measurement system to another. Although gases and vapors usually act in a predictable manner, it is necessary to recognize deviations and to apply appropriate corrections.

The purpose of this chapter is to derive, organize, and correlate the important fundamental relationships that exist in both pure and mixed gases. The mathematical expressions described in this chapter will appear throughout the remaining chapters, and sample calculations will be given as each equation is used in its own specialized manner. Since this book is intended to be primarily a guide for the working chemist, mathematical derivations will be sparse. The mathematical formulas will be presented to provide a sound basis for dealing with practical problems in a practical way.

Table 1.1 Characteristics of Methods for Producing Gas and Vapor Mixtures

Methods	Usual Range of Concentration	Volume of Test Mixture (L)	Average Best Accuracy (%)	Operating Pressure (atm)	Degree of Wall Loss
Static systems					
Single vessel	10 ppm to 5%	10% of vessel	3–5, 1	≤1	Medium
Multiple vessel	10 ppm to 5%	Vessel volume[a]	5–10, 3	≤1	Medium
Nonrigid chamber	10 ppm to 5%	1–1000	3, 1	1	High
Gravimetric	Low ppm to 50%	1–5000	0.1–0.002	1–150	Medium
Partial pressure	100 ppm to 50%	0.5–5000	5–10, 1	1–150	Medium
Volumetric	Low ppm to 50%	0.5–5000	3–5, 2	1–150	Medium
Partial evacuation	Low ppm to 50%	0.1–2	1–5, 1	0.001–1	Low to medium
Dynamic systems					
Gas stream mixing					
Single dilution	100 ppm to 50%	>5000	1–3, 1	≥1	Low
Double dilution	Low ppb to 1%	>5000	5–10, 2	≥1	Low
Injection					
Syringe drive	1 ppm to 0.1%	5–1000	2–5, 1	≥1	Low
Liquid pumps	1 ppm to 5%	Very large	5, 1	1	Low
Gravity feed	100 ppm to 5%	Large	5, 3	≥1	Low
Pulse diluter	0.1 ppb to 0.1%	Very large	3–5, 1	≥1	Low to medium
Diffusion	0.1 to 500 ppm	Very large	2–6, 1	1	Low
Permeation	0.05 to 200 ppm	Large	1–3, 0.5	1	Low
Evaporation	50 ppm to 10%	Large	3–15, 2	≥1	Low
Electrolysis	Low ppm to 1%	Ltd. by soln. vol.	2–5, 1	1	Low
Chemical reaction	Low ppm to 1%	Ltd. by react vol.	3–10, 1	1	Low to medium

Introduction and General Principles

Static systems				
Single vessel	Low	Low	Yes[b]	No
Multiple vessel	Low	Low	Yes[b]	No
Nonrigid chamber	Very low	Low	Yes	Not always
Gravimetric	Medium to high	Low	Yes[c]	Usually
Partial pressure	High	Medium	Yes[c]	Usually
Volumetric	Medium	Medium	Yes[c]	Usually
Partial evacuation	Medium	Low	No	Not always
Dynamic systems				
Gas stream mixing				
Single dilution	Low to medium	Low	No	Yes
Double dilution	Medium	Low to medium	No	Usually
Injection				
Syringe drive	Medium to high	Medium to high	No	Usually
Liquid pumps	Medium	Low to medium	No	No
Gravity feed	Low	Low	No	No
Pulse diluter	High	High	No	Yes
Diffusion	Medium	Low to medium	No	No
Permeation	Low to medium	Low to medium	No	Not always
Evaporation	Low to medium	Low	No	No
Electrolysis	Medium	Medium	No	Yes
Chemical reaction	Medium	Medium	No	Yes

[a] Three vessels in series.
[b] Dilution occurs when test mixture is removed from vessel.
[c] No, if downstream dilution is not available.
[d] Best for gases only.
[e] Best for liquids only.

1.1.1 Ideal Gases

1.1.1.1 Pressure, Temperature, and Volume

From the distant past of your high school and freshman college chemistry, the names of Boyles and Charles can be vaguely remembered as giants in the world of gas. We will again summon their work as a quick review of the gas laws.

The relationship between the pressure and volume of a gas at a constant temperature is described by Boyle's law, which states that the volume, V, is inversely proportional to the pressure, P:

$$V = K\left(\frac{1}{P}\right) \quad (1.1)$$

where K is a proportionality constant, or

$$P_1 V_1 = P_2 V_2 \quad (1.2)$$

The relationship between the temperature and the volume of a gas at a constant pressure is described by Charles's law, which states that the volume is directly proportional to the absolute temperature, T:

$$V = KT \quad (1.3)$$

A more general equation involving all three variables can be obtained by combining Equations 1.1 and 1.3:

$$PV = KT \quad (1.4)$$

or

$$\frac{P_1 V_1}{T_1} = \frac{P_2 V_2}{T_2} \quad (1.5)$$

If K is proportional to the number of moles of gas, n, then

$$PV = nRT \quad (1.6)$$

which is the well-known ideal gas law. The most commonly used values of the molar gas constant, R, are listed in Appendix C.

1.1.1.2 Density

The calculation of gas density is important for correcting flow rates and for determining how well gases conform to the ideal gas law. If W is the weight and

Introduction and General Principles

M is the molecular weight (sometimes referred to as molar mass and formula weight), then

$$n = \frac{W}{M} \tag{1.7}$$

Equation 1.6 then becomes

$$PV = \frac{WRT}{M} \tag{1.8}$$

or

$$\frac{W}{V} = \frac{PM}{RT} \tag{1.9}$$

where W/V is the ideal gas density. The calculated densities of some of the more common gases are given in Appendix E.

1.1.1.3 Concentration

The concentrations of gas and vapor mixtures are usually expressed in units of percent or parts per million by volume. For concentrations above 0.1%, the concentration, C, is usually expressed in percent. This can be calculated for ideal gases by

$$C_\% = \frac{10^2 v_a}{v_a + v_b + \ldots + v_n} = \frac{10^2 p_a}{p_a + p_b + \ldots + p_n} \tag{1.10}$$

where $v_{a,b,\ldots,n}$ and $p_{a,b,\ldots,n}$ are the volumes and partial pressures of components a, b, ..., n at a constant temperature. For concentrations below 0.1%, the concentration is usually expressed in parts per million by volume (ppm; i.e., the number of parts by volume of trace gas in a million parts of gas mixture). This can be calculated for ideal gases by

$$C_{ppm} = \frac{10^6 v_a}{v_D + v_a} = \frac{10^6 p_a}{p_D + p_a} \tag{1.11}$$

where v_D and p_D are the volume and pressure of the diluent gas. The v_a and p_a terms in the denominator are usually neglected when dealing with concentrations below 5000 ppm, since their contribution to the total volume is usually insignifi-

Table 1.2   Comparison of Concentration Calculations

\multicolumn{2}{c	}{Concentration from $v_a/(v_D+v_a)$}	\multicolumn{2}{c}{Concentration from v_a/v_D}
%	ppm	%	ppm
0.0001	1	0.0001000001	1.000001
0.001	10	0.00100001	10.0001
0.01	100	0.010001	100.01
0.1	1,000	0.1001	1,001
0.5	5,000	0.5025	5,025
1	10,000	1.010	10,101
10	100,000	11.11	111,111
20	200,000	25.0	250,000
50	500,000	100	1,000,000
75	750,000	300	3,000,000
95	950,000	1900	19,000,000
100	1,000,000	—	—

cant. However, above 5000 ppm, the contaminant gas becomes more dominant and its volume must be included in the total gas volume. Table 1.2 illustrates how the numerical value of the concentration is altered depending on whether or not the trace gas volume is included in the denominator of Equation 1.11.

Concentration can also be calculated when a known weight or volume of liquid is vaporized in a known volume of a diluent gas. If the liquid has a weight W and a molecular weight M, then the volume of vapor produced is given by

$$v_a = \frac{WRT}{MP} \tag{1.12}$$

If v_a is diluted with v_D, then the resulting concentration on a volume percent basis is determined by the equation

$$C_\% = \frac{10^2 v_a}{v_a + v_D} = \frac{10^2 \frac{WRT}{MP}}{\frac{WRT}{MP} + v_D} = \frac{10^2}{1 + \frac{v_D MP}{WRT}} \tag{1.13}$$

Introduction and General Principles

Similarly, the concentration in parts per million by volume is determined by

$$C_{ppm} = \frac{10^6 v_a}{v_D} = \frac{10^6 \frac{WRT}{MP}}{v_D} \quad (1.14)$$

Usually W is expressed in terms of the volume of liquid, v_L, and density, ρ_L. Equation 1.14 then becomes

$$C_{ppm} = \frac{10^6 v_L \rho_L RT}{V_D MP} \quad (1.15)$$

At room temperature (25°C) and pressure (760 mmHg), this reduces to

$$C_{ppm} = \frac{24.5 \times 10^6 v_L \rho_L}{v_D M} \quad (1.16)$$

In some air pollution, industrial hygiene, and animal toxicology work, the most convenient unit of concentration is typically weight per unit volume. This is especially true when dealing metal fumes, mists, or dusts, such as copper, lead, or iron fumes. However, gases and vapors in air are also expressed in these terms. Such a concentration is calculated by

$$C_{\frac{W}{V}} = \frac{W}{v_D} \quad (1.17)$$

which is related to concentration in parts per million by

$$C_{\frac{W}{V}} = \frac{10^{-6} C_{ppm} MP}{RT} \quad (1.18)$$

and to vapor pressure by

$$C_{\frac{W}{V}} = \frac{p_a MP}{p_D RT} \quad (1.19)$$

Usually, however, P_D closely approximates P. Equation 1.19 therefore simplifies to

$$C_{\frac{w}{v}} = \frac{p_a M}{RT} \qquad (1.20)$$

Since the concentration in weight per unit volume is expressed in milligrams per cubic meter, Equation 1.18 becomes

$$C_{\frac{mg}{M^3}} = \frac{C_{ppm} M}{24.5} \qquad (1.21)$$

at 25°C and 760 mmHg.

1.1.2 Nonideal Gases

1.1.2.1 Atmospheric Pressure

Most gases adhere closely to the ideal gas law at room temperature and atmospheric pressure. However, for the best accuracy with mixed gases, deviations from ideality must be taken into account, especially at high pressures. This is accomplished by comparing the actual gas density, ρ_a, with the ideal volume, v_I, under the same conditions. The actual gas volume, v_A, can then be corrected to the ideal volume, v_i, and can be used to make precise gas mixtures by the equation

$$v_I = K v_A = \frac{\rho_A}{\rho_I} v_A \qquad (1.22)$$

At 0°C and 760 mmHg, p_i can be found from

$$\rho_I = \frac{M}{22,414} \qquad (1.23)$$

Values for ρ_A, ρ_I, and ρ_A/ρ_I under standard conditions are tabulated in Appendix E for easy reference.

1.1.2.2 High Pressure

If pressurized systems are used, the deviation from ideality becomes more pronounced and large correction factors must be applied if even moderate accuracies (±10 to 20%) are to be achieved (Figure 1.1). The correction factor, compressibility, Z, is defined by

$$Z = \frac{PV}{RT} \qquad (1.24)$$

Introduction and General Principles

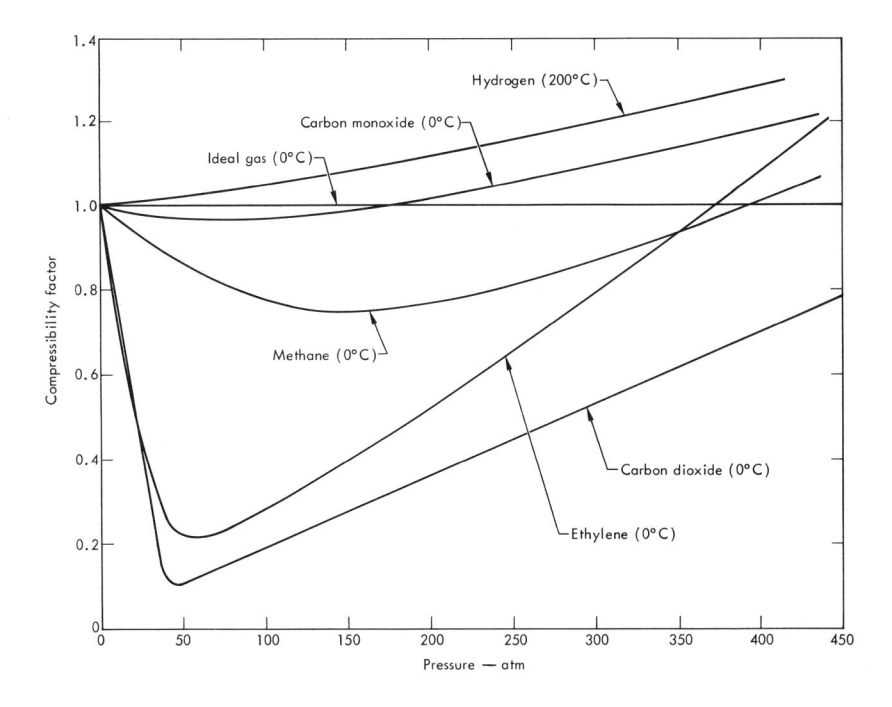

Figure 1.1 Compressibility vs pressure for six gases. Note the large deviations at 50 atm for ethylene and carbon dioxide.

where P, V, and T are measured under the conditions of interest. Equation 1.10 then becomes

$$C_\% = \frac{10^2 \dfrac{P_a}{Z_a}}{\dfrac{P_a}{Z_a} + \dfrac{P_b}{Z_b} + \ldots + \dfrac{P_n}{Z_n}} \quad (1.25)$$

The values of Z are known for pure gases but are not always found for mixed gases.

1.2 SAFETY CONSIDERATIONS

Most of the materials discussed in this book are hazardous in nature. They are often under extreme pressure, explosive, carcinogenic, highly reactive, and potentially dangerous when inhaled or adsorbed through the skin. Be sure to consult the Material Safety Data Sheets and industrial hygiene safety professionals for the proper handling techniques and precautions. It is essential that the user wear proper clothing and use safety glasses, gloves, and shields when required. All potentially dangerous operations should be carried out in a well-lighted laboratory inside a hood operating at the proper face velocity.

1.3 SCIENTIFIC NOTATION AND UNITS OF MEASUREMENT

Many sample problems will be presented throughout this book to illustrate the use of various equations. In general, all answers will be expressed to three significant figures. Assume that all the input given numbers are known to three significant figures. For example, a 1-L flask is actually known to 1.00 L. Therefore, for convenience, some significant figures have not been shown.

The problems in this book were set up and calculated using the Lotus 1-2-3 spread sheet software. This technique requires that exponents be shown in scientific notation. For example, 22.4×10^6 and 10^{-6} are expressed as 22.4E+06 and E-06 respectively. Hence, all the problems which require 10 raised to either a positive or negative power will use this E notation.

Before 1960, the dominant units used in science were based on the metric system. In 1960, the General Conference on Weights and Measures proposed an updated system called the International System of Units (abbreviated SI after the French Le Systeme International d'Unites). This had most of the elements of the metric system but did present a few new twists. For example, the pressure unit *pascal* (Pa) was born. This particular unit just did not catch the fancy of the scientific community. Most pressure units today are still given in psi, mmHg, or atmospheres. Many of the examples given throughout this book will represent mostly the metric system, with some English units mixed in. This is not an oversight but the way problems are encountered and solved in the real world. The precise implementation of the SI system will be left to those zealots who follow.

CHAPTER 2

AIR PURIFICATION

Laboratory compressed air is the most common source of diluent gas for low-concentration, high-volume standard gas mixtures. It is continuously supplied as needed, usually by diesel or electric compressors at pressures of 80 to 125 psi, and it is stored in holding tanks. Several undesirable contaminants can be introduced during compression and storage. Oil mists are a common by-product, as are substantial amounts of condensed water vapor. Carbon dioxide, nitrogen oxides, aldehydes, carbon monoxide, unburned hydrocarbons, pipe scale, and airborne dust are also potential problems. Even if air is claimed to be 99.9% pure, it can still contain up to 1000 ppm of undesirable materials. Before any quality low-concentration work can be done, the air supply system must be scrupulously cleaned to prevent contamination and possible chemical reactions. The composition of clean, dry air is given in Table 2.1.

This chapter describes the basic methods of removing contaminants from flowing air streams. General multipurpose filtering devices are discussed, as are methods for removing excess water vapor, oil mists, extraneous gases, and particulate matter. The air-purification procedures that are described can also be applied to such relatively stable gases as nitrogen, oxygen, and inert gases.

2.1 REMOVAL OF WATER VAPOR

Moisture can be removed from gases by a variety of methods. The most common methods are chemical, compression, cooling, and permeation dryers. Most laboratory systems will use one or a combination of these techniques.

2.1.1 Solid Desiccants

Solid desiccants constitute the most conventional method for removing water vapor in small-scale laboratory operations. They remove moisture either by chemical reaction (absorption) or by capillary condensation (adsorption).[1] Solid absorbing agents include calcium chloride, calcium sulfate, and magnesium perchlorate; solid adsorbing agents include activated alumina and silica gel. Solid

Table 2.1 Composition of Clean, Dry Air

Component	Composition
Nitrogen	78.08%
Oxygen	20.95%
Argon	0.934%
Carbon dioxide	0.033%
Neon	18.2 ppm
Helium	5.24 ppm
Methane	2.0 ppm
Krypton	1.14 ppm
Hydrogen	0.5 ppm
Nitrous oxide	0.5 ppm
Xenon	0.087 ppm

Source: Weast.[28]

desiccants are one of the most practical tools for drying gases because they are commercially available, they are easy to store, they can be regenerated by heating, and they often indicate their condition by their color.

Solid desiccants are generally evaluated by comparing their drying efficiencies and capacities. The efficiencies of a drying agent (i.e., the degree of dryness achieved) can be compared by measuring the water vapor remaining in a gas after it passes through the desiccant at the equilibrium velocity.[2] The drying efficiencies of several desiccants are compared in Table 2.2. Barium oxide and magnesium perchlorate are the most efficient desiccants of those compared, whereas copper sulfate and granular calcium chloride are the least efficient. An extensive investigation by Trusell and Diehl evaluated the efficiencies of 21 desiccants in drying a stream of nitrogen.[3] Their results, shown in Table 2.3, indicate that the most efficient desiccant is anhydrous magnesium perchlorate.

The capacity of a desiccant is the amount of water it is able to remove per unit of the desiccant's dry weight. Often a drying agent is efficient but is unsuitable for drying large quantities of gas because of its limited capacity. Anhydrous calcium chloride is an example of such a desiccant. The capacity of a desiccant depends not only on the kind of material of which it is composed, but also on the size of the grains, the amount of surface area exposed to the gas, and the thickness through which the gas flows. Additional factors include the type of gas being dried, as well as its velocity, temperature, pressure, and moisture content.[7] The capacities of several desiccants as a function of relative humidity are shown in Figure 2.1. The relative capacities of the most common drying agents can be determined from Table 2.3 by comparing the volumes of gas each desiccant is able to dry. The materials with the highest capacities are anhydrous magnesium perchlorate, calcium sulfate, and phosphorous pentoxide.

One should not indiscriminately choose a desiccant simply because it has the proper efficiency and capacity. The geometry and size of the drying train must be considered to allow enough residence time to achieve equilibrium. In addition, many drying agents heat up violently when they are exposed to too much moisture

Air Purification

Table 2.2 Comparative Efficiencies of Various Solid Desiccants Used in Drying Air

Desiccant	Composition	Granular Form	Residual Water After Drying mg/L
Barium oxide	BaO		0.00065
Magnesium perchlorate	Mg(ClO$_4$)$_2$	Anhydrous	0.002
Calcium oxide	CaO		0.003
Calcium sulfate	CaSO$_4$	Anhydrous	0.005
Aluminum oxide	Al$_2$O$_3$		0.005
Potassium hydroxide	KOH	Sticks	0.014
Silica gel			0.030
Magnesium perchlorate	Mg(ClO$_4$)$_2$·3H$_2$O		0.031
Calcium chloride	CaCl$_2$	Dehydrated	0.36
Sodium hydroxide	NaOH	Sticks	0.80
Barium perchlorate	Ba(ClO$_4$)$_2$		0.82
Zinc chloride	ZnCl$_2$	Sticks	0.98
Calcium chloride	CaCl$_2$	Anhydrous	1.25
Calcium chloride	CaCl$_2$	Granular	1.5
Copper sulfate	CuSO$_4$	Anhydrous	2.8

Source: Data taken from Hammond.[2]

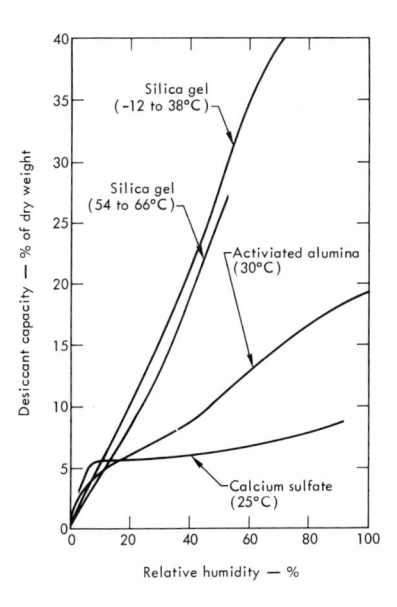

Figure 2.1 Desiccant capacity vs relative humidity for silica gel, activated alumina, and calcium sulfate (Drierite). The calcium sulfate temperature is estimated. Data taken from Hougen, O. A. and F. W. Dodge. *The Drying of Gases* (Ann Arbor, MI: Edwards Brothers Inc., 1947) with permission.

Table 2.3 Comparative Efficiencies and Capacities of Various Solid Desiccants[a]

Desiccant	Initial Composition	Regeneration Drying Time (hr)	Requirements Drying Temp (°C)	Average Efficiency[b] (mg/L)	Relative Capacity[c] (L)
Magnesium perchlorate[d,l]	Mg(ClO$_4$)$_2$ · 0.12 H$_2$O	48[e]	245[e]	0.0002	1168
Anhydrone[d,f]	Mg(ClO$_4$)$_2$ · 1.48 H$_2$O		240[e]	0.0015	1157
Barium oxide	96.2% BaO		1000[h]	0.0028	244
Activated alumina	Al$_2$O$_3$	6–8[i]	175, 400[i]	0.0029	263
Phosphorous pentoxide[j]	P$_2$O$_5$			0.0035	566
Molecular sieve 5A[f]	Ca, Al, silicate			0.0039	215
Ind. magnesium perchlorate[d,l]	88% Mg(ClO$_4$)$_2$, 1% KMnO$_4$	48[e]	240[e,g]	0.0044	435
Lithium perchlorate[j,l]	LiClO$_4$	12,[e] 12	70,[e] 110	0.013	267
Calcium chloride[j,l]	CaCl$_2$ · 0.18 H$_2$O	16[e]	127[e]	0.067	33
Drierite[f]	CaSO$_4$ · 0.02 H$_2$O	1–2	200–225[k]	0.067	232
Silica gel		12	118–127[h]	0.070	317
Ascarite[f]	91% NaOH			0.093	44
Calcium chloride[j]	CaCl$_2$ · 0.28 H$_2$O		200[e]	0.099	57
Calcium chloride[j,l]	CaCl$_2$	16[e]	245[e]	0.137	31
Anhydrocel[f]	CaSO$_4$ · 0.21 H$_2$O	1–2	200–225[k]	0.207	683
Sodium hydroxide	NaOH · 0.03 H$_2$O			0.513	178
Barium perchlorate[l]	Ba(ClO$_4$)$_2$	16	127	0.599	28
Calcium oxide	CaO	6	500, 900[g]	0.656	51
Magnesium oxide	MgO	6	800	0.735	22
Potassium hydroxide[j]	KOH · 0.52 H$_2$O			0.939	18
Mekohbite[f,j]	69% NaOH			1.378	68

[a] Source: Data taken from Trusell and Diehl.[3]
[b] The amount of water remaining in the nitrogen after it was dried to equilibrium.
[c] The maximum volume of nitrogen dried to the specified efficiency for a given volume of desiccant.
[d] Hygroscopic.
[e] Dried in a vacuum.
[f] Trade name.
[g] Source: Trusell and Diehl.[3]
[h] Source: Hougen and Dodge.[5]
[i] Source: Morton.[6]
[j] Deliquescent.
[k] Source: Hammond.[2]
[l] Anhydrous.

over too short a time, and the pressure drop through the desiccant can be a problem at high flow rates. The most acceptable desiccants are anhydrous magnesium perchlorate, calcium sulfate, silica gel, activated alumina, and molecular sieve. These are briefly described below.

2.1.1.1 Anhydrous Magnesium Perchlorate

Anhydrous magnesium perchlorate (Anhydrone or Dehydrite) has the highest efficiency as well as the greatest capacity.[3] It is hygroscopic but not deliquescent, and it can absorb up to 35% of its own weight without evolving corrosive fumes, as phosphorous pentoxide does. Since the monohydrate does not dissociate to liberate water until 134°C, it can be used to dry gases at high temperatures. Hydration continues until the hexahydrate, which has a theoretical capacity of 48.4%, is formed. Magnesium perchlorate is available in either the regular or the indicating form, the latter containing about 1% potassium permanganate.

The chief disadvantages of anhydrous magnesium perchlorate are its relatively high cost (roughly four times the cost of the other three desiccants, or about $50 a pound) and the difficulty of regenerating it. The temperature must be raised slowly while the perchlorate is dried in a vacuum in order to prevent the crystals from fusing. A final temperature of about 245°C is recommended to return the perchlorate to its anhydrous state.[3] A further disadvantage of this and other perchlorate desiccants is the tendency to form explosive compounds in the presence of organic materials, especially when they are heated. Oil mists and other organic vapors must therefore be removed before such desiccants are used.

2.1.1.2 Calcium Sulfate

Calcium sulfate (Drierite or Anhydrocel) has an average efficiency of about 0.1 mg/L and a capacity of 7 to 14% at 25°C. It is stable, inert, and not deliquescent, even at peak capacity. It is easily regenerated (1 to 2 h at 200°C), and it operates at an almost constant efficiency over a wide range of temperatures. However, continued regeneration is difficult, because the constant formation and destruction of the hemihydrate breaks down the grains and forms a dusty residue. Calcium sulfate is available in sizes from 4 to 20 mesh and in either the regular or indicating form.

2.1.1.3 Silica Gel

Silica gel has a moderately high efficiency and capacity because of its large number of capillary pores which occupy about 50% of the gel's specific volume.[1] The capacity of the gel varies from batch to batch because of differences in the size and shape of the pores. The gel maintains its efficiency until it has absorbed 20% of its weight and it can be regenerated indefinitely at 120°C. At this relatively low regeneration temperature, however, the gel cannot be used for high-temperature drying. If it is regenerated above 260°C, it loses some of its capacity. Silica gel is available in sizes from 2 to 300 mesh. The addition of cobalt chloride to the surface of the gel provides an indicating ability.

2.1.1.4 Activated Alumina

Activated alumina has a higher efficiency than silica gel but offers less capacity (12 to 14%), especially at high humidities. It can be regenerated between 180 and

400°C without losing much of its capacity. It is available in sizes from 14 mesh to 1 in. and in either the regular or indicating form.

2.1.1.5 Molecular Sieve

Molecular sieves are synthetic crystalline alkali metal aluminosilicates that have a highly developed porous structure. About one half of their volume is comprised of a series of interconnected cavities of precisely uniform size. They can withstand temperatures as high as 600°C, with no effect on their crystallinity or adsorptive properties. Molecular sieves have a relatively high efficiency and a capacity of 20 to 29%. Table 2.4 compares four types of material.

2.1.2 Liquid Desiccants

If solid desiccants are not practical, then liquid desiccants can be used. Liquid desiccants have a much higher capacity than their solid counterparts (Figure 2.2) and can be continuously regenerated via spraying, pumping, or recirculating. On the other hand, their efficiencies are normally very low unless the anhydrous forms are used, and they usually cannot produce the relative humidities below 20%.[5] Some of the more common liquid desiccants are described in Table 2.5.

Note that although the strong acids and bases achieve the best efficiencies, they also emit corrosive vapors. In general, these techniques are not practical for small-laboratory air supplies.

2.1.3 Cooling

Cooling is the most efficient laboratory method for removing water from a gas stream. The gas is directed through a vessel in a low-temperature bath, and the excess water condenses on the cold walls of the vessel. For example, a bath of dry ice and acetone at $-70°C$ removes all but about 0.01 mg/L of water in air at equilibrium. A liquid nitrogen bath at $-194°C$ removes all but about 1×10^{-23} mg/L at equilibrium. This is about 19 orders of magnitude more efficient than anhydrous magnesium perchlorate, the best solid desiccant. Although extremely efficient, these types of cooling systems are appropriate for only a few liters per minute or less due to the ice buildup in the air lines.

Another type of cooling system that can effectively dry larger flows of air is the refrigerated compressed-air dryer. Water is removed by chilling the air and allowing the condensed water to drain out of the system automatically. These systems are available commercially and have the capacity to process 10 to 300 ft^3/min at inlet pressures of 100 psi.[8]

2.2 REMOVAL OF PARTICULATES

For relatively low-flow systems near 1 ft^3/min, there are a number of filters that can remove micron-sized particles. One example is a sintered bronze mesh, 1 in. in diameter and 2.5 inches long, that can filter out 2-µm particles at a flow rate of 1 ft^3/min and an operating pressure of 110 psi.[9] Filters made from metal fibers (5 to 750 µm pore size),[10] sintered stainless (2 to 150 µm pore size),[11] and foamed

Air Purification

Table 2.4 Characteristics of Molecular Sieves

Type	Common Form	Nominal Pore Diameter (Å)	Density (g/mL)	Equilibrium Capacity (wt %)
3A	Powder	3	0.48	23
3A	1/8-in. pellets	3	0.75	20
3A	8 × 12 beads	3	0.71	20
4A	Powder	4	0.48	28.5
4A	1/8-in. pellets	4	0.72	22
4A	8 × 12 beads	4	0.72	22
4A	14 × 30 mesh	4	0.71	22
5A	Powder	5	0.48	28
5A	1/8-in. pellets	5	0.69	21.5
13X	Powder	10	0.48	36
13X	1/8-in. pellets	10	0.64	28.5
13X	8 × 12 beads	10	0.64	28.5

Source: Union Carbide Corp.[29]

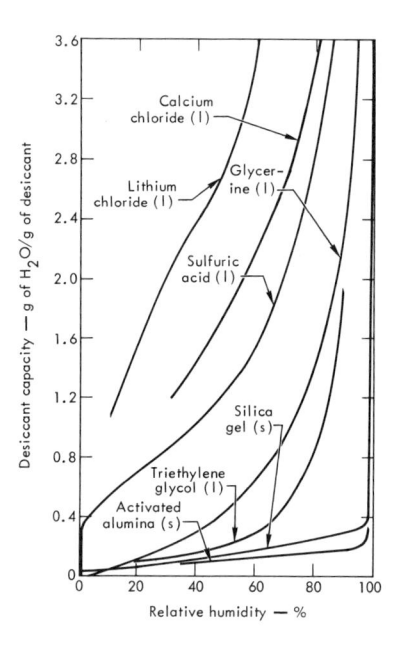

Figure 2.2 Desiccant capacity vs relative humidity for two solid desiccants and five liquid desiccants. Data taken from Hougen, O. A. and F. W. Dodge. *The Drying of Gases* (Ann Arbor, MI: Edwards Brothers Inc., 1947) with permission.

Table 2.5 Comparative Properties of Eight Liquid Desiccants

Desiccant	Relative Humidity at 21°C (%)	Solution Conc. (%)	Operating Temperature Range (°C)	Remarks
Calcium chloride	20–25	40–50	32–49	Regenerated at 150°C
Diethylene glycol	5–10	70–95	16–43	Oxidizes and decomposes at high temps.; regenerated with vacuum evaporation
Glycerol	30–40	70–80	21–38	
Lithium chloride	10–20	30–45	21–38	Corrosive; fumes during drying process; does not fume during regeneration
Phosphoric acid	5–20	80–95	16–38	
Sodium and potassium hydroxides	10–20	Saturated	29–49	Corrosive; frequently used to remove carbon dioxide and water simultaneously
Sulfuric acid	5–20	60–70	21–49	Corrosive; most efficient liquid desiccant
Triethylene glycol	5–10	70–95	16–43	

Source: Perry.[1]

Air Purification

metals (5 μm to 0.1 in. pore size)[12] are available. Porous Teflon® and Kel-F® filters that can remove particles as small as 2 μm are also available.[13]

If extremely pure air is required and a certain particle size must be removed, then membrane filters are often useful. These are available in sizes from 13 to 293 mm with pores from 7.5 nm to 8 μm.[14-16] These filters are constructed from a wide selection of microporous materials (e.g., regenerated cellulose, polyvinyl chloride, glass fibers, polypropylene, nylon, and Teflon) whose characteristics are accurately known.[14] The flow rate per unit of filter area at a given temperature is a function of the pore size and the upstream pressure. The relationship between pressure, pore size, and flow rate for a typical membrane filter is shown in Figure 2.3.

Laboratory air supplies, because of the flow requirements, must use filters that can operate at lower pressure drops and retain a higher mass of contaminants. Excess water from the drying process and any compressor oil must be removed. Most companies that supply compressed air equipment can furnish the appropriate filtration devices.

2.3 REMOVAL OF ORGANIC VAPORS

Organic vapors are often present in an air stream after the compression stage. Particulate filters will remove the aerosols, but the vapor remains as a possible contaminant problem.

Activated carbon has long been the most popular material for removing organic vapors. The carbon has internal submicroscopic capillaries that are just slightly larger than the molecules they trap. Even though water vapor may be present, the carbon will selectively remove the organic contaminants. Not all organic vapors are completely adsorbed in carbon, in spite of using large filter areas and low flow rates. For example, such low molecular weight compounds as acetylene, ethane, ethylene, methane, hydrogen, carbon monoxide, and carbon dioxide have almost no affinity for activated carbon.[17] Other organic materials such as methanol, methyl chloride, and vinyl chloride, are adsorbed in limited amounts.[18]

Another more vigorous laboratory method of removing organic vapors was devised by Kusnetz et al. Contaminated air is passed through a 2-in. diameter Mullite tube, which is filled with copper shavings and heated to 1250°C with a combustion furnace. The existing gas is cooled and the combustion products are removed by passing the gas through a dichromate solution, as well as Ascarite, activated carbon, and glass wool filters.[19]

2.4 REMOVAL OF MISCELLANEOUS CONTAMINANTS

A host of filters are available for treating compressed gases in the laboratory. Most compressed-gas filters operate via a two-stage separation that involves centrifugal or inertial separation followed by diffusion through a filter[20-22] or special sorptive material.[23] The filters can remove condensed water, oil mists, and particulate matter as small as 1 μm. Drain plugs are provided for periodically

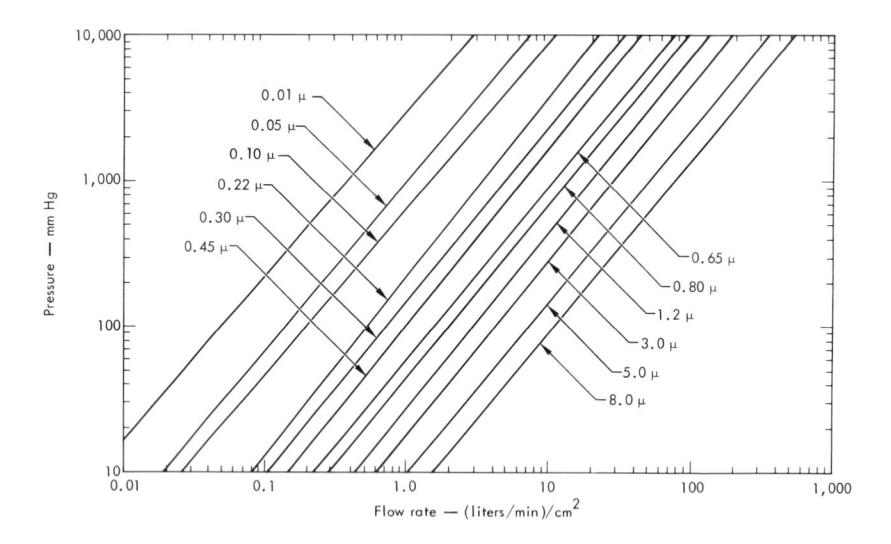

Figure 2.3 Pressure vs flow rate for 12 mean pore sizes at 25°C. Data taken from catalog no. MF-64, Millipore Corporation, Bedford, MA, with permission.

removing the collected liquid, and the filters themselves can be exchanged when they become clogged or excessively loaded. Automatic drain traps are also available.[24]

Compressed gases may also contain acid gases (hydrogen cyanide, sulfur dioxide, chlorine, and hydrogen chloride), carbon monoxide, and carbon dioxide. Acid gases and carbon dioxide can be removed with soda lime, a mixture of calcium and sodium hydroxides. Soda lime with various moisture contents is available in sizes of 4 to 14 mesh, in either the regular or the indicating form, and can absorb up to 25% of its own weight in carbon dioxide. The rate at which acid gases are absorbed depends on the condition of the lime. As the lime becomes spent, a thin film of calcium carbonate covers the surface of the soda-lime particles and cannot be removed by regeneration. Acid gases can be removed by several types of treated carbons. The usual impregnation materials are metal salts of copper and chrome. These materials are also commonly used in air-purifying, respiratory-protective cartridges and canisters.

Carbon monoxide is relatively unaffected by its passage through soda lime, activated carbon, or any of the previously described desiccants. Special materials must be used for its removal. Hopcalite, a mixture of copper and manganese oxides, has traditionally been the most practical agent for removing carbon monoxide. It operates as a catalytic oxidizing agent, converting the carbon monoxide to carbon dioxide. The main requirement for using Hopcalite is that it be kept scrupulously dry, for it loses its catalytic ability in the presence of water. Currently, other Zeolite-type materials are being developed that will also catalytically remove carbon monoxide in humid gas streams.

There are a number of pure air generators that operate from room or com-

Air Purification

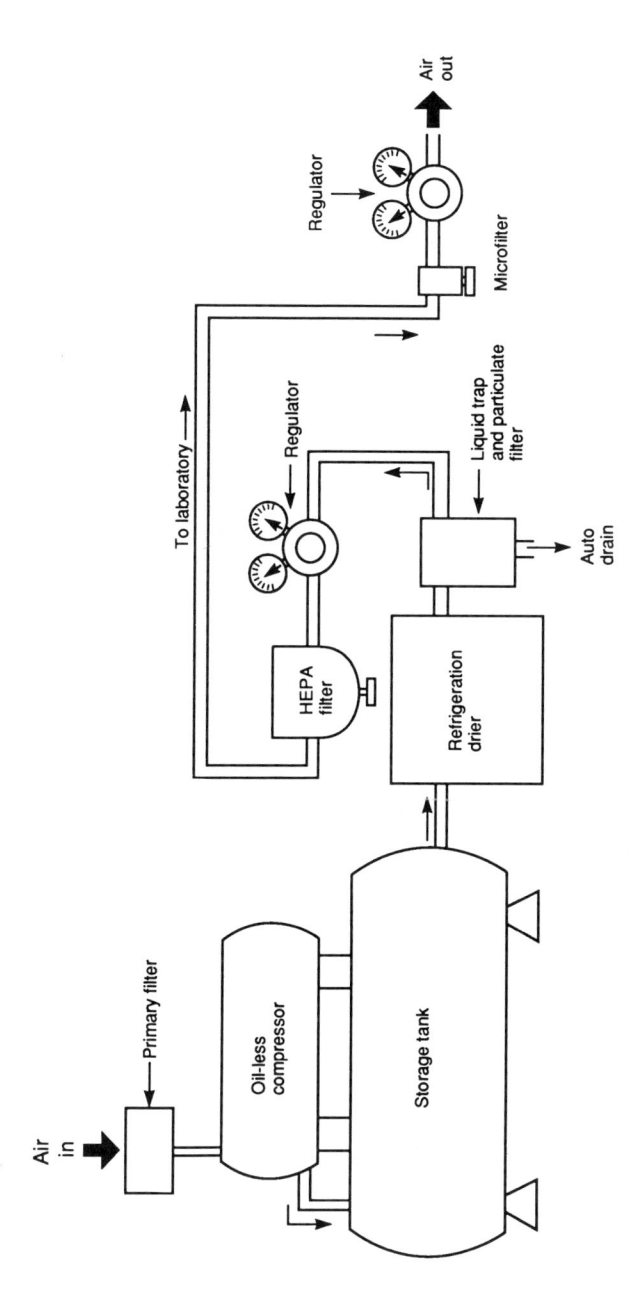

Figure 2.4 A typical laboratory compressed-air system with appropriate purification devices.

pressed air sources. Hydrocarbons, even methane, and carbon monoxide are converted to carbon dioxide and water.[25] Output flows are generally 1 to 20 L/min.

2.5 A TYPICAL LABORATORY COMPRESSED-AIR SYSTEM

Figure 2.4 describes a typical system for the small laboratory. Let us assume the requirements call for a maximum of 150 L/min at relative humidity of less than 10% at 25°C. The heart of the system is a 1.5 to 2 hp oilless piston air compressor (part no. 4Z707).[8] Since the pistons contain lap-jointed Teflon piston rings with stainless steel seals, no oil vapor is generated. Supply air is first passed through a filter, which removes atmospheric dust, acid gases, and higher molecular weight organic vapors. An acid gas-organic vapor respirator canister equipped with a high-efficiency particulate air (HEPA) filter does an outstanding job in this capacity. Next the air is compressed and exits through a refrigeration dryer, which will operate at pressures 175 psi or below and has an automatic drain (part no. 3Z529).[8] This type of dryer is capable of bringing 15 ft^3/min of air to a dew point of 35°F. Next the air passes through an air-line filter, which removes any water droplets or particles down to 40 μm (part no. 2Z763).[8] The final compressor filter is an HEPA filter and air-purifying chemical cartridge (part no. 81857).[26] This removes 99% of the particles 0.3 μm and larger, and any organic vapors or acid gas formed during the compression or drying process.

The cleansed air now travels, under pressure, to the laboratory use site. Just before exiting through the regulator, it passes through a final microfiber filter, which removes 99.95% of the 0.6-μm particulate matter resulting from loose flux or impurities in the conducting pipe system (part no. A944-BX).[27] The air supply system is now ready to supply clean, dry air, with the exception of ambient levels of methane and carbon monoxide.

REFERENCES

1. Perry, J. H. *Chemical Engineers Handbook* (New York: McGraw-Hill Book Company, Inc., 1984).
2. Hammond, W. A. *Drierite, the Versatile Desiccant, and its Applications in the Drying of Solids, Liquids, and Gases* (Columbus, OH: The Stoneman Press, 1958).
3. Trusell, F. and H. Diehl. Anal. Chem., 35: 674 (1963).
4. Skoog, D. A. and D. M. West. *Fundamentals of Analytical Chemistry* (New York: Holt, Rinehart & Winston, Inc., 1963).
5. Hougen, O. A. and F. W. Dodge. *The Drying of Gases* (Ann Arbor, MI: Edwards Brother, Inc., 1947).
6. Morton, A. A. *Laboratory Techniques in Organic Chemistry* (New York: McGraw-Hill Book Company, Inc., 1938).
7. Nonhebel, G. *Gas Purification Processes* (London: George Newnes, Ltd., 1964).
8. Catalog No. 375 Spring 1989, W. W. Grainger Co, Salinas, CA.

Air Purification

9. Catalog No. 200, Permanent Filter Corporation, Compton, CA.
10. Bulletin Nos. FM-1000, FM-1200, and FM-1300, Huyck Metals Company, Milford, CT.
11. Bulletin No. 1, Sintered Specialties, Janesville, WI.
12. Brochure, General Electric Company, Detroit, MI.
13. Brochure, Pall Corporation, Glen Cove, NY.
14. Catalog, 1969, Pure Aire Corporation of America, Van Nuys, CA.
15. Brochure, Arthur H. Thomas Company, Philadelphia, PA.
16. Catalog No. MF-64, Millipore Corporation, Bedford, MA.
17. Brochure, The Dexter Corporation, Windsor Locks, CT.
18. Nelson, G. O. and C. A. Harder. *Am. Ind. Hyg. Assoc. J.,* 35: 391 (1974).
19. Kusnetz, H. L., B. E. Saltzman, and M. E. Lanier. *Am. Ind. Hyg. Assoc. J.,* 21: 361 (1960).
20. Bulletin No. 118, R.P. Adams Company, Inc., Buffalo, NY.
21. Circular No. 1066, Wilkerson Corporation, Englewood, CO.
22. Bulletin No. 200, Dollinger Corporation, Rochester, NY.
23. Form No. 101-E, Deltech Engineering, Inc., New Castle, DE.
24. Catalog No. 600, King Engineering Corporation, Ann Arbor, MI.
25. Bulletin A-200, Aadco Company, Rockville, MD.
26. Data Sheet 01-01-01, Mine Safety Appliances Company, Pittsburgh, PA.
27. Bulletin 101-D, Balston, Inc., Lexington, MA.
28. Weast, R. C. *Handbook of Chemistry and Physics* (Boca Raton, FL: CRC Press, Inc., 1987).
29. Bulletin Nos. XF-21, XF-22, F 37, and XF-23, Union Carbide Corporation, Linde Division, New York, NY.

CHAPTER 3

FLOW RATE AND VOLUME MEASUREMENTS

Flow rate and volume measurements play an important role in the production of both static and dynamic gas mixtures. The accuracy with which gases are mixed is directly dependent on the accuracy of such measurements. Hence, in order to minimize errors in gas measuring systems, one must thoroughly understand the methods and characteristics of flow rate and volume measurements.

This chapter deals with flow rates between a fraction of a milliliter per minute and several cubic feet per minute and does not cover the higher flow rates (100 ft^3/min or more) that are often encountered in industrial ventilation systems. Primary, intermediate, and secondary laboratory standards are discussed, and a complete spectrum of devices for measuring flow rates and volumes in the laboratory is evaluated.

3.1 PRIMARY STANDARDS

A primary standard is usually some type of volume-measuring device. The volume under consideration is measured by direct measurement or weight. It is not dependent on any physical property of the gas involved.

3.1.1 Spirometers

Spirometers or bell provers are the most accurate standards for flow rate and volume measurements. They come in a variety of sizes; 9 L and 2, 5, 10, and 20 ft^3 are a few of the available volumes.[1,2] A representative spirometer is shown in Figure 3.1.

The spirometer functions as follows. When a gas enters the inlet, the movable bell of precisely known dimensions rises. The bell is supported by a chain, balanced by a counterweight, and separated from the stationary tank by a liquid interface, often a light oil. A volume scale, fixed to the side of the bell, and a pointer, attached to the stationary tank, indicate the total volume entering the tank over a given time interval. A pressure differential of 2 inches of water is usually

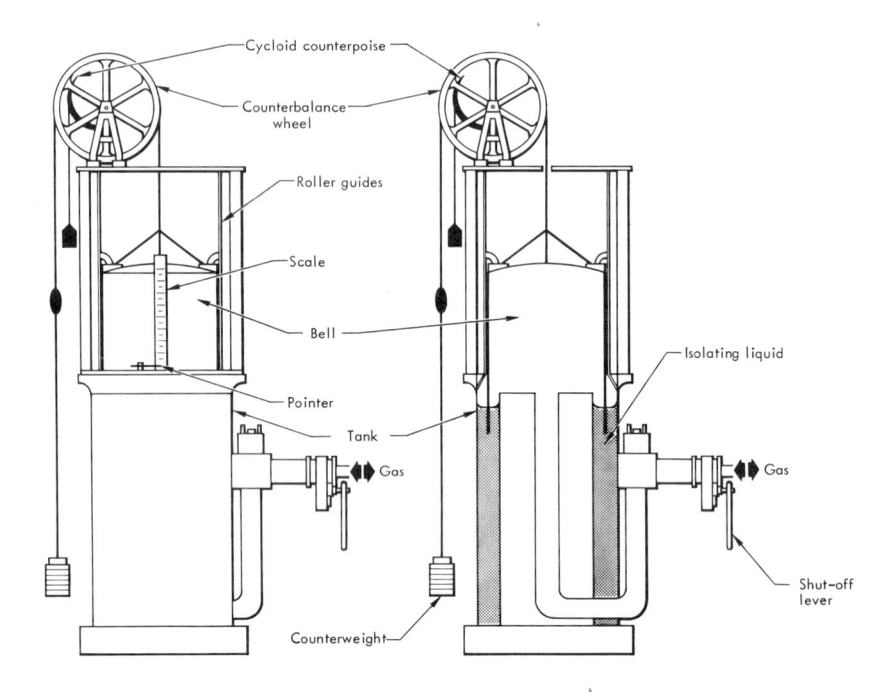

Figure 3.1 Orthographic and cross-sectional view of a 5 ft³ spirometer.

all that is required to raise or lower the bell. The downward pressure exerted by the bell, regardless of the depth it is submerged in the liquid, is kept essentially constant by a cycloid counterpoise that automatically compensates for buoyancy changes exerted by the liquid medium.[3] Thus, the bell requires the same pressure to raise or lower it over its entire working range. The temperatures of the liquid and the ambient air, as well as differences in the working pressure, are measured with the attached thermometers and oil manometer.

Most spirometers are individually calibrated against "cubic foot" bottles (a vessel that is certified by the U.S. Bureau of Standards to have a volume of exactly 1 ft³) at the factory,[4] but can be rechecked if minor inaccuracies are suspected. Before a spirometer is checked, it must be properly aligned, the liquid level must be correct, all leaks must be eliminated, and the difference between the temperature of the isolating liquid and the temperature of the ambient air must not be more than 0.3°C.[5]

The first step in calibrating a spirometer is to see if the bell drifts when all valves are open. If it does, the counterweight should be adjusted accordingly. In models that do not have a cycloid counterpoise, the bell should drift toward the geometric center of the fully raised and the fully lowered positions.[6] When admitting air to the spirometer, make sure all valves are open; otherwise, manometer oil can be suddenly discharged upward and permanently decorate the ceiling with a red stain. Once the lines are open, the spirometer can be checked with a cubic foot bottle or by means of the strapping procedure.

Flow Rate and Volume Measurements

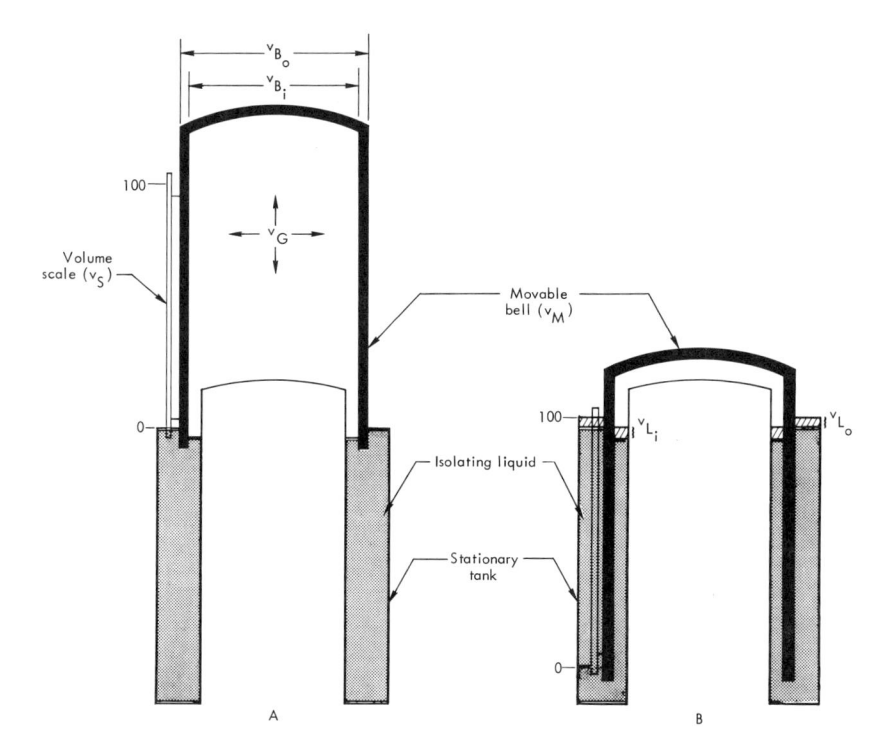

Figure 3.2 Simplified cross section of a spirometer with the bell fully raised (A) and fully lowered (B).

Figure 3.2 gives an illustration of the strapping technique. Basically this is a method of measuring the dimensions of the bell with a steel tape and calculating the volume.[5] The bell is shown both in fully raised or zero scale position (Figure 3.2A) and the empty or 100 scale position (Figure 3.2B). Volume measurements at both bell positions are made at the same pressure differential between the inside and outside of the bell. The gas volume, v_G, displaced when the bell is moved from the raised to the empty position, is equal to the volume of the bell interior, v_{B_i}, between the reference marks plus the liquid volume, v_{L_i}, which is caused by bell displacement. The interior bell volume, v_{B_i}, is in turn equal to the exterior bell volume, v_{B_o}, minus the volume of the metal bell, v_M. The volume of the metal bell plus the scale volume, v_S, is equal to the liquid that rises inside, v_{L_i}, and outside, v_{L_o}, the bell. These relationships can also be shown as

$$v_G = v_{B_i} + v_{L_i} \tag{3.1}$$

$$v_{B_i} = v_{B_o} - v_M \tag{3.2}$$

and

$$v_M + v_S = v_{L_o} + v_{L_i} \tag{3.3}$$

28 Gas Mixtures: Preparation and Control

By combining equations 3.1, 3.2, and 3.3, and solving for v_G, we obtain

$$v_G = v_{B_o} + v_S - v_{L_o} \tag{3.4}$$

The strapping method is often as good as a cubic foot bottle, and it routinely yields accuracies of ±0.2% when performed by an experienced person.

Example 3.1 Using the strapping technique, calculate the percentage error in the volume scale on a 5 ft³ spirometer. [5]

Measurements
Outside diameter of bell   66.047 in.
Length of volume scale   25.031 in.
Width of volume scale   1.125 in.
Thickness of volume scale   0.117 in.
Thickness of measuring tape   0.006 in.
Distance between outer surface
 of the bell and the inner surface
 of the stationary tank   1.942 in.
Rise in the level of the isolating
 liquid for full travel of bell   0.345 in.
Number of in.³/ft³   1728 in.³/ft³

Calculations
Corrected outside diameter of bell:

$$66.047 - (3.141)(0.006) = 66.028 \text{ in.}$$

Actual outside diameter of bell:

$$\frac{(66.028)}{(3.1416)} = 21.017 \text{ in.}$$

Volume of the bell as calculated from the actual outside diameter:

$$\frac{(3.1416)\left(\frac{21.017}{2}\right)^2 (25.031)}{1728} = 5.0255 \text{ ft}^3$$

Volume of the volume scale:

$$\frac{(25.031)(1.125)(0.117)}{(1728)} = 0.0019 \text{ ft}^3$$

Diameter of the stationary tank at the level of the liquid:

Flow Rate and Volume Measurements

$$21.017 + (2)(1.942) = 24.901 \text{ in.}$$

Difference in the apparent volume of the liquid:

$$\frac{(3.1416)\left[\left(\frac{24.901}{2}\right)^2 - \left(\frac{21.017}{2}\right)^2\right](0.345)}{1728} = 0.0280 \text{ ft}^3$$

Volume of displaced gas:

$$5.0255 + 0.019 - 0.0280 = 4.9995 \text{ ft}^3$$

Volume scale too short by

$$\left[\frac{(5.0000 - 4.9995)}{5.000}\right](100) = 0.011\%$$

The spirometer is primarily used to determine the flow rates of intermediate and secondary standards. The flow rate at standard conditions (25°C and 760 mmHg) can be calculated from

$$Q = \left(\frac{V}{t}\right)\left(\frac{P+p}{760}\right)\left(\frac{298}{T}\right) \tag{3.5}$$

where Q is the flow rate (L/min), V is the displaced volume (L), t is the time interval to observe V (min), P is the atmospheric pressure (mmHg), p is the internal manometer pressure (mmHg), and T is the ambient temperature (°K).

3.1.2 Pitot Tubes

The primary standard for measuring gas velocities is the Pitot tube. This device is particularly useful for measuring high flows in large ducts. It is primarily used in ventilation work, but it does have some laboratory applications. Pitot tubes are extensively described in the literature;[7,8] thus, the following section presents only the most basic concepts.

Basically, a Pitot tube is a pressure-sensing instrument. The standard Pitot tube adopted by the American Conference of Governmental Industrial Hygienists is shown in Figure 3.3. It consists of two concentric tubes — an inner tube that senses the impact pressure of a flowing gas and an outer tube that senses the static pressure. In order for a Pitot tube to function properly, the orifice in the upstream end of the tube must face the flow squarely. The impact and static pressures can be measured with an upright U-tube and an inclined manometer, or some other suitable pressure measuring device, depending on the magnitude of the pressure

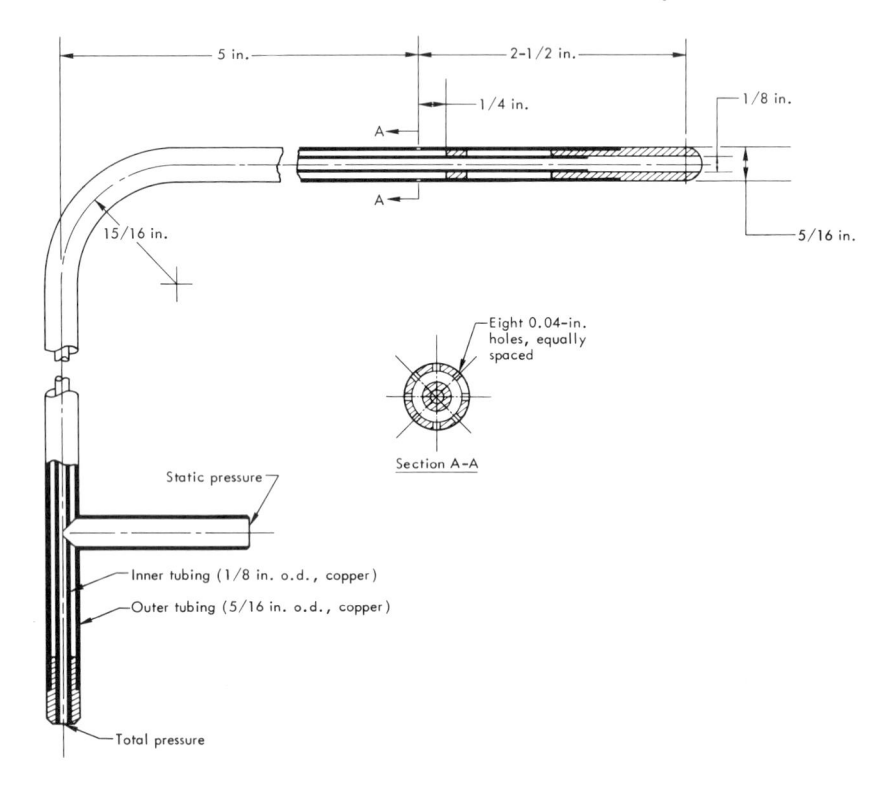

Figure 3.3 The standard Pitot tube. Data taken from Powell, C. H. and A. D. Hosey, Eds. *The Industrial Environment—Its Evaluation and Control* (Washington, D.C.: U. S. Government Printing Office, 1965).

and velocity. When a U-tube is employed, acceptable accuracies can be achieved only at velocities above 2500 ft/min. On the other hand, a carefully made and accurately leveled inclined manometer can retain its accuracy at velocities as low as 600 ft/min.[8,9] Some typical velocity pressures and velocities are 0.062 inches of water at 1000 ft/min, 0.14 inches at 1500 ft/min, 0.25 inches at 2000 ft/min and 0.56 inches at 3000 ft/min.[6]

Since Pitot tubes contain no moving parts, they can be used in almost any position or location. If they are made of stainless steel, that can be used at high temperatures and pressures, and even in corrosive atmospheres. Gases that contain particles, aerosols, or fumes must be avoided because such contaminants may foul the small, carefully machined orifices.

The amount of air flowing through a duct can be calculated from

$$Q = A\bar{u} \tag{3.6}$$

where Q is the flow rate of the gas, A is the cross-sectional area of the duct, and u is the average linear velocity of the gas. For a Pitot tube, $\bar{u}$ is determined by

Flow Rate and Volume Measurements

$$\bar{u} = \sqrt{\frac{2p}{\rho}} \qquad (3.7)$$

where p is the difference between the impact and static pressures, and ρ is the density of the gas.

Since the velocity of a gas in a given duct varies across any given cross section because of frictional losses and the shape of the duct, many measurements must be made to determine the most accurate average velocity.[8] However, one can closely approximate the average velocity by assuming that it is roughly 90% of the velocity in the center of the duct. Equation 3.5 combined with Equation 3.6 then becomes

$$Q = 0.9 \, A \sqrt{\frac{2p}{\rho}} \qquad (3.8)$$

If greater accuracy is desired, the traversing technique described in *Industrial Ventilation*[8] can be used.

3.1.3 Frictionless Pistons

Although spirometers and wet test meters are useful standards, their accuracy diminishes at flow rates of 1 L/min or less. To fill this gap, soap bubble meters[10] and mercury-sealed pistons[11] have been devised. These instruments measure flow rates of 1 mL/min to several L/min with reasonable accuracy and are used to check the accuracy of secondary standards, particularly rotameters, mass flow meters, critical orifices, and porous plug devices.

Two kinds of soap-bubble meters are shown in Figure 3.4. Gas from the instrument being calibrated enters the inlet and travels up the buret. It usually has a capacity of 10, 50, 250, or 1000 mL, depending on the needed range. Soap bubbles are introduced by squeezing the rubber bulb and raising the soap solution above the inlet. The gas rises through the solution, forming bubbles, and the progress of the bubbles between selected volume markings is timed with a stop watch. As the bubbles rise, they act essentially as frictionless pistons — a pressure of 0.02 inches of water is all that is required to move the bubbles at a uniform rate.[12]

Under average laboratory conditions, soap bubble meters are usually accurate to within 1%. However, with carefully controlled conditions and for relatively nonreactive and insoluble gases, accuracies of ±0.25% have been reported.[13] The accuracy of soap bubble meters can be improved slightly[14] by adding photoelectric devices[15,16] or electrical contacts[17] that actuate relays and timers. As the flow rate approaches 1 or 2 mL/min, the accuracy declines because of gas permeation through the soap film and water vapor contributions to the gas volume. For example, Czubryt and Gesser report errors of 3 and 7%, respectively, for carbon

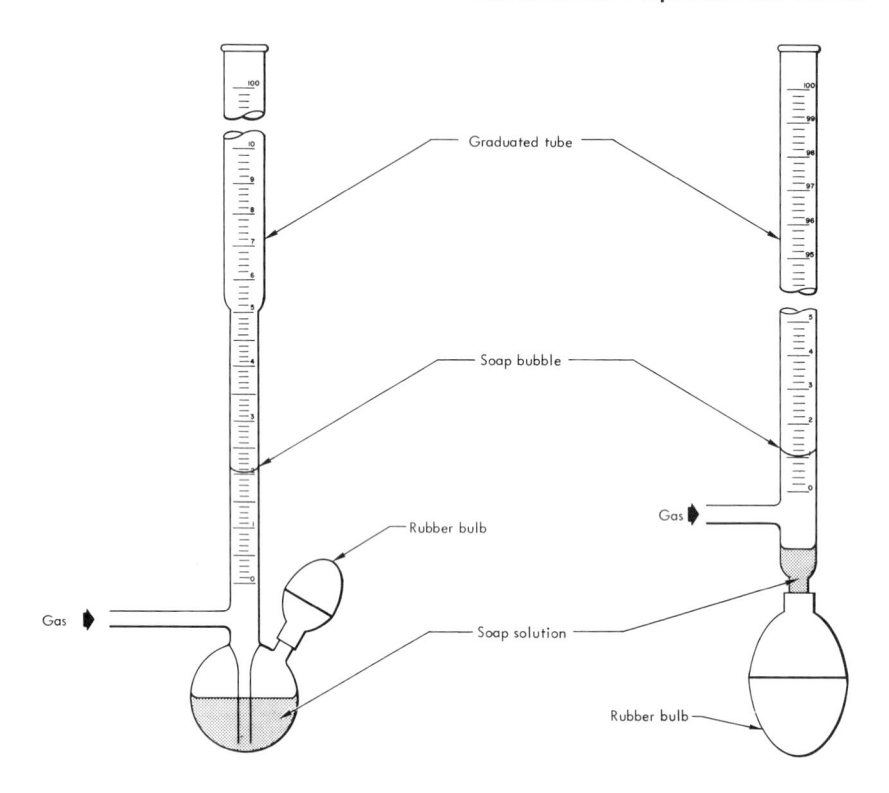

Figure 3.4 Two kinds of soap bubble meters for measuring gas flows.

dioxide and argon flowing at 1.6 and 1.2 mL/min.[18] It must be emphasized that bubble meters should not be used to calibrate any device with reactive gases, such as ammonia, hydrogen chloride, and sulfur dioxide. Such gases lead to the destruction of the soap bubble and a rather spectacularly unsuccessful calibration effort.

If a soap-bubble meter is undesirable, then a precisely measured and calibrated electronically actuated piston may be used.[19] Noble reports that such a device has an accuracy of ±0.03% at flow rates between 1 and 100 mL/min.[14]

A commercially available instrument for calibrating low-flow meters is the mercury sealed piston shown in Figure 3.5. The gas enters the precision-bore glass cylinder and displaces the polyvinyl chloride piston, which is made gas tight by the mercury O-ring seal. This instrument is virtually frictionless, but the weight of the piston must be compensated for in order to achieve the best accuracy. The capacity of this instrument ranges from 1 to 24,000 mL/min, and its accuracy is ±0.2% for timing intervals of 30 sec or more.[11] This technique has been extended to larger bore cylinders using counterweights to offset the weight of the piston.[20] Using automatic timing circuits, accuracies of 0.001% have been reported for flow rates of 50 L/min.

The mercury-sealed piston, while compatible with most gases, must not be used with any gas that reacts with mercury. Hydrogen chloride, chlorine, and

Flow Rate and Volume Measurements

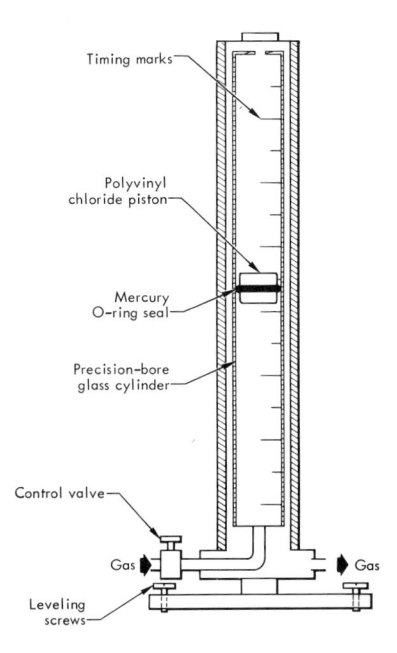

Figure 3.5 Mercury-sealed piston for measuring gas flows.

phosgene, for example, will combine with the mercury to produce a salt rather than a liquid seal. Another gas to avoid is ethylene oxide. While inert to mercury, it will dissolve slowly into the polyvinyl chloride piston and yield erroneous flow rate data. If a reactive gas must be used, then a chamber of inert gas is placed between the mercury-sealed piston and the device to be calibrated.[46] The reactive gas then pushes only the inert gas into direct contact with the liquid mercury.

3.1.4 Syringe Drive Systems

Many gases that are corrosive or that are soluble in water cannot be used in the previously mentioned calibration devices. One technique that circumvents the material interaction problem is the syringe drive system shown in Figure 3.6. Instead of flowing the gas of interest through the device to be calibrated and into the standard system, the calibration of the flow device is carried out in somewhat the reverse fashion. First the syringe is filled with the gas of interest. Knowing the injection rate and the volume per unit length of the gas-filled syringe, a calibration curve can be constructed using

$$Q = \frac{LK}{\tau} \tag{3.9}$$

where Q is the flow rate (mL/min), L is the length of plunger displacement (mm), K is the syringe constant (mL/mm), and τ is the time required to displace the plunger a distance L. For best results, use syringes that use a plunger and Luer-

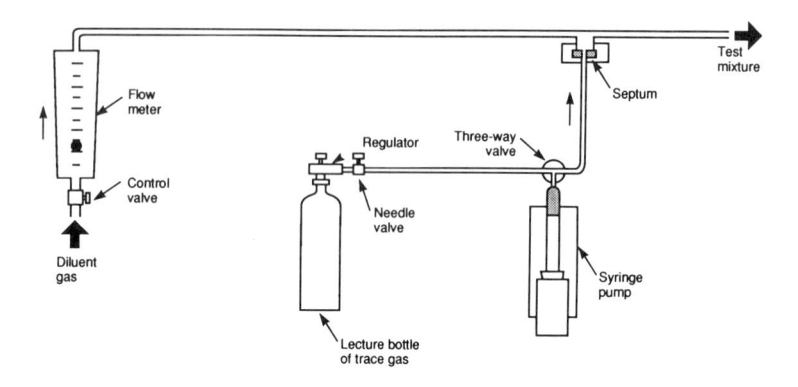

Figure 3.6 Syringe-drive system for calibrating flow devices with corrosive gases.

tip fitting contructed of Teflon®, and avoid metal connections and rubber-tipped syringes.

3.1.5 Aspirator Bottles

Flow rates of 10 mL/min to 10 L/min can be measured with an aspirator bottle like that shown in Figure 3.7. Gas from the secondary standard being calibrated flows into the aspirator bottle, which is filled with water. The water is then displaced into the graduated cylinder. The time required to displace a set volume of water yields the flow rate. The flow rate should be corrected to standard conditions, or the temperature and pressure should be noted. Aspirator bottles are more of an historical apparatus and do not have as wide an appeal as soap bubble meters because they have less accuracy (±1 to 3%), convenience, and range. If one wishes to maximize the accuracy of the device, an article by Christian discusses how to correct for the effects of water vapor pressure, solubility, and gas nonideality.[47] If all these effects are taken into account, accuracies of 0.1 to 1% can be achieved.

3.2 INTERMEDIATE STANDARDS

Reasonable accuracy in volume or flow measurements does not necessarily require the use of primary standards, for there are several intermediate standards that are almost as effective. Although the internal volumes of intermediate standards cannot always be obtained by dimensional measurements (hence the intermediate rating), such standards still exhibit accuracies of ±1% or better, and can therefore be used with confidence. Intermediate standards include the wet test meter and the dry gas meter.

3.2.1 Wet Test Meters

Wet test meters are usually used to calibrate secondary standards, but they can also be used to meter gases directly.[21] A wet test meter is shown in Figure 3.8. A container houses a revolving drum that is about two-thirds submerged in water.

Flow Rate and Volume Measurements

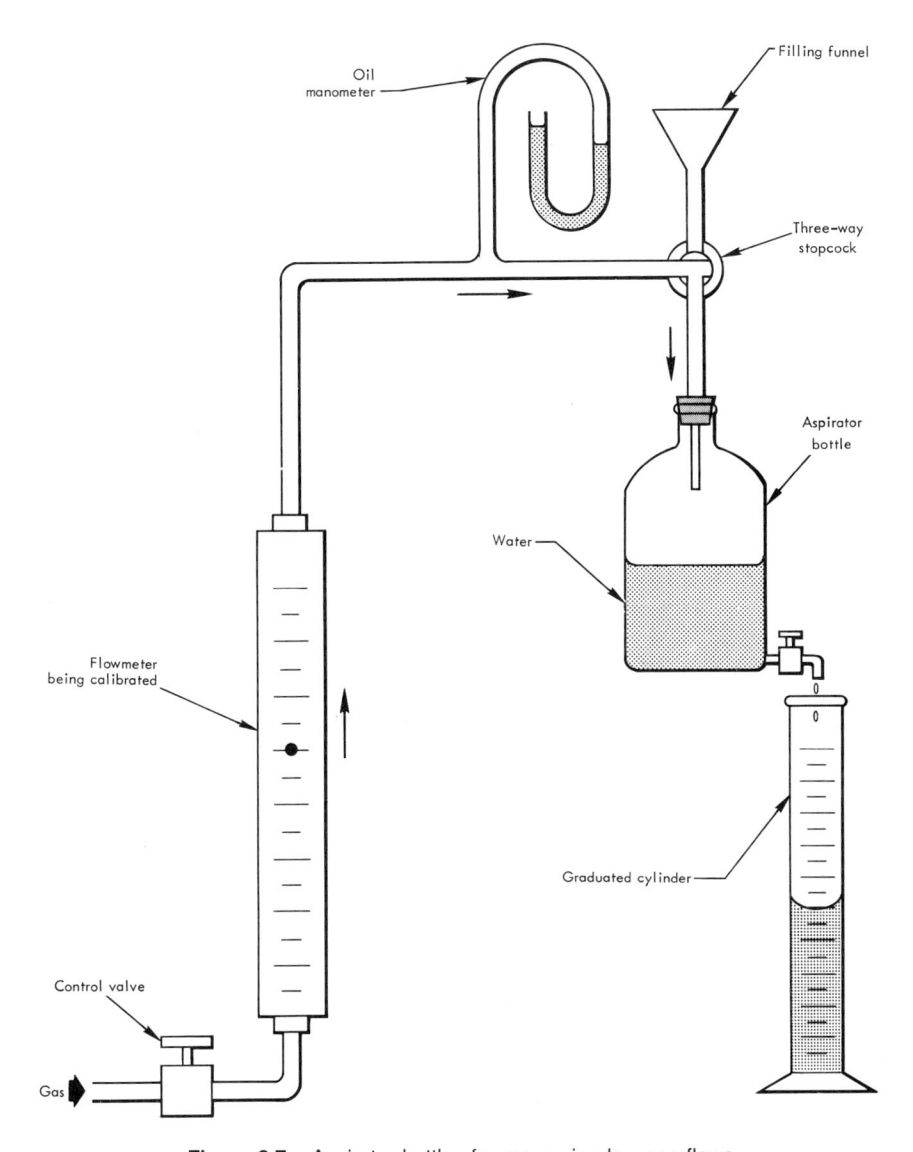

Figure 3.7 Aspirator bottles for measuring low gas flows.

The drum is divided into four sections, each of which has an inlet and outlet. When a gas enters the meter, it exerts a buoyant force that turns the drum clockwise. As one quadrant becomes full, a new quadrant rotates into the filling position, and the full one begins to expel gas through its outlet. The drum rotates smoothly if the gas is fed to the meter at a steady rate. The total volume for a given time interval is recorded by the decade dials provided. Attached to the meter are a water manometer for measuring the internal and external pressure differential and a thermometer for measuring the temperature of the incoming gas. This thermometer is commonly missing due to an errant flying elbow from the last calibration

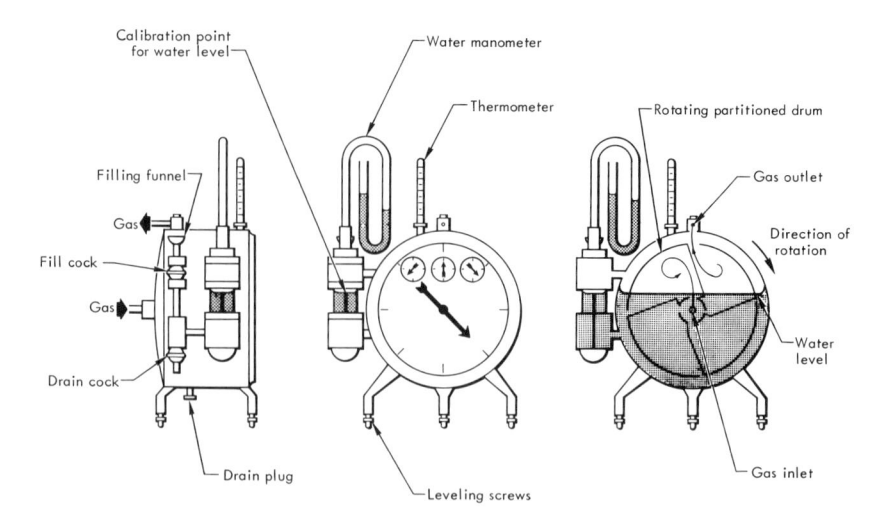

Figure 3.8 Side, front, and cross-sectional views of a wet test meter.

effort. A filling funnel and drain-and-fill cocks are used to keep the water level at the precise calibration point. A toggle switch controls an electrical circuit that is closed when the pointer strikes a contact prong at the zero point. Binding posts are provided for connection to an external signal alarm or event monitor.

Liter and cubic foot models are available, and their specifications are shown in Table 3.1. If the meter is used at lower flow rates than the minimum specified, unacceptable friction losses can occur. Excessively high flows cause unwanted pressure differentials.

Before a wet test meter can be checked for accuracy, it must be leveled and filled with water to the calibration point. Then the gas should be run through the meter for several hours in order to saturate the water and to allow the meter to equilibrate.[6] The meter can then be checked against a 0.1 ft^3 bottle[22,23] or a spirometer, as shown in Figure 3.9. Example 3.2 gives a sample calculation using a spirometer as a primary standard. Enough gas is withdrawn to significantly displace the bell in the spirometer and to make the drum in the wet test meter revolve three times. The initial and final volume, atmospheric pressure, the temperature of the gas, and internal pressures inside the devices are all noted. It should be noted that since a vacuum source is the primary air mover, the internal pressures are less than atmospheric and therefore are reported as negative values. If compressed air were to be used, and air flowed in the other direction under a slight positive pressure, the internal pressures would be a positive contribution to the overall system pressure.

Each calibration point is repeated three or more times at various flow rates so that a curve relating flow rate to error can be drawn. The volume measurements

Flow Rate and Volume Measurements

Table 3.1 Wet Test Meter Specifications

Specifications	Wet test meter type	
	Cubic foot	Liter
Capacity/hr	20 ft^3/hr	680 L/hr
Capacity/min	0.33 ft^3/min	11.3 L/min
Vol per revolution	0.1 ft^3	3 L
Dial subdivisions	0.001 ft^3	0.01 L
Counting dial max	100 ft^3	3000 L
Pressure range	0.3–6 in. water	0.3–6 in. water
Minimum air flow/hr	2 ft^3/hr	68 L/hr
Minimum air flow/min	0.03 ft^3/min	1.1 L/min
Maximum air flow/hr	24 ft^3/hr	800 L/hr
Maximum air flow/min	0.40 ft^3/min	13.3 L/min
Maximum vacuum	2 in. water	2 in. water
Accuracy	±1/2%	±1/2%
L × W × H	22 × 30 × 36 cm	22 × 30 × 36 cm

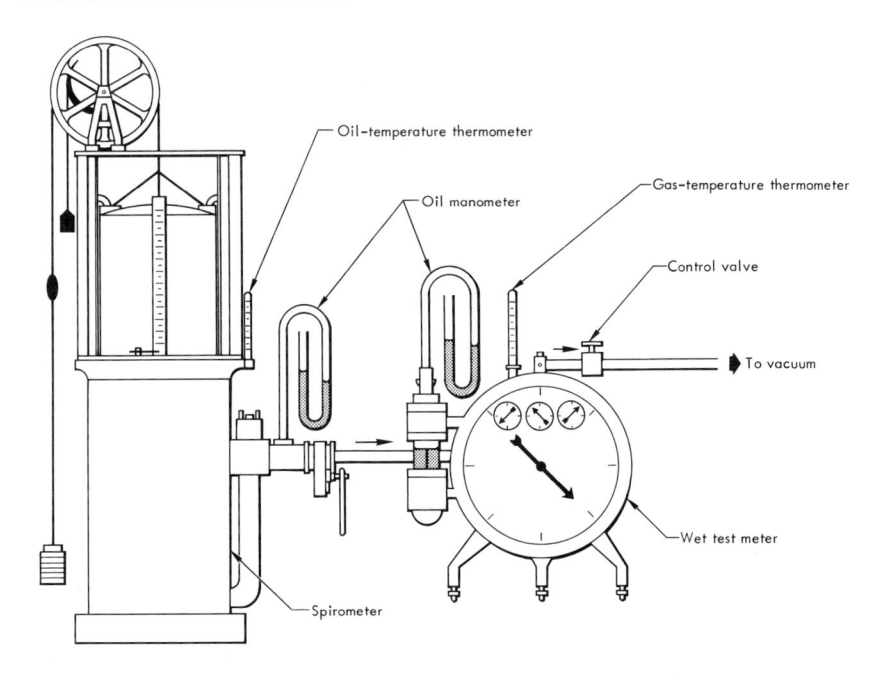

Figure 3.9 Setup for calibrating a wet test meter against a spirometer.

are corrected to standard conditions, but the water vapor effects are usually ignored.[6] Wet test meters should check out to within 0.5% of the values indicated by the volume dials, and new instruments are often within 0.25%.[24]

Example 3.2. Calculate the percent error between the wet test meter and the 5 ft³ spirometer using the following information:

Ambient conditions	27°C	740 mmHg
Standard conditions	25°C	760 mmHg

	Trial Number			
	1	2	3	4
Spirometer data				
Initial reading, ft³	2.000	2.500	2.000	2.000
Final reading, ft³	2.692	3.262	2.746	2.828
Total volume, L	19.60	21.58	21.13	23.45
Manometer reading, in. water	-0.015	-0.025	-0.040	-0.095
Manometer reading, mmHg	-0.028	-0.047	-0.075	-0.178
Duration of run, min	18.65	10.50	5.63	3.00
Flow rate from Eq. 3.5, Qs, L/min	1.016	1.988	3.629	7.558
Wet test meter data				
Initial reading, ft³	2.500	3.000	2.000	2.500
Final reading, ft³	3.194	3.763	2.746	3.327
Total volume, L	19.65	21.61	21.13	23.42
Manometer reading, in. water	-0.200	-0.500	-1.200	-2.300
Manometer reading, mmHg	-0.374	-0.935	-2.244	-4.301
Duration of run, min	18.65	10.50	5.63	3.00
Flow rate from Eq. 3.5, Qw, L/min	1.019	1.988	3.618	7.507
Error [(Qs - Qw)/Qs] × 100, %	-0.24	-0.01	0.29	0.68

3.2.2 Dry Gas Meters

The dry gas meter is a common industrial device used to monitor the use volume of our natural gas supply. Such meters can measure flow rates of 5 to 5000 L/min at pressures up to 250 psi with an accuracy of a few percent.[4,25] The specifications of dry gas meters suitable to our range of interest are summarized in Table 3.2.

Figure 3.10 shows a cross section of a dry gas meter. The meter consists of a sealed outer case, two sliding valves, two bellows, several volume dials, and a linkage system. In Figure 3.10A, the gas enters the meter and passes the left-hand sliding valve, filling the left-hand bellows and forcing gas out of the chamber on the left. When this bellows is fully extended, the sliding valves shift and direct the incoming gas to the right-hand bellows. This forces the gas from the right-hand chamber, as shown in Figure 3.10B. When both bellows are fully extended, the sliding valves again shift and the bellows empty separately, as shown in Figures 3.10C and 3.10D. This cycle of alternately filling and emptying the bellows is linked to the volume dials, which register the corresponding volume changes. If

Flow Rate and Volume Measurements

Table 3.2 Dry Gas Meter Specifications

Model	Volume per Revolution	Capacity	Dial Subdivisions	Max Working Press. (psi)
DTM-115-1	0.1 ft^3	115 ft^3/hr	0.001 ft^3	5
DTM-115-2	1 ft^3	115 ft^3/hr	0.01 ft^3	5
DTM-115-3	1 L	3200 L/hr	0.01 L	5
DTM-115-4	10 L	3200 L/hr	0.10 L	5
DTM-200-1	0.1 ft^3	200 ft^3/hr	0.001 ft^3	5
DTM-200-2	1 ft^3	200 ft^3/hr	0.01 ft^3	5
DTM-200-3	10 L	5600 L/hr	0.10 L	5
DTM-200-4	100 L	5600 L/hr	1.00 L	5
DTM-325-2	1 ft^3	325 ft^3/hr	0.01 ft^3	10
DTM-325-4	10 L	9200 L/hr	0.10 L	10
DTM-325-5	100 L	9200 L/hr	1.00 L	10

Data taken from American Dry Test Meters, American Meter Division Bulletin 500.3.

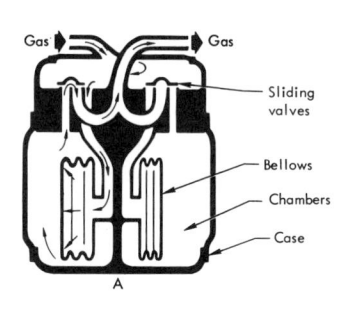

Figure 3.10 Idealized cross section of a dry gas meter. The operation of this instrument is described in text.

a steady supply of gas is supplied to the meter, the main indicating dial turns in a rather uneven fashion due to the nature of the linkage system. To minimize errors, at least one full revolution must be timed. When calibrating low-range dry gas meters (1 L per revolution), groups of four revolutions must be timed to avoid these linkage-related errors.

Dry gas meters can be calibrated in the same manner as wet test meters, but their usual accuracies are about ±1%. If the recorded flow rate differs from the actual flow rate by more than 2%, it can be made to correspond more closely by means of the tangential adjusting weights, which influence the motion of the linkage to the volume dials. Although dry gas meters are less accurate, they are sometimes preferred over wet test meters because they are about one-third lighter, cost half as much, occupy less space, and do not have any water vapor interaction problems.

3.3 SECONDARY STANDARDS

Most flow-measurement devices used in the laboratory are classified as secondary standards; that is, they have been compared against a primary standard at known conditions of gas type, pressure, and temperature, and have a calibration curve relating the meter reading to the actual flow rate.

Secondary standards, usually not as accurate as their primary counterparts, are nevertheless useful because of their convenient size, weight, cost, and general availability. These standards include rotameters, mass flow meters, orifice meters, critical orifices, and controlled leaks and plugs. Well over 100 types of flow meters are available, and periodic tabulations of their specifications appear in the literature.[26-29,90]

3.3.1 Rotameters

The most widely used laboratory method for measuring gas or liquid flow rates is the rotameter (also spelled rotometer) or variable-area flow meter. It is usually a round glass tube of increasing diameter that houses one or more floats that are free to move vertically up and down the tube axis. The float inside the tube is engineered so that the diameter is nearly the same as the tube's inlet diameter. On the sides are permanently engraved reference marks that may be either linearly (by adjusting the taper[40]) or exponentially inscribed on the tube. Figure 3.11 gives an example. As the gas flows up the tube, the float is displaced and continues to move upward until equilibrium is reached. This occurs when the downward force of the float just equals the buoyant force of the moving gas stream. Any change in the flow rate will cause the float to occupy a proportionally different position.

The flow rate range of a full set of rotameters is enormous and spans from 1 mL/min to over 300 ft^3/min with reasonable accuracy. The range of each individual rotameter, however, is considerably less and is usually one and a half to two orders of magnitude. The range can be extended by using multiple floats, and many commercial types are equipped with dual floats of glass and stainless steel. Table 3.3 gives the flow rate characteristics of different float materials over a wide range of flows.

Flow Rate and Volume Measurements

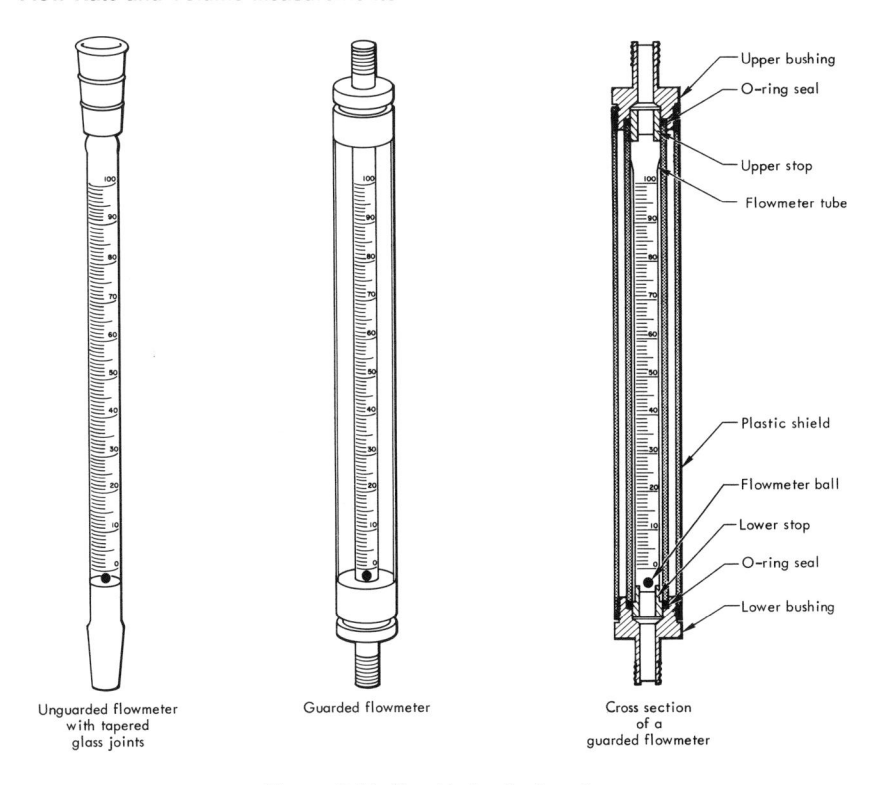

Figure 3.11 Two kinds of rotameters.

Float design will vary depending on the manufacturer and the flow rate desired. Most lower range rotameters will use spherical floats, but a wide variety of configurations exist, especially at higher flow rates. Figure 3.12 shows several types of floats in current use. No matter which type is used, the rotameter reading is conventionally taken at the widest point of the float.[30] For example, spherical types are usually read at the center and not at the top or bottom of the float.

Some commercial flow meters have not only a dual range but also have the capacity to measure total volume through a system over a given time interval. A volume counter is linked to a turbine wheel, whose spin is proportional to the amount of gas turning it. This works well when the the flow rate changes are gradual, but significant errors arise when sharp pulses and fast rate changes are encountered. Another type of flow meter uses a light and photocell on either side of the float to electronically measure the float position.[48]

Rotameters are available with a wide variety of inlet and outlet connections. Hose barbs, pipe threads, compression fittings, and tapered glass joints are commercially supplied. Usually it is best to obtain rotameters that have a protective outer shield, such as the one shown in Figure 3.11. Multiple guarded units are also available. This outside protection is quite useful in preventing accidental breakage from an errant piece of laboratory apparatus. These jacketed flow meters can be connected in series or parallel to extend the desired range, as shown in Figure 3.13.[31] The parallel configuration is preferred because it produces less pressure

Table 3.3 Flow Rate Characteristics of Different Float Materials

Model Number	Float Material	Airflow mL/min	ft³/hr	Model Number	Float Material	Airflow L/min	ft³/hr
B-125-6	Glass	36	0.076	B-250-1	Glass	2.40	5.08
	Sapphire	57	0.121		Sapphire	3.08	6.53
	St. steel	108	0.229		St. steel	4.60	9.75
	Carboloy	203	0.430		Carboloy	6.75	14.3
	Tantalum	251	0.532		Tantalum	7.30	15.5
B-125-10	Glass	50	0.106	B-250-2	Glass	5.20	11.0
	Sapphire	81	0.172		Sapphire	6.60	14.0
	St. steel	150	0.318		St. steel	9.90	21.0
	Carboloy	263	0.557		Carboloy	14.1	29.9
	Tantalum	300	0.636		Tantalum	15.4	32.6
B-125-20	Glass	89	0.189	B-250-3	Glass	7.78	16.5
	Sapphire	144	0.305		Sapphire	9.60	20.3
	St. steel	268	0.568		St. steel	14.3	30.3
	Carboloy	473	1.00		Carboloy	20.0	42.4
	Tantalum	560	1.19		Tantalum	22.0	46.6
B-125-30	Glass	396	0.839	B-250-4	Glass	12.0	25.4
	Sapphire	521	1.10		Sapphire	15.3	32.4
	St. steel	835	1.77		St. steel	22.6	47.9
	Carboloy	1,250	2.65		Carboloy	32.4	68.6
	Tantalum	1,370	2.90		Tantalum	34.2	72.5
B-125-40	Glass	850	1.801	B-250-5	Glass	16.3	34.5
	Sapphire	1,110	2.35		Sapphire	19.5	41.3
	St. steel	1,700	3.60		St. steel	30.8	65.3
	Carboloy	2,400	5.08		Carboloy	40.2	85.2
	Tantalum	2,600	5.51		Tantalum	44.1	93.4
B-125-50	Glass	2,340	4.96	B-250-6	Glass	20.7	43.9
	Sapphire	3,100	6.57		Sapphire	23.2	49.2
	St. steel	4,600	9.75		St. steel	40.5	85.8
	Carboloy	6,600	14.0		Carboloy	50.0	105.9
	Tantalum	7,000	14.8		Tantalum	53.3	112.9
B-125-60	Glass	3,800	8.05	B-250-7	Glass	8.7	18.4
	Sapphire	5,000	10.6		Sapphire	11.2	23.7
	St. steel	7,500	15.9		St. steel	16.6	35.2
	Carboloy	10,700	22.7		Carboloy	22.9	48.5
	Tantalum	11,500	24.4		Tantalum	24.3	51.5
B-125-70	Glass	5,600	11.9	B-250-8	Glass	22.4	47.5
	Sapphire	7,400	15.7		Sapphire	29.1	61.7
	St. steel	10,990	23.3		St. steel	43.4	91.9
	Carboloy	15,500	32.8		Carboloy	62.0	131
	Tantalum	16,600	35.2		Tantalum	67.4	143

Data taken from Porter Instrument Co.

Flow Rate and Volume Measurements

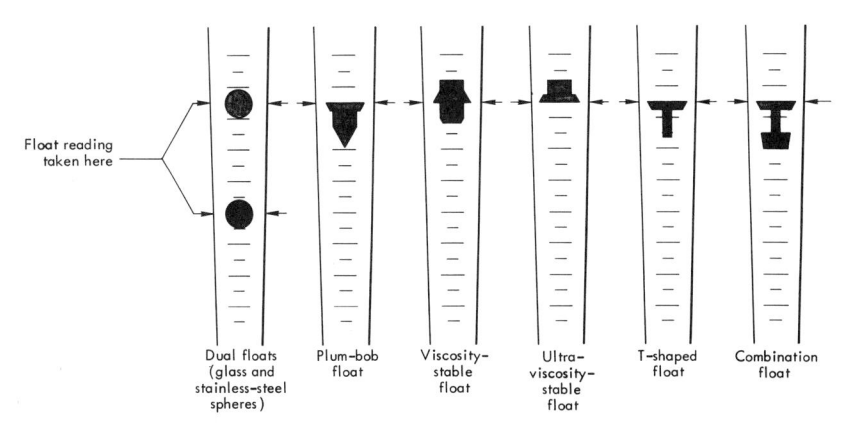

Figure 3.12 Six kinds of rotameter floats. Readings are conventionally taken at the widest point of the float.

Float reading taken here — Dual floats (glass and stainless-steel spheres) | Plum-bob float | Viscosity-stable float | Ultra-viscosity-stable float | T-shaped float | Combination float

drop across the measurement system. Gas flow control is usually done with a valve just before the rotameter. However, a valve downstream can be used if the supply-gas pressure is kept constant.[43]

Most rotameters are calibrated at room temperature and pressure, but they may be corrected to experimental conditions by using the appropriate equations. Pressures of 0.5 to 5 atm and temperatures from 0 to 150°C are not unreasonable conditions of use.

The flow rate of a gas through a rotameter is seldom calculated from tube diameters and float dimensions, although the development and use of such equations are adequately covered in the literature.[32-37,41,45] Numerous experiments have concluded that mathematical formulas do not provide accurate information, particularly at flow rates below 50 L/min.[46] Instead, curves relating meter reading to flow rate are derived from a calibration against one of the primary or intermediate standards discussed, usually a bubble meter, spirometer, or wet or dry gas meter. Although the manufacturer generally provides reasonably accurate calibration curves, rotameters should be recalibrated at the experimental conditions of interest, with the primary standard on either the inlet or the outlet and open to the atmosphere, as shown in Figure 3.14. In this manner, the flow conditions at a known temperature and pressure are documented, while the actual conditions need not be known. Once this basic information is available, correction factors can be more realistically applied.

This next section will discuss rotameter correction factors. The work of Caplan will be extensively referenced, since he is one of the few individuals to have unraveled the mysteries of this seemingly simple device.[38]

The basic rotameter flow equation is given as[39]

$$W = KD_f \sqrt{\frac{W_f(\rho_f - \rho)\rho}{\rho_f}} \qquad (3.10)$$

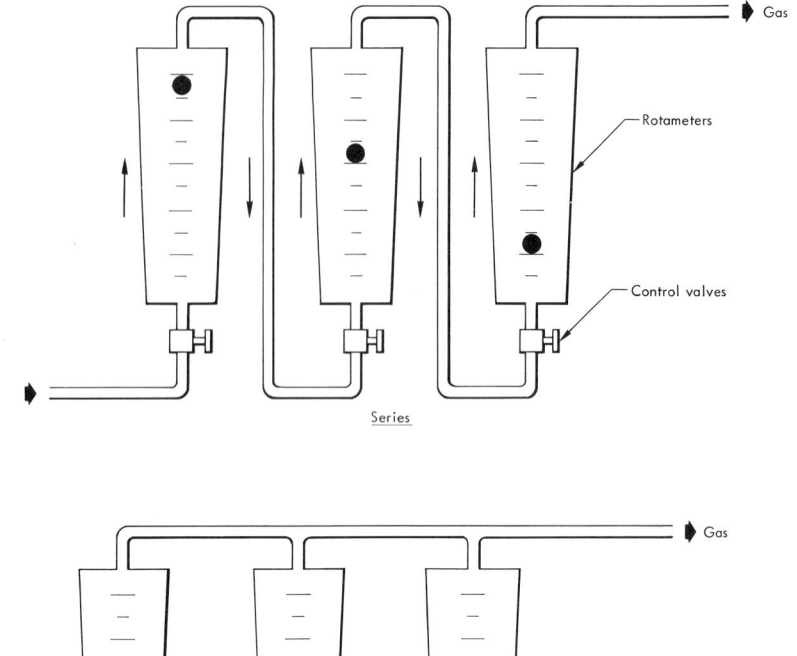

Figure 3.13 Rotameters arranged in series and parallel.

where W is the weight rate of flow, K is the rotameter coefficient (for a given design), D_f is the float diameter, W_f is the float weight, ρ_f is the density of the float, and ρ is the density of the fluid. The main function of this equation is to determine the capacity of a rotameter for a given fluid. It contains no scale reading or pressure correction capacity.

The complete equation for a rotameter is given also by Equation 3.11:

$$Q = C\sqrt{\frac{\pi g}{2}\left(\frac{D_t^2 - D_f^2}{D_f}\right)}\sqrt{\frac{V_f(\rho_f - \rho)}{\rho}} \tag{3.11}$$

Flow Rate and Volume Measurements

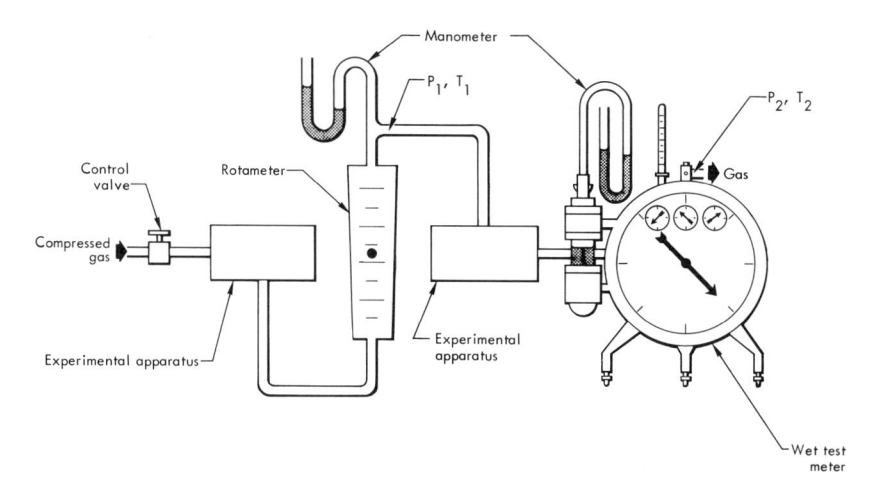

Figure 3.14 Setup for calibrating a rotameter against a wet test meter.

where Q is the volume flow rate, C is the coefficient of discharge, g is the gravitational acceleration, D_t is the tube diameter, and V_f is the volume of the float. C is a function of the Reynold's number and viscosity. This, however, is negligible at ordinary pressures and can be considered a constant for a given rotameter. The terms π, g, D_f, V_f, and ρ_f are also all constant. The term $(\rho_f - \rho)$ is essentially equal to ρ_f since ρ is so small in comparison. Thus we can take all of these constants and lump them into a new constant K_1. Equation 3.11 then becomes

$$Q = K_1 \left(D_t^2 - D_f^2 \right) \sqrt{\frac{1}{\rho}} \tag{3.12}$$

Since D_t varies with the scale reading and D_f is constant, then the term $(D_t^2 - D_f^2)$ is proportional to the flow area between the float and the tube, and therefore to the scale reading, which is designated as R. Substituting and solving for R, Equation 3.12 then becomes

$$R = K_2 Q \sqrt{\rho} \tag{3.13}$$

The equation is now in a form that relates the basic parameters of interest — the rotameter reading at a given flow and density condition. When relating two sets of conditions, the constant cancels and Equation 3.13 takes the form

$$\frac{R_1}{R_2} = \frac{Q_1 \sqrt{\rho_1}}{Q_2 \sqrt{\rho_2}}$$

$$\tag{3.14}$$

The next exercise will be to use Equation 3.14 to correct the basic rotameter for actual volumetric flow. In this condition Q_1 equals Q_2 and Equation 3.14 rearranges to

$$R_2 = R_1 \sqrt{\frac{\rho_2}{\rho_1}} \qquad (3.15)$$

Conditions will be simplified to assume that temperature is always constant and the density term is changed only by pressure. This pressure change can come from an internal apparatus configuration or from a change in altitude (see Appendix M). Assume that the rotameter has been calibrated and that the flow rate is read from a calibration curve showing flow as a function of float setpoint, R.

Example 3.3. Rotameter A is calibrated at sea level and the resulting data are shown in the left-hand column in Table 3.4. What would be the rotameter calibration curve correction factor at Denver (5,000 ft) and Mildred Lake (10,000 ft)? In other words, what would be the new settings to yield the same actual flow at altitude using the sea-level calibration?

The air densities in g/L are

1.205 at sea level
1.003 at 5,000 ft
0.829 at 10,000 ft

Using Equation 3.14, at 5,000 ft

$$R_2 = \sqrt{\frac{1.003}{1.205}} = 0.912$$

and at 10,000 ft

$$R_2 = \sqrt{\frac{0.829}{1.205}} = 0.829$$

The calculated set point correction factor is less than 1, indicating that the setting is lower at higher elevations to achieve the same flows. This is because the gas velocity is the same at any elevation. However, since the density is less at higher elevations, the force exerted on the float is less because the density decreases as the elevation increases. Table 3.4 (rotameter A) shows how the readings of a typical flow meter will vary at 5,000 and 10,000 ft if the sea level calibration is known.

Flow Rate and Volume Measurements

Table 3.4 Rotameter Readings at Various Altitudes for Actual Flow Rates

Actual Flow Rate (L/min)	Rotameter A Reading Sea Level	5,000 Feet	10,000 Feet	Actual Flow Rate (L/min)	Rotameter B Reading Sea Level	5,000 Feet	10,000 Feet
0.0	0.0	0.00	0.00	0.0	0.00	0.0	0.00
1.0	2.0	1.82	1.66	1.0	3.29	3.0	2.73
2.0	4.0	3.65	3.32	2.0	6.58	6.0	5.45
3.0	6.0	5.47	4.98	3.0	9.86	9.0	8.18
4.0	8.0	7.30	6.64	4.0	13.15	12.0	10.91
5.0	10.0	9.12	8.29	5.0	16.44	15.0	13.64

Air densities in g/L: 1.205 Sea level, 1.003 at 5,000 ft, 0.829 at 10,000 ft.

If the calibration is done at an altitude other than sea level, the correction process is similar.

Example 3.4. Rotameter B is calibrated at the 5,000 ft and the resulting data are shown in the right-hand columns in Table 3.4. What would be the rotameter calibration curve correction factor at sea level and 10,000 ft? In other words, what would be the new settings to yield the same actual flow at altitude using the 5,000 ft calibration?

The air densities in g/L are

$$1.205 \text{ at sea level}$$
$$1.003 \text{ at } 5{,}000 \text{ ft}$$
$$0.829 \text{ at } 10{,}000 \text{ ft}$$

Using Equation 3.14, at sea level

$$R_2 = \sqrt{\frac{1.205}{1.003}} = 1.096$$

and at 10,000 ft

$$R_2 = \sqrt{\frac{0.829}{1.003}} = 0.909$$

Note that the reading to attain a given flow is higher at sea level due to the denser air.

The two preceding examples pertain to changes in actual flow rate with changing conditions of density. How does one correct for the change in reading at different densities when the same weight (mass) flow is desired? Substituting W/ρ, where W is the weight rate of flow, for Q in Equation 3.13, R becomes

$$R = K_2 \frac{W}{\rho} \sqrt{\rho} = \frac{K_2 W}{\sqrt{\rho}} \tag{3.16}$$

Using the proportion

$$\frac{R_1}{R_2} = \frac{\dfrac{K_2 W_1}{\sqrt{\rho_1}}}{\dfrac{K_2 W_2}{\sqrt{\rho_2}}} \tag{3.17}$$

and setting W_1 to equal W_2, and cancelling the constants,

$$\frac{R_1}{R_2} = \sqrt{\frac{\rho_2}{\rho_1}} \tag{3.18}$$

becomes

$$R_2 = R_1 \sqrt{\frac{\rho_1}{\rho_2}} \tag{3.19}$$

Note that the density correction is the inverse of Equation 3.15. The following example uses Equation 3.19 to correct standard sea-level flow rates to flow rates at different altitudes. This is also shown for a hypothetical rotameter in Table 3.5

Example 3.5. A rotameter, calibrated at sea level, is to be used at 10,000 ft. The flow rate desired is 2.0 standard (sea level) L/min at 10,000 ft. What is the flow rate reading at 10,000 ft to achieve this condition?

The air densities in g/L are
1.205 at sea level
1.003 at 5,000 ft
0.829 at 10,000 ft

The given flow rate at sea level is 2.0 L/min

To obtain the setting at 10,000 ft corresponding to 2.0 standard L/min, use Equation 3.19:

$$R_2 = 2.0 \sqrt{\frac{1.205}{0.829}} = 2.41 \quad L/min$$

Flow Rate and Volume Measurements

Table 3.5 Rotameter Readings at Various Altitudes for Standard Flow Rates

Standard Flow Rate (L/min)	Rotameter reading		
	Sea Level	5,000 Feet	10,000 Feet
0.00	0.00	0.00	0.00
1.00	2.00	2.19	2.41
2.00	4.00	4.38	4.82
3.00	6.00	6.58	7.23
4.00	8.00	8.77	9.65
5.00	10.00	10.96	12.06

Air densities in g/L: 1.205 sea level, 1.003 at 5,000 ft, 0.829 at 10,000 ft.

The final example addresses the problem of what corrections are needed when a flow meter is calibrated at other than sea level yet standard sea-level mass flow conditions are needed when the device is used at yet another altitude.

Example 3.6. A rotameter, calibrated at 5,000-ft, is to be used at 10,000 ft. The flow rate desired is 2.0 standard (sea level) L/min at 10,000 ft. What is the flow rate reading to achieve this condition?

The air densities in g/L are

$$1.205 \text{ at sea level}$$
$$1.003 \text{ at } 5,000 \text{ ft}$$
$$0.829 \text{ at } 10,000 \text{ ft}$$

The given flow rate at 5,000 ft is 2.0 L/min

The sea-level flow, based on the 5,000-ft calibration, would be

$$Q_s = 2.0 \left(\frac{1.003}{1.205} \right) = 1.66 \text{ standard L / min}$$

To obtain the setting at 5,000 ft corresponding to 2.0 standard L/min, use Equation 3.19:

$$R_2 = 2.0 \sqrt{\frac{1.205}{1.003}} = 2.19 \text{ L / min}$$

Next, use Equation 3.19 to get 2.0 standard L/min at 10,000 ft from the 5,000-ft calibration.

$$R_2 = 2.19 \sqrt{\frac{1.205}{0.829}} = 2.64 \text{ L / min}$$

Another important question often arises. Can a flow meter calibration for one gas be applied to other gases? Many references say that this can be done by using Equation 3.14 and using the appropriate densities. However, it is the experience of the author that this equation fails in many instances. Chlorine and hydrogen chloride have densities of 3.2 and 1.6 g/L, respectively, yet their calibration curves for the 10 to 100 ml/min range are virtually identical. This discrepancy is probably due to other factors, such as viscosity.

As can be seen, the correction factor calculation can be somewhat confusing depending on the situation. Such corrections can be ignored when using the electronic mass flow meters discussed in the following section.

3.3.2 Mass Flow Meters

A large number of air velocity and flow meters depend on the rate of cooling or heat transfer. This transfer of heat depends on three factors: the amount of heat added to the gas, the number of molecules (mass flow) passing the heat source, and the heat capacity of each molecule.[49] This type of flow meter, unlike the variable-area volumetric type, has a visual output that is dependent only on the mass flow of gas. Therefore temperature or pressure (i.e., altitude) variations do not influence the amount of mass flowing through a given system.

A prime example is the mass flow meter shown in Figure 3.15. Here the gas divides into two streams. The bulk goes through a laminar-flow bypass, while a small portion is allowed to travel through the sensor section. The sensor section contains two resistance temperature detector coils around the sensor tube, which direct a constant amount of heat into the gas stream.[50] If no gas is flowing, the heat reaching each sensing coil is equal. However, in the flow condition heat is carried from the upstream sensor toward the downstream sensor, as shown in Figure 3.16.[51] The temperature difference is proportional to the flow of gas. This temperature difference is normally converted to a 0 to 5 V output signal and is viewed on a meter readout.

Quite often flow control, as well as knowledge of the gas flow, are required. Control is normally achieved by using an electromagnetic throttling solenoid or a thermal expansion valve. The desired flow is set using a potentiometer, which generates a command signal. This signal is compared with the signal generated by the sensor. Any difference between the signals causes the valve to open or close until the signals are comparable in strength. The valve will readjust automatically any time there is a change in the potentiometer setting, gas supply pressure, back pressure, or temperature.[49]

Although no correction factors are required to compensate for changes in gas density from temperature and pressure, there are corrections that must be made when using different gases. Quite often a reference gas will be used for calibration instead of the toxic, flammable, or corrosive gas for which the meter is intended. In addition, a flow meter calibrated for one gas will quite often be used to measure

Flow Rate and Volume Measurements

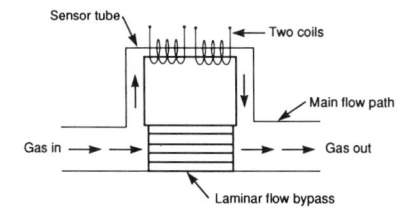

Figure 3.15 Cross section of a typical mass flow meter.

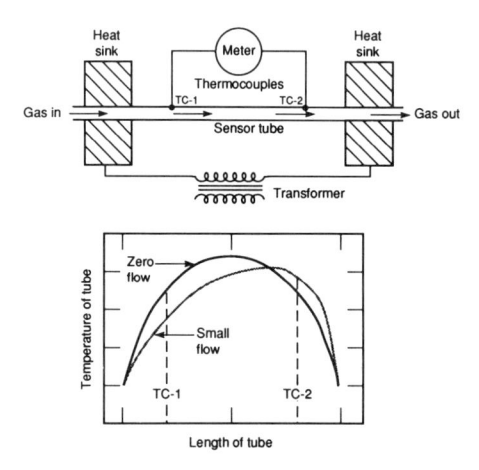

Figure 3.16 Mass flow meter schematic showing the temperature distribution under static (zero flow) and flowing conditions.

the flow of several gases. In this case, the correction factors shown in Appendix K must be applied. These gas conversion factors are dimensionless ratios relating the sensitivity of a mass flow meter to two different gases.[52] The conversion factor for nitrogen is usually set to unity, and all other gases are referenced to nitrogen. This is stated mathematically by

$$\frac{Q_1}{Q_2} = \frac{K_1}{K_2} \qquad (3.20)$$

where Q is the volumetric flow rate at standard conditions (usually 0°C and 1 atm) and K is the conversion factor. The subscript 1 is the unknown gas and the subscript 2 is the reference gas. The conversion factor is derived from the first law of thermodynamics. Given

$$Q_m = \frac{NH}{C_p \Delta T} \qquad (3.21)$$

where Q_m is the mass flow rate (g/min), N is the correction factor for the molecular structure (Table 3.6), H is the constant amount of heat applied to the sensor tube (cal), C_p is the specific heat (cal/g), and ΔT is the temperature difference between the downstream and upstream sensor coils.

The mass flow rate can also be written as

$$Q_m = \rho Q \qquad (3.22)$$

where ρ is the gas density at standard conditions (Appendix E) (g/L). The temperature difference, ΔT, is proportional to

$$\Delta T = aE \qquad (3.23)$$

where a is a constant and E is the output voltage. If Equations 3.22 and 3.23 are combined and inserted into Equation 3.21, Q then becomes

$$Q = \frac{bN}{\rho C_p} \qquad (3.24)$$

where b = H/aE at a constant output voltage. For our work, we want the ratio of the flow rate, Q_1, for an actual gas to the flow rate of a reference gas, Q_2, which will produce the same output voltage in a given mass flow meter. This is done by combining Equations 3.20 and 3.24, and canceling out b. This yields

$$\frac{Q_1}{Q_2} = \frac{K_1}{K_2} = \frac{(N_1 \rho_1 C_{p1})}{(N_2 \rho_2 C_{p2})} \qquad (3.25)$$

which is the fundamental relationship used in Appendix K.[50] K factors can be calculated directly from Equation 3.25 using Table 3.6, gas densities, and the heat capacity. However, the values given in Appendix K may be somewhat different, since they are quite often derived from experiment.

The next four examples illustrate how conversion factors are used. Example 3.10 is the most common, since it yields the exact flow setting when the actual gas used is different than the calibration gas.

Flow Rate and Volume Measurements

Table 3.6 Values for Molecular Correction Factor

Gas	Value of N
Monatomic, i.e., argon, helium, xenon	1.01
Diatomic, i.e., nitrogen, oxygen, nitric oxide	1.00
Triatomic, i.e., carbon dioxide, nitrous oxide	0.94
Polyatomic, i.e., ammonia, arsine, diborane	0.88

Example 3.7. A mass flow meter is calibrated with nitrogen and the flow rate is 1000 standard mL/min for a 5.00 VDC full-scale output signal. What would be the flow rate at full scale if carbon dioxide were used?

From Appendix K,

$$K(N_2) = 1.00$$
$$K(CO_2) = 0.74$$

N_2 flow at full scale $Q(N_2) = 1000$ mL/min

Using Equation 3.25:

$$Q(CO_2) = \left(\frac{0.74}{1.00}\right)1000 = 740 \text{ mL/min at standard conditions}$$

Example 3.8. A mass flow meter is calibrated with argon and the flow rate is 100 standard mL/min for a 5.00 VDC full-scale output signal. What would be the flow rate at full scale if nitrous oxide were used?

From Appendix K,

$$K(Ar) = 1.42$$
$$K(N_2O) = 0.71$$

Air flow at full scale $Q(Ar) = 100$ mL/min

Using Equation 3.25:

$$Q(N_2O) = \left(\frac{0.71}{1.42}\right)100 = 50 \text{ mL/min at standard conditions}$$

Example 3.9. A mass flow meter is to be calibrated with ammonia at 100 mL/min full-scale flow. The preferred reference gas, nitrous oxide, will be used. What flow of nitrous oxide must be generated to do the calibration?

From Appendix K,
$$K(NH_3)/K(NO_2) = 1.03$$

NH_3 flow at full scale $Q(NH_3) = 100$ mL/min

Using Equation 3.25:

$$Q(N_2O) = \frac{100}{1.03} = 97.1 \text{ mL/min at standard conditions}$$

Example 3.10. A mass flow controller has been calibrated with air at 200 mL/min full-scale flow. The controller is to be used with sulfur dioxide. What percent of the full-scale flow must be set to achieve 120 mL/min sulfur dioxide?

From Appendix K,
$$K(SO_2) = 0.69$$
$$K(air) = 1.00$$

Air flow at full scale $Q(air) = 200$ mL/min
Desired sulfur dioxide flow = 120 mL/min
Using equation 3.25:

$$Q(SO_2) = \left(\frac{0.69}{1.00}\right)200 = 138 \text{ mL/min at standard conditions}$$

The percent of full scale that must be set on the controller is

$$\% \text{ full scale} = \left(\frac{120}{138}\right)100 = 87.0\%$$

3.3.3 Orifice Meters

One of the oldest devices for flow rate measurement is the orifice meter. It is presently probably more of historical significance due to the current popularity of variable-area and mass flow meters. Orifice meters, however, have been successfully used to meter moving gas streams at flow rates from a maximum of 50 L/min to several milliliters per minute if special techniques are employed.

A typical orifice meter is shown in Figure 3.17. A restrictive orifice, usually a capillary, is placed in the moving gas stream. The pressure differential created

Flow Rate and Volume Measurements

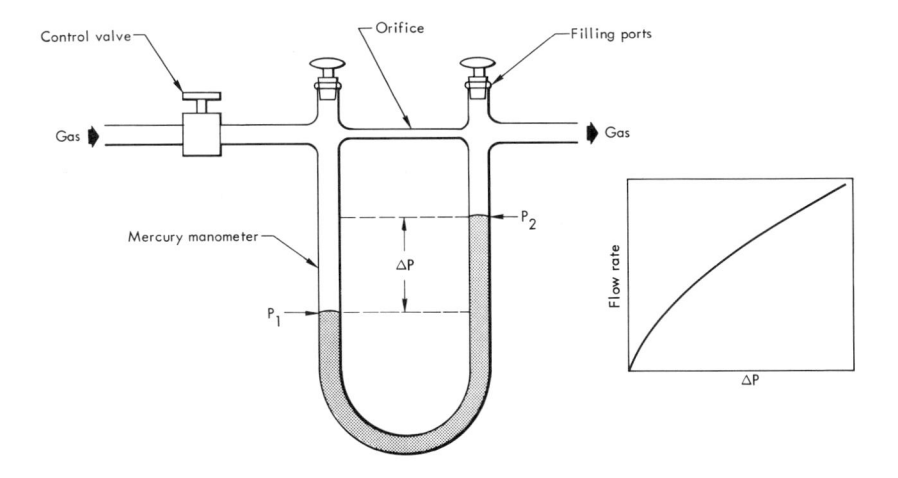

Figure 3.17 Sketch of a typical orifice meter and a calibration curve correlating the meter reading with the flow rate.

by the movement of gas through the orifice is related almost directly to the flow rate over a specified pressure range. The pressure can be measured with any of the normal devices available; U-tubes containing mercury, water, or light oil, and inclined manometers are popular devices. The flow rate pressure relationship is usually determined experimentally and the calibration is displayed graphically, as shown in Figure 3.17.

Flow rates can also be estimated from Equation 3.26. According to Poiseuille's law, the flow of gas through a tube, neglecting the effects at the entrance and exit of the capillary, is [53,54]

$$Q = \frac{(P_1 - P_2)\left(1 + \frac{P_1 - P_2}{2P_2}\right)}{r} \tag{3.26}$$

where

$$r = \frac{128 \, \eta L}{\pi d^4} \tag{3.27}$$

Here, Q is the flow rate (cm^3/sec), P_1 is the upstream pressure (dyne/cm^2), P_2 is the downstream pressure (dyne/cm^2), r is the capillary flow resistance (g/cm^4-sec), L is the capillary length (cm), d is the internal capillary diameter (cm), and η is the viscosity of the gas (poise or g/cm-sec). The viscosities at ambient temperatures of the most common gases are listed in Appendix Q.

Example 3.11. Calculate the flow rate of nitric oxide in liters per minute through a glass capillary tube. The upstream and downstream pressures are 20.0 and 14.7 psi, respectively. The capillary is 2.54 cm long and has a 0.1-mm inside diameter. The room temperature is 20°C.

Pressure, upstream	20.0 psi or 1.38E + 06 dynes/cm^2
Pressure, downstream	14.7 psi or 1.01E + 06 dynes/cm^2
Viscosity	1.86E – 04 g/cm-sec
Tube length	2.54 cm
Tube i.d.	0.01 cm
Dynes/cm^2/psi	6.90E + 04
cm^3/L	1000
sec/min	60

From Equation 3.27, the capillary flow resistance is

$$r = \frac{(128)(1.86E-04)(2.54)}{(3.14)(0.01)^4} = 1.93E+06 \ g/cm^4-sec$$

From Equation 3.26, the flow is

$$Q = \frac{(1.38E+06 - 1.01E+06)\left(1 + \frac{(1.38E+06)-(1.01E+06)}{2(1.01E+06)}\right)}{1.93E+06}$$

$$Q = 0.224 \ mL/sec \ or \ 0.0134 \ L/min$$

Orifice meters are available to everyone with even the most rudimentary laboratory equipment. They can be constructed and easily calibrated against a wet test meter or soap bubble meter. Care must be exercised not to plug up the orifice with particles or to exceed the maximum pressure and force the manometer indicating material out of the U-tube. These meters are useful but have the disadvantage of suffering pressure losses of up to 90%.[55]

3.3.4 Critical Orifices

Critical orifices have been used extensively for maintaining and controlling gas streams at predetermined flow rates.[56-60] Sections of glass tubing, capillary tubing,[61,66] hypodermic needles,[59,62-64] watch jewels,[65] and plastic pipet tips[67] have all been used to achieve a remarkably uniform flow. The driving force for the flow device is usually atmospheric pressure or a constant pressure source upstream and a vacuum source downstream of the orifice.

An orifice is shown in Figure 3.18 and is said to be critical when the ratio of the downstream absolute pressure, P_2, to the upstream absolute pressure, P_1, produces a sonic velocity at the orifice gas exit. For air, the critical ratio, R_{cr}, is normally considered to be

Flow Rate and Volume Measurements

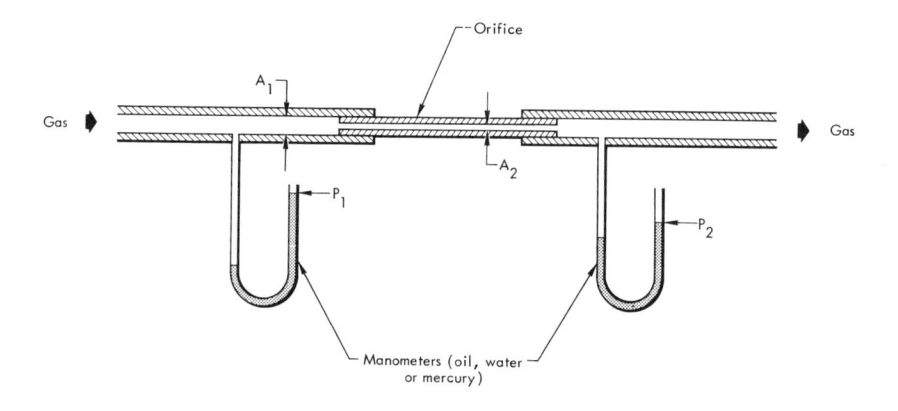

Figure 3.18 Sketch of a typical critical orifice meter using a pressurized gas source. Critical flow is achieved when the pressure drop is sufficient to sustain sonic velocity.

$$R_{cr} = \frac{P_2}{P_1} \leq 0.53 \qquad (3.28)$$

if the ratio of A_1/A_2 is greater than 25.[55,68,69] The 0.53, however, is only an approximation, and it can vary from 0.2 to 0.8 depending on the orifice configuration.[67] Table 3.7 lists several different orifice designs that increase R_{cr} to as high as 0.87. Even higher values (0.92) can be achieved by radiusing the tip of the pipette into a bell-shaped entry. The implications of increasing the R_{cr} value mean that the energy requirements to achieve sonic velocity are greatly reduced.

The calculations for the maximum flow rate, Q_{max}, are complex and are obtained from William et al.[70] and Anderson and Friedman.[71] For a perfect gas through rounded orifices and nozzles,

$$Q_{max} = C_D A_2 P_1 \sqrt{\frac{gK_H M}{RT}\left(\frac{2}{K_H+1}\right)^{\frac{(K_H+1)}{(K_H-1)}}} \qquad (3.29)$$

where Q_{max} is the mass flow rate (g/sec), A_1 is the orifice area (cm^2), P_1 is the absolute upstream presssure (dynes/cm^2), K_H is the ratio of the specific heat at a constant pressure to the specific heat at a constant volume, M is the molecular

Table 3.7 Critical Ratio Values for Different Orifice Configurations

Orifice Type	Throat dia. (mm)	Critical Ratio
Blunt entry, elongated parallel sides	0.8	0.44
Square edged	0.8	0.44
Sharp edged	0.5	0.46
Converging	0.95	0.47
Converging to elongated parallel sides	1.0	0.47
Short converging, included angle 0.7°	1.0	0.56
Short converging, included angle 16.6°	1.1	0.60
Short converging, included angle 7.8°	1.2	0.76
Short converging, included angle 4.4°	1.0	0.85
Short converging, included angle 2.5°	1.0	0.85
Short converging, included angle 4.4°	2.4	0.87

Data taken from Zimmerman, N. J. and P. C. Reist. *Am. Ind. Hyg. Assoc. J.* 45:340 (1984), with permission.

weight (g/mol), T is the ambient temperature (°K), R is the gas constant (8.31 × 10^{-7} erg/mol-°K), and C_D is the discharge coefficient (usually 0.8 to 0.95). For air, Equation 3.29 can be reduced to the more workable expression,

$$Q_{max} = \frac{0.388 \, C_D A_2 P_1}{\sqrt{T}} \quad (3.30)$$

where P_1 is the absolute pressure (g/cm^2).

Example 3.12. At 1 atm, critical flow can be induced in a 2.6-mm diameter orifice using a vacuum source at 640 mm absolute pressure. Calculate the maximum flow at 25°C assuming a coefficient of discharge of 1.

Coefficient of discharge	1
Throat diameter	2.6 mm
Throat area	0.0531 cm²
Temperature	25°C or 298°K
Pressure, upstream	1 atm or 1034 g/cm²
Air density at 25°C	1.18 g/L
sec/min	60

From Equation 3.30, the maximum flow is

$$Q_{max} = \frac{0.388(1)(0.0531)(1034)}{\sqrt{298}} = 1.23 \text{ g/sec}$$

$$Q_{max} = 62.7 \text{ L/min}$$

Example 3.13. For a 1-mm² orifice the critical ratio is 0.75. If the upstream and downstream pressures are 1.00 and 0.46 atm, respectively, does

Flow Rate and Volume Measurements

the gas flow reach sonic velocity? If so, what is the gas flow in g/sec if the discharge coefficient is 0.85 and the temperature is 100°C?

Coefficient of discharge	0.85
Throat area	0.0100 cm²
Temperature	100°C or 373°K
Pressure, upstream	1.00 atm or 1034 g/cm²
Pressure, downstream	0.46 atm or 476 g/cm²

$$R_{cr} = \frac{1.00}{0.46} = 0.460$$

This is less than 0.75. Therefore critical flow exits.
From Equation 3.30, the maximum flow is

$$Q_{max} = \frac{0.388(0.85)(0.01)(1034)}{\sqrt{373}} = 0.18 \text{ g/sec}$$

Although Equation 3.30 would lead one to believe that the length of the orifice has no effect on the existing gas flow rate, this is not the case in practice. Table 3.8 shows data on flow rate variance with length under the same laboratory conditions. The table shows that the longer the needle, the smaller the critical flow. The theoretical calculated value for Q_{max} is also shown for comparison.

Critical orifices have the advantages of being inexpensive, convenient, and easy to operate. They are ideal for sampling or mixing gas streams at a single constant rate for long periods if reasonable care is taken to protect the orifice from debris or wear.[72] By using high-efficiency particulate air, membrane, or glass-wool filters placed upstream from the orifice,[59] reasonable accuracies (±3% for hypodermic needles[63]) can be maintained.

The disadvantages are few but significant. Only one critical flow rate is possible from each orifice, and the pressure differential required to maintain it is relatively high. Each orifice must be individually calibrated for even reasonable accuracy, since the flow rate from calculations can differ from the experimental flow by as much as 300%.[63]

3.3.5 Porous Plugs

A system regulating gas flows using a packed plug of asbestos was first devised by Saltzman and Gilbert,[73] and it was later improved upon by Avera[74] and Hales.[75] The device was used to regulate flows to 0.01 mm/min. It consists of a porous, packed, or asbestos plug flow meter, which has no moving parts and can be built from basic materials in the laboratory. This flow meter is particularly useful for producing gas mixtures in the parts per million range without double dilution or the use of excessive volumes of gas for a single dilution. Ethylene oxide, chlorine, chlorine dioxide, ammonia, phosgene,[76] nitrogen dioxide (initially diluted to 0.5%),[77] nitric oxide,[73] and radioactive sulfur dioxide[79] are examples of gases that

Table 3.8 Critical Orifice Flow Rate as a Function of Needle Size

Needle Gauge	Inner Diameter in.	Inner Diameter mm	Calc Flow[b] L/min	Actual Flow at Specified Needle Size[a] 0.5 in. L/min	1.0 in. L/min	1.5 in. L/min	2.0 in. L/min	3.0 in. L/min
10	0.106	2.69	55.7					
11	0.094	2.39	43.9					
12	0.085	2.16	35.9					
13	0.071	1.80	24.9					
14	0.063	1.60	19.7					
15	0.054	1.37	14.4					
16	0.047	1.19	10.9					
17	0.042	1.07	8.81				5.89	
18	0.033	0.84	5.43		4.45	4.1	3.77	
19	0.027	0.69	3.66			2.43	2.37	2.15
20	0.024	0.60	2.77		1.97	1.83	1.75	
21	0.020	0.51	2.00		1.47	1.32	1.23	
22	0.016	0.41	1.29		0.89	0.8	0.68	0.59
23	0.013	0.34	0.889	0.63	0.54	0.49	0.42	
24	0.012	0.31	0.739	0.51	0.42	0.29		
25	0.010	0.26	0.520	0.34	0.26		0.19	
26	0.010	0.26	0.520					
27	0.008	0.21	0.339					
28	0.006	0.16	0.197					
29	0.007	0.18	0.249					
30	0.006	0.16	0.197					
31	0.005	0.13	0.130					
32	0.004	0.11	0.093					
33	0.004	0.11	0.093					

[a] Measured at 630 mmHg. Data taken from Lodge, J.P. et al., *J. Air Pollut. Control Assoc.* 16:197 (1969), with permission.
[b] Assume temperature is 25°C, the air density is 1.18 g/L, and the discharge coefficient is unity.

have been metered through such devices. Porous-plug flow meters provide a steady concentration for many hours of operation. They are compact, inexpensive, and rapid changes in concentration can be made. These devices were more popular in the 1970s, but with advances in producing low-concentration gas standards in cylinders, their use has greatly diminished.

A porous-plug system is shown in Figure 3.19. The gas of interest passes into the side arm of a 1 to 4 mm bore of a three-way capillary stopcock, where it is split into two gas streams. The bulk of the gas proceeds through a droplet trap and escapes into a graduated vessel containing some liquid. The balance of the stream proceeds through the packed arm of the stopcock, which contains medium-length, acid-washed asbestos fibers of the type used in Gouch crucibles.[77] The stopcock, which is always maintained in the same position, is filled by tamping the fibers with a blunt piece of wire of appropriated diameter from both sides of the T. The amount of tamping required will depend on the flow rates desired. A more tightly packed stopcock will yield a proportionally lower flow rate through the plug. The pack must be kept scrupulously free from grease or moisture, and it should be

Flow Rate and Volume Measurements

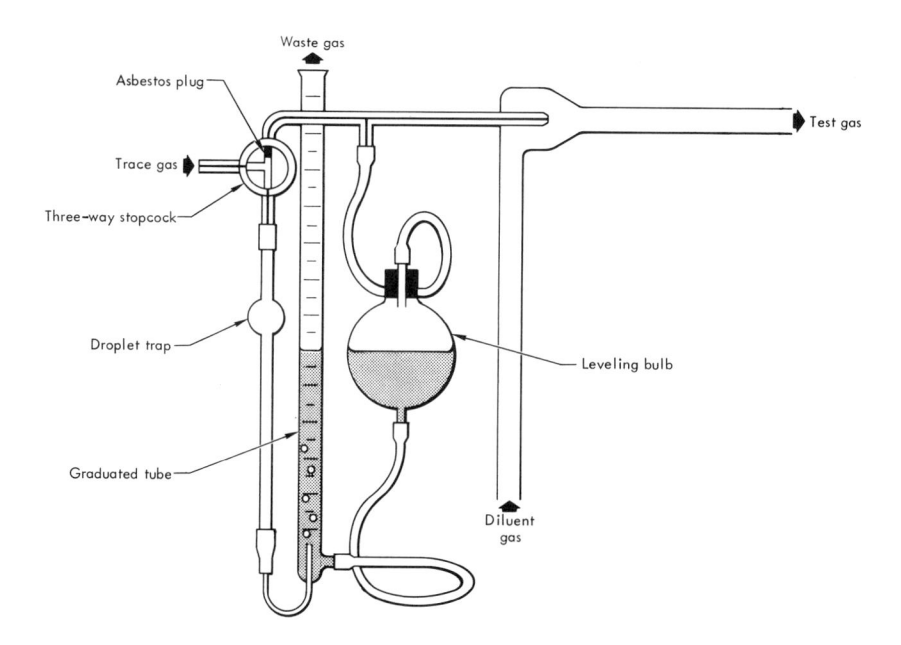

Figure 3.19 Sketch of a typical porous-plug flow meter. The flow is adjusted by changing the height of the leveling bulb.

dried and cleaned with trichloroethylene or chloroform before tamping. The stopcock plug should be lightly coated with fluorocarbon or fluorosilicon grease to prevent gas leaks, but excess lubricant must be kept out of the bore.

The flow rate is proportional to the pressure drop across the compacted plug, and the upstream pressure is maintained by discharging the excess gas into a liquid column of known height. By varying the liquid height between the surface and the point of gas admission, a linear calibration plot of height vs flow rate can be constructed.

Losses from the waste gas outlet can minimized if low or microflow needle valves are employed. If appreciable downstream pressures are encountered, a leveling bulb can be added to the system, as shown, to automatically compensate for either positive of negative output pressures.[74] By raising or lowering the bulb, the flow rate can be altered without adding water to the system. The leveling bulb is connected between the outlet and the graduated vessel with flexible tubing, and it can be raised and lowered freely. The area of the bulb should be 80 to 100 times the area of the gas escape vessel so that it will change only a small amount compared to the receiving vessel if downstream pressure changes occur.[74]

The outlet bore of the flow meter should be as small as possible, and capillary stopcocks are preferred. The small bore reduces the time to reach equilibrium and is especially important at low flow rates. It is also best to employ a large-diameter waste tube so that the pressure drop is negligible, even at the higher waste escape flow rates.[76] A droplet trap is provided in case the gas supply is disconnected too rapidly, for the inertia of the waste liquid could rise and wet the plug, rendering it useless.

Calibration is accomplished by timing the rate of movement of water or oil drop in a graduated volumetric pipet attached to the delivery end of the capillary stopcock.[79]

3.3.6 Controlled Leaks

Any material that offers resistance to gas flow can be used to control the flow rate. Just as tamped asbestos or ceramic frits,[80] with their convolutions and tortuous channels, inhibit the progress of gaseous material, so also does a wide variety of other materials. For example, a flattened copper tube containing a stainless steel wire will allow a small but perceptible and reproducible amount of gas to flow.[81]

The most widely used and commercially available restrictors are constructed from porous metal. A powdered metal alloy is compressed and cold-welded in an appropriate coupling. The particle size of the powder determines the pore openings and hence the flow rate. Gas flows when a pressure differential exists between the inlet and outlet. This occurs either from an inlet pressure or an outlet vacuum.[82,83]

Controlled leaks and restrictors can perform the triple functions of gas flow regulation, filtration, and surge dampening. Small flows on the order of 10^{-3} to 10^{-8} mL/sec at specified pressure differentials are said to be controlled leaks, while those from 10^{-3} to 10^{-2} are said to be flow restrictors.[84] They are usually constructed from stainless steel, but others are made of gold, silver, Monel, nickel, Inconel, and various other alloys. Several shapes and sizes can be obtained, and a few are shown in Figure 3.20.

Standard flow tolerances as low as ±1% are also available, but uniform permeability is usually ±5 to 10%. The flow rate, of course, varies with the porosity, and a typical performance chart for a 1/16-in. element is shown in Figure 3.21.

3.3.7 Miscellaneous Devices

There are a number of techniques that can measure flow rate that have not been previously mentioned in this chapter. They include additional thermal transfer, mechanical, and electromechanical methods. Their flow rates cover almost the entire spectrum from less than 0.1 mL/min to over 1000 ft^3/min. Such techniques involve heated wire, rotating and deflecting vane, drag body, turbine, Venturi tube, vortex shedding, acoustic, ionization, and magnetically activated flow meters to name a few. For additional information, refer to Owen and Pankhurst,[7] Arya and Plate,[85] and several other journal summaries.[86-89]

Flow Rate and Volume Measurements 63

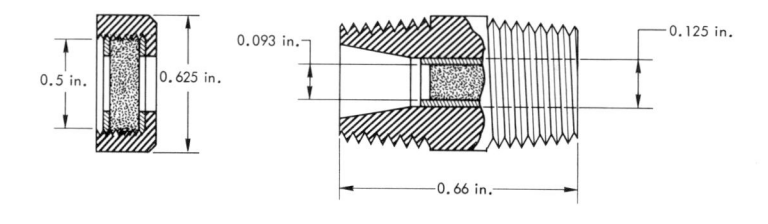

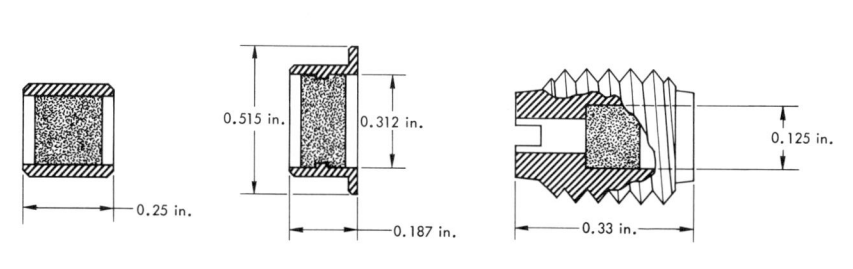

Figure 3.20 Cross sections of five controlled leaks and restrictors. Data taken from Mott Metallurgical Corporation Brochure, Farmington, CT, with permission.

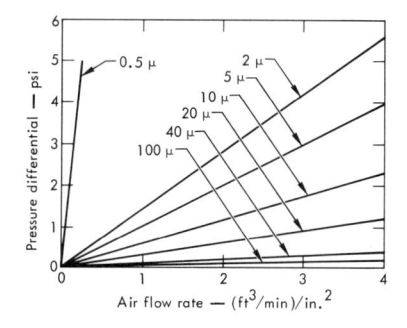

Figure 3.21 Pressure differential vs the flow rate of air for seven restrictor porosities. Data taken from Mott Metallurgical Corporation Brochure, Farmington, CT, with permission.

REFERENCES

1. Catalog No. P-900, Warren E. Collins, Inc., Braintree, MA.
2. Bulletin No. 207.1, American Meter Controls, Philadelphia, PA.
3. Catalog AG-1, American Meter Controls, Philadelphia, PA.
4. Catalog LPG-4A, American Meter Controls, Philadelphia, PA.
5. Bulletin AIM-207, American Meter Controls, Philadelphia, PA.
6. Powell, C. H. and A. D. Hosey, Eds. *The Industrial Environment—Its Evaluation and Control* (Washington, D.C.: U.S. Government Printing Office, 1965).
7. Owen, E. and R. C. Pankhurst. *The Measurement of Air Flow*, 4th ed. (Elmsford, NY: Pergamon Publishing Co., 1966).
8. *Industrial Ventilation*, 10th ed. (Ann Arbor, MI: Edwards Brothers, Inc., 1968).
9. Brandt, A. D. *Industrial Health Engineering* (New York: John Wiley & Sons, Inc., 1947).
10. Barr, G. J. *J. Sci. Inst.* 11:321 (1934).
11. 1969 Catalog, Emerson Electric Company, Hatfield, PA.
12. Kusnetz, H. *Am. Ind. Hyg. Assoc. J.* 21:340 (1960).
13. Levy, A. *J. Sci. Instr.* 41:449 (1964).
14. Noble, F. W., K. Abel, and P. W. Cook. *Anal. Chem.* 37:1631 (1965).
15. "Instrument Calibrates Low Gas-Rate Flowmeters," Technical Brief No. 65-10137, Houston, TX: National Aeronautics and Space Administration (1965).
16. Hunter, J. J. *J. Sci. Instr.* 42:175 (1965).
17. Frisone, G. J. *Chemist-Analyst* 54:56 (1965).
18. Czubryt, J. J. and H. D. Gesser. *J. Gas Chromatog.* 6:528 (1968).
19. Van Swaay, M. V. *J. Chromatog.* 12:99 (1963).
20. Bulletin Cal 4/84, MKS Instruments, Burlington, ME.
21. Brief, R. S. and R. G. Confer. *Am. Ind. Hyg. Assoc. J.* 30:576 (1969).
22. *Standard Methods for Measurement of Gaseous Fuel Samples*, publication No. ASTM-D-1071-55, (Philadelphia, PA: American Society for Testing and Materials, 1963).
23. Bulletin No. TS-63110-3, Precision Scientific Company, Chicago, IL.
24. Data Sheet No. LM.1, Parkinson-Cowan Measurement, Stretford, Manchester, England.
25. Catalog No. EG-40, American Meter Controls, Philadelphia, PA.
26. *Instr. Control Systems* 27:101 (1970).
27. *Anal. Chem.* 41:176LG (1969).
28. *Instr. Control Systems* 42:100 (1969).
29. *Instr. Control Systems* 42:115 (1969).
30. Fairchild, E. J. and H. E. Stokinger. *Am. Ind. Hyg. Assoc. J.* 19:171 (1958).
31. Nash, D. L. *Appl. Spectr.* 21:126 (1967).
32. Martin, J. J. *Chem. Eng. Progr.* 45:338 (1949).
33. Haenel, R. D. *Instr. Control Systems* 41:127 (1968).
34. Hougen, J. O. *Instruments* 26:1716 (1953).
35. Gilmont, R. and B. Roccanova. *Instr. Control Systems* 39:89 (1966).
36. Polentz, L. M. *Instr. Control Systems* 34:1048 (1961).
37. Coleman, M. C. *Trans. Inst. Chem. Engrs. (London)* 34:339 (1956).
38. Caplan, K. J. *Am. Ind. Hyg. Assoc. J.* 46:B-10 (1985).
39. *Variable Area Flowmeter Handbook*, Vols. I, II, (Warminster, PA: Fischer and Porter Co., 1970, 1977).
40. Karusek, F. W. *Res. Develop.*, 27:40 (1976).

41. Weber, H. L. *ISA,* ISBN 87664-362-4 (1977).
42. Gilmont, R. and B. Roccanova. *Instr. Control Systems* 39:35 (1966).
43. Veillon, C. and J. Y. Park. *Anal. Chem.* 42:684 (1970).
44. *Variable Area Flowmeter Handbook,* Vol II (Warminster, PA:Fischer and Porter Co., Catalog 10a1022, 1968).
45. Specification Guide 6, *Inst.Control Systems* 49:63 (1976).
46. Okladek, J. *Am. Laboratory* 20:84 (1988).
47. Christian, A. *Anal. Chem.* 45:698 (1973).
48. "Electronic Rotameter", Product Bulletin, Quantachrome Corp., Greenvale, NY.
49. Okladek, J. *Am. Laboratory* 20:92 (1988).
50. Instruction Manual, Mass Flow Meters and Controllers, 830/840, Sierra Instruments, Carmel Valley, CA (1987).
51. Benson, J. M., W. C. Baker, and E. Easter. *Instr. Control Systems,* 43:85 (1970).
52. Instruction Manual, Mass Flowmeters and Controllers, FC-260, 261, 262, FM-360, 361, 362, P/N 900457-001, Tylan Corporation, Torrance, CA (Sept 1989).
53. Hersch, P. A. *J. Air Pollut. Control Assoc.* 19:164 (1969).
54. Jentzsch, D. *J. Gas Chromatogr.* 5:226 (1967).
55. Powell, C. H. and A. D. Hosey, Eds. "The Industrial Environment—Its Valuation and Control" (Washington, D.C.: U.S. Government Printing Office, 1965).
56. Hatch, T., H. Warren, and P. Drinker. *J. Ind. Hyg. Tox.* 14:301 (1932).
57. Page, R. T. *Ind. Eng. Chem.* 7:355 (1935).
58. Collins, W. T. "A Gravimetric Standard for Primary Gas Flow Measurements," Report No. K-L-6181, Oak Ridge National Laboratory, Oak Ridge, TN, (1967).
59. Lodge, J. P., J. B. Pate, B. E. Ammons, and G. A. Swanson. *J. Air Pollut. Control Assoc.* 16:197 (1966).
60. Pinkerton, M. K., J. M. Lauer, P. Diamond, and A. A. Tamas. *Am. Ind. Hyg. Assoc. J.* 24:239 (1963).
61. Malkova, E. M. and Z. T. Kezina. *Meas. Tech.* (USSR) (English transl.) 921 (1961).
62. Axelrod, H. D., J. H. Carey, J. E. Bonelli, and J. P. Lodge. *Anal. Chem.* 41:1856 (1969).
63. Corn, M. and W. Bell. *Am. Ind. Hyg. Assoc. J.* 24:502 (1963).
64. Imada, M. R. and E. Jeung. "Integrated Gas Sampling Techniques Utilizing Critical Orifices," paper presented at the 8th Conference on Methods in Air Pollution and Industrial Hygiene Studies, Kaiser Center, Oakland, CA, February 6–8, 1967.
65. Brenchley, D. L. *J. Air Pollut. Control Assoc.* 22:967 (1972).
66. Huygen, C. *J. Air Pollut Control. Assoc.* 20:675 (1970).
67. Zimmerman, N. J. and P. C. Reist. *Am. Ind. Hyg. Assoc. J.* 45:340 (1984).
68. Perry, J. H. *Chemical Engineers Handbook* (New York: McGraw-Hill Book Company, Inc., 1950).
69. Wartburg, A. F., J. B. Pate, and J. P. Lodge. *Environ. Sci. Technol.* 3:767 (1969).
70. William, F. W., J. P. Stone, and H. G. Eaton. *Anal. Chem.* 48:442 (1976).
71. Anderson, J. W. and R. Friedman. *Rev. Sci. Instrum.* 20:61 (1949).
72. DiNardi, S. R. and C. L. Sacco. *J. Air Pollut. Control Assoc.* 28:603 (1978).
73. Saltzman, B. E. and N. Gilbert. *Am. Ind. Hyg. Assoc. J.* 20:379 (1959).
74. Avera, C. B. *Rev. Sci. Instr.* 32:985 (1961).
75. Hales, J. W., D. A. Snyder, and J. L. York. *Am. Ind. Hyg. Assoc. J.* 32:299 (1971).
76. Saltzman, B. E. *Anal. Chem.* 33:1100 (1961).
77. Saltzman, B. E. and A. F. Wartburg. *Anal. Chem.* 37:1261 (1965).
78. Bostrom, C. E. *Int. J. Air Water Pollut.* 9:333 (1965).

79. Kusnetz, H. L., B. E. Saltzman, and M. E. Lanier. *Am. Ind. Hyg. Assoc. J.* 21:361 (1960).
80. Axelrod, H. D., R. J. Teck, and J. P. Lodge. *Anal. Chem.* 43:496 (1971).
81. Hersch, P. A. *J. Air Pollut. Control Assoc.* 19:164 (1969).
82. Paty, L. *Vacuum* 7–8:80 (1957–58).
83. Guthrie, D. G. "A Gas Blender for Minor Components at Concentrations Between 10 and 1000 v.p.m.," Report No. 270(D), United Kingdom Atomic Energy Authority, London (1962).
84. Brochure, Mott Metallurgical Corporation, Farmington, CT.
85. Arya, S. P. S. and E. J. Plate. *Instr. Control Systems* 42:87 (1969).
86. Flow Measurement Reference Book, Instr. Control Systems, 69-05.
87. Bailey, S. J. "Flow Measurement '74," *Control Eng.* 5:35 (1974).
88. McShane, J. L. and F. G. Geil. "Measuring Flow," Res. Develop. 30: (1975).
89. Baker, W. C. and J. F. Pouchot. "The Measurement of Gas Flow, Part 1," *J. Air Pollut. Control Assoc.* 33:66 (1983).
90. Wasserman, R. and H. Grant. *Instrum. Control Systems* 46:59 (1973).

CHAPTER 4

STATIC SYSTEMS FOR PRODUCING GAS MIXTURES

The static or batch system is commonly used for producing standard gas and vapor mixtures. This involves the introduction of a known weight or volume of contaminant into a container of fixed dimensions.[1-3] Nonrigid containers holding metered volumes of gases are also used. Static devices may be used at any desirable pressure, but laboratory applications are generally limited to systems at atmospheric pressure or slightly below. Static systems are usually employed when comparatively small volumes are required, and they are used extensively for instrument calibration and the production of gas phase standards for gas chromatography, mass spectrometry, and infrared spectrophotometry.

4.1 SYSTEMS AT ATMOSPHERIC PRESSURE

4.1.1 Single Rigid Chambers

One of the most convenient methods of making a gas standard is based on admitting a predetermined amount of solvent or gas into a single rigid vessel of known dimensions, as shown in Figure 4.1. The contaminant is vaporized if necessary, mixed with the diluting gas, and sampled. Another inlet provides for replacement gas in order to keep the system at atmospheric pressure. A wide variety of chamber sizes has been used — from small syringes,[4,5] plastic containers,[6-10] flasks,[11] and glass bottles,[12-14] 20-m gas cells,[15] to large chambers in the cubic meter range[16-18] and even large rooms.[19,20] The usual size is on the order of 20 to 40 L.[6,21-23] This allows for the removal of enough useful gas without causing excessive dilution by the replacement gas.

The materials chosen for construction should be such that the walls are smooth and do not cause excessive adsorption or reaction. Glass bottles are generally used, but phenolic,[7] epoxy,[18] Teflon, stainless steel,[17,24,25] and polyvinyl chloride[26] materials have been used.

The volume of the system must be easily determined, no matter what the size,

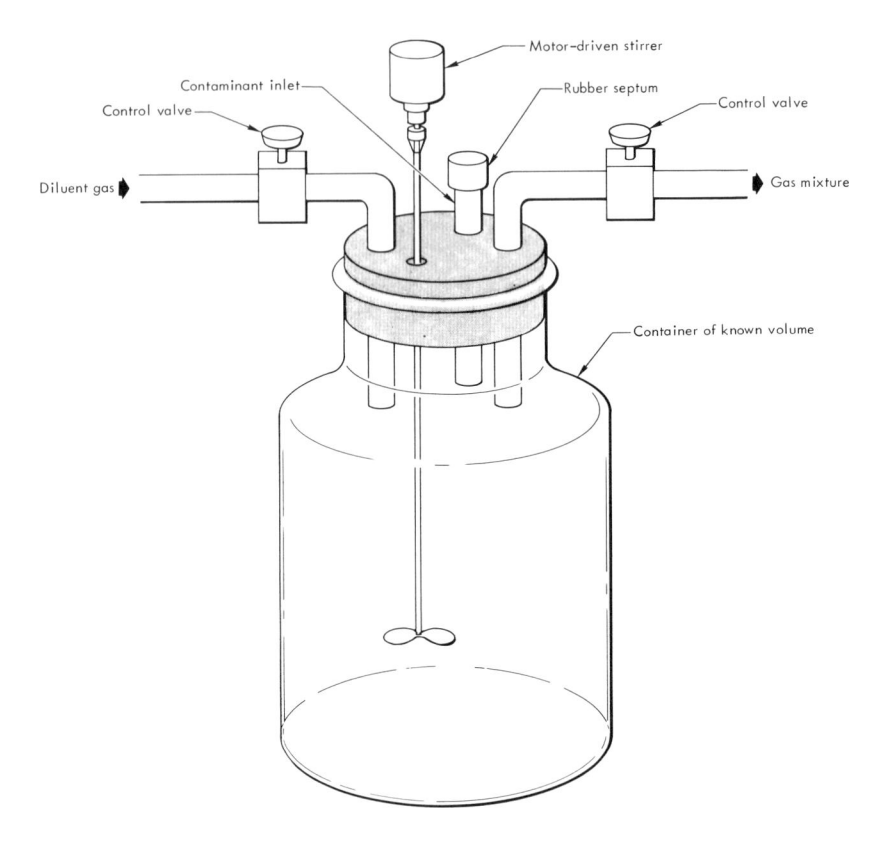

Figure 4.1 Sketch of a rigid static system for producing gas and vapor mixtures.

shape, or material chosen for construction. The volume is found by measuring the chamber boundaries directly or by filling the chamber with water and determining the volume either volumetrically or gravimetrically. An alternative method is to evacuate the vessel and meter the diluent gas into it through a wet test meter.

4.1.1.1 Sample Introduction

Introduction of the contaminant gas or liquid into the system can be accomplished in a number of ways. Some current methods are summarized in Table 4.1 and are shown in Figure 4.2. One of the best devices is a glass syringe. Syringes for liquids in the microliter range are reported to have accuracies and reproducibilities of ±1%. Gas-tight syringes fitted with Teflon plungers in the 0.01- to 50-ml volume range may be used for both gases and liquids. Huge "super syringes" (up to 2.0 L capacity) and Teflon syringes for reactive materials are also commercially available. Ground-glass syringes should be avoided because of leakage, especially when dispensing small quantities of liquids.

If no pipets or syringes are available, precision-bore tubing can be used. The pollutant gas is allowed to pass through the tubing and the ends are sealed with Scotch® tape.[27] Another approach is to place sealed ampoules containing weighed

Static Systems for Producing Gas Mixtures

Table 4.1 Characteristics of Solvent and Gas Injectors

Injector	Phase Injected	Plunger Material	Capacity	Accuracy at Capacity, %	Source
Microliter syringe	Liquid	Wire	1–5 µL	2	Hamilton
Microliter syringe	Liquid	Wire	5–500 µL	1	Hamilton
Gas-tight syringe	Liquid, gas	Teflon tip	0.05–2.5 mL	1	Hamilton
Gas-tight syringe	Liquid, gas	Teflon tip	5–50 mL	1	Hamilton.
Micropipet	Liquid	Air	2–4 µL	1.8	Am. Sci. Prod.
Micropipet	Liquid	Air	5–25 µL	0.5	Am. Sci. Prod.
Micropipet	Liquid	Air	35–150 µL	0.3	Am. Sci. Prod.
Microm buret	Liquid	Teflon	0.2–2 mL	0.5	Cole-Parmer
Microm buret	Liquid	Glass	0.25–2.5 mL	0.04	Cole-Parmer
Volumetric pipet	Liquid	Air	0.5–200 mL	1	

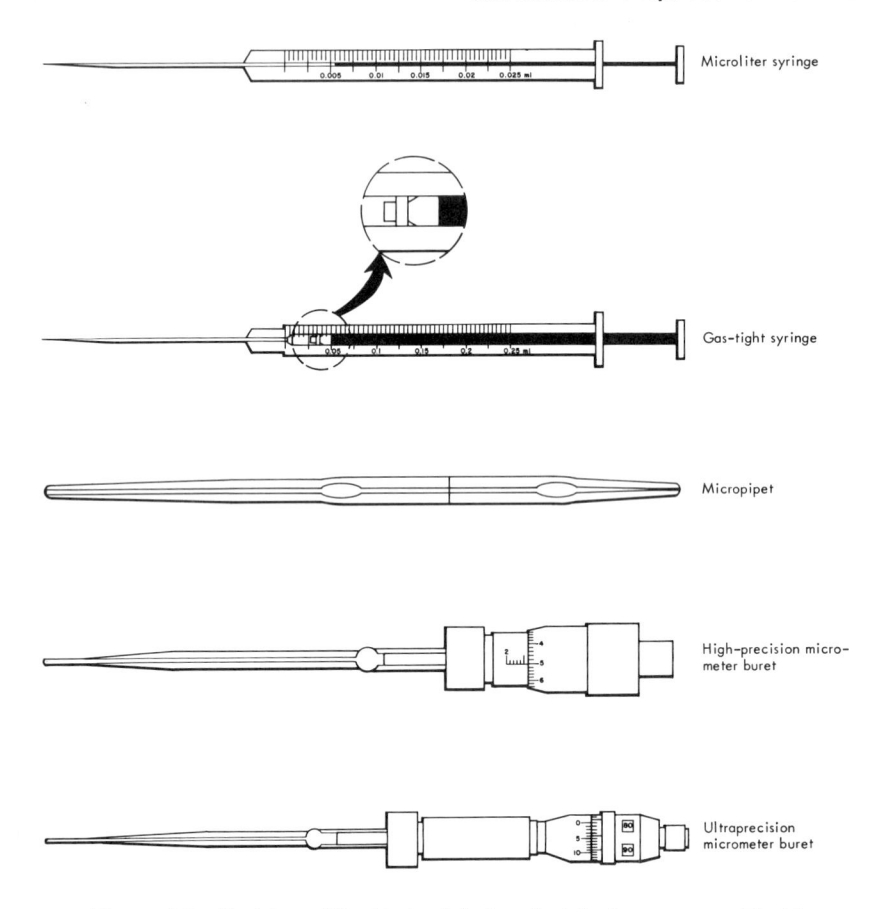

Figure 4.2 Sketches of five kinds of devices for injecting gases and liquids.

quantities of the test material.[1,28,29] The ampoule is then broken inside the container, and the contents escape and mix with the dilution air. Care should be taken to break the ampoule without destroying the main container.

There are several methods used to fill syringes with gases,[16,29,30] one of which is illustrated in Figure 4.3. Pure gas from a small lecture bottle flows slowly into a liquid reservoir through a section of rubber tubing. The valve is turned just enough to start the gas bubbling in the liquid. The syringe needle pierces the tubing, and after several flushes the syringe is filled and withdrawn. Small lecture bottles of pure gas (usually 99+%) are convenient for this purpose.[31,32] Care must be taken to make sure the gas does not react with or dissolve into the liquid in the reservoir. Ammonia and ethylene oxide, for example, will cause water to migrate up the Tygon tubing right into the lecture bottle. Never allow this to occur.

The contaminant gas or liquid is usually injected into the vessel through a soft material, which allows the syringe needle to penetrate and be withdrawn without significant leakage. Rubber stoppers, serum caps,[33] and various rubberized septa[22] have been used for this purpose. The syringe should be depressed smoothly and

Static Systems for Producing Gas Mixtures

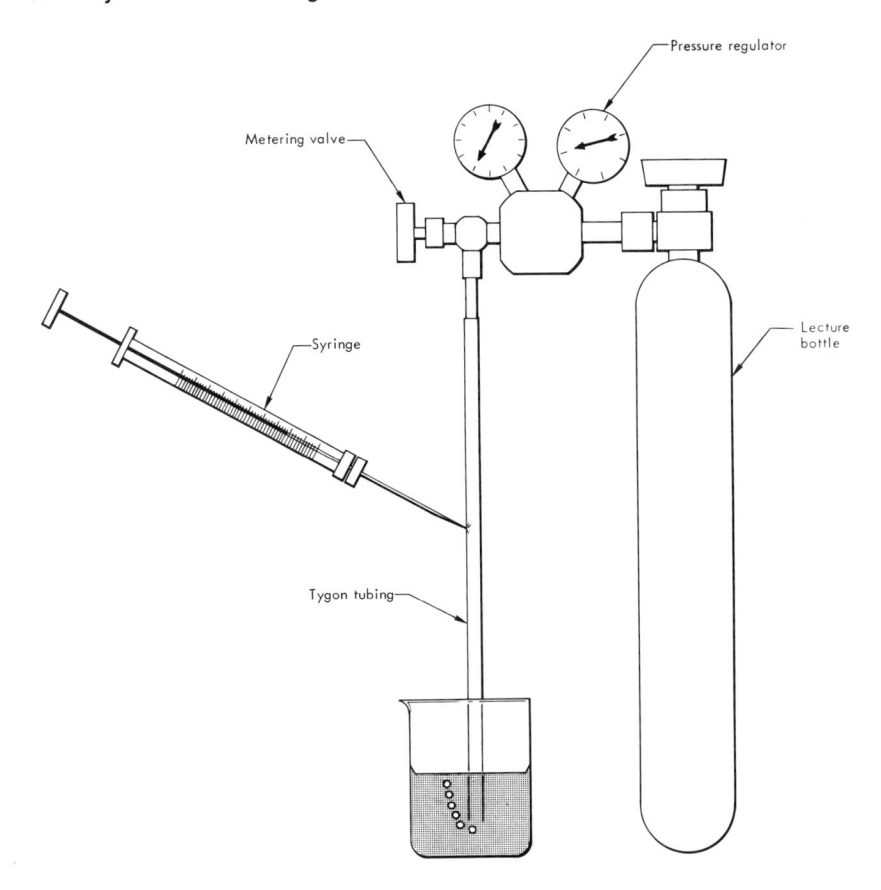

Figure 4.3 Setup for filling a syringe with a gas.

evenly only once when injecting into the calibration vessel. If the dead volume in the needle is injected, it could cause high readings, especially when microliter quantities are being used. A slight vacuum in the sample container is sometimes needed to assist entry of the contaminant into the container, especially when working with moderately sized steel cylinders, which have no internal mixing device.[34]

Measured volumes of liquids are also dispensed with micro or lambda pipets, as shown in Figure 4.2. They have the advantage of no dead volume, but unlike a syringe, which can dispense a wide range of volumes, a pipet yields only one measured solvent volume. Variable-volume precision liquid dispensers are available and are also shown in Figure 4.2. Micrometer syringes and burets have a reported accuracy of 0.5% and can dispense volumes up to 2 mL.[35] Ultraprecision micrometer syringes and burets made of glass and Teflon attain an accuracy of 0.02 to 0.04% when delivering volumes up to 2.5 mL. Variable volume (0.030 to 3.000 mL) gas dispensers are not generally available but have been designed and can yield accuracies in the neighborhood of 0.2%.[36]

4.1.1.2 Gas Blenders

When the contaminant concentration of a mixture is relatively large (in the percent range), gas burets are commonly used to measure, move, and mix the various gaseous components.[37-40] Figure 4.4 illustrates an example of this type. The gas buret, a vessel of known volume, moves measured volumes of gas to the main chamber by means of an advancing front of mercury or water.[41] Various volumes may be preselected by using calibrated scribe marks on the buret.[42,43] Volumetric burets are often used to measure these gas analyzers. When extreme care is taken, certain types of gas blenders have reported blending accuracies of 0.01%,[44] but this is unusual.

4.1.1.3 Miscellaneous Dispensers

Small volumes of gases can be swept into the test chamber with calibrated bypasses, as shown in Figure 4.5. The main test chamber is evacuated, then the material of interest flows through and fills the bypass. The stopcocks are readjusted, and the contaminant is swept into the main calibration vessel with diluent gas.

Liquids, as previously mentioned, can be introduced into test chambers in small ampoules. The solvent is added to a tared ampoule, sealed with a flame, weighed, and added to the vessel. The ampoule can be broken either with a metal plunger[31] or by shaking it with another metal object in an unbreakable container.[6] Another technique for the quantitative transfer and injection of small samples (0.001 to 10 µmol) of gases and vapors has been accomplished using closed glass tubes equipped with hemispheric break seals.[45]

The use of chemical reactions in preparing standard mixtures is especially important for some gases. Hydrogen cyanide, for example, is especially difficult to handle, but it can be handled more easily if small quantities are generated by reacting known quantities of potassium cyanide in an excess of concentrated sulfuric acid (see Chapter 6). Nitrogen dioxide is similarly prepared from the decomposition of lead nitrite.

4.1.1.4 Mixing Devices

Once the sample has been added to the vessel, means must be provided to mix the gas or to evaporate the liquid and mix the vapors. Several of these methods are summarized in Figure 4.6. One of the most common and probably the best all-around method is the use of some type of fan or stirrer. For very large chambers that approach room size, an overhead circulation fan is sufficient.[18,46] Smaller systems in the 20 to 40 L size use internal propellers driven by external motors. Small squirrel cage motors placed inside the vessel have been used for mixtures below the explosive range.[22] For smaller systems on the order of 1 L or so, a magnetic stirrer will suffice.[40]

Before mixing can occur, some source of heat maybe required to vaporize the higher boiling solvents. In large systems, the sample is simply evaporated on a hot plate inside the vessel. Volatilization with infrared lamps,[47] sand baths,[18] small light bulbs,[46] and Nichrome wire[16] have also been used. These evaporation techniques are sometimes needed for volatilization of gases if they are admitted to the vessel dissolved in a liquid.[48]

Static Systems for Producing Gas Mixtures

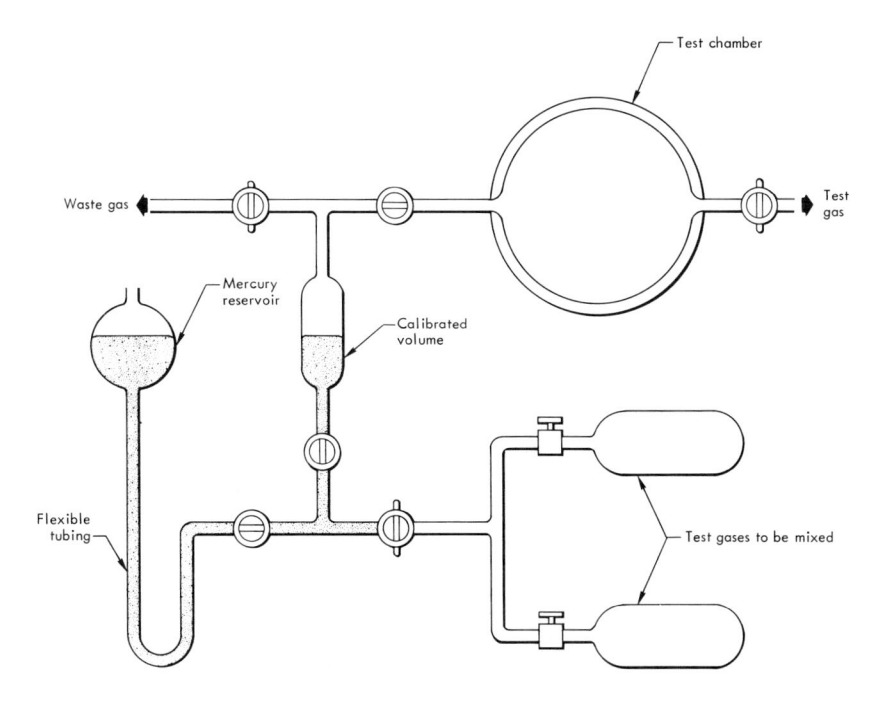

Figure 4.4 Setup for mixing measured volumes of gases via an advancing mercury or water front.

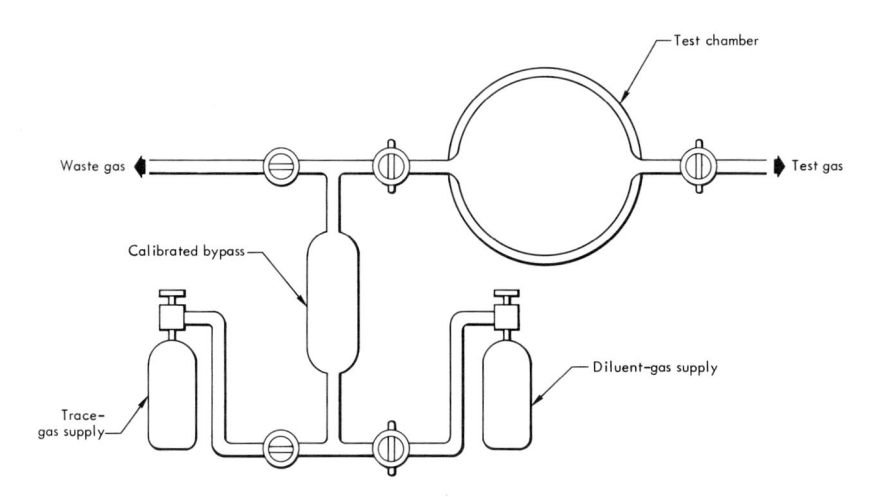

Figure 4.5 Setup for mixing volumes of gases with a calibrated bypass.

If no stirrer is available, the liquid sample can be injected on to adsorbent paper, which increases the surface area and leads to faster evaporation. Foil strips, moved about by container inversion, will greatly increase the speed of evaporation.[6]

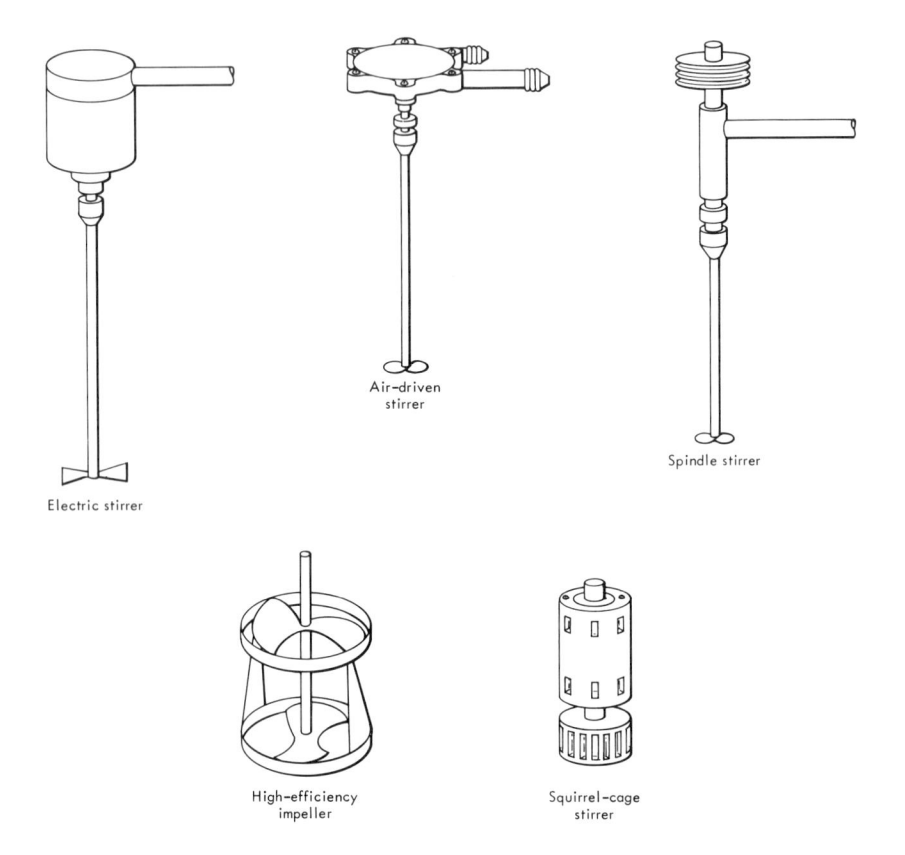

Figure 4.6 Sketches of five devices for mixing gases and vapors in a rigid chamber.

4.1.1.5 Concentration Calculations

The concentration by volume of a contaminant gas or vapor in a gas mixture can be calculated either in percent or in parts per million, depending on the magnitude of the concentrations desired. For a gas mixture in the parts per million range, the concentration is

$$C_{ppm} = \frac{10^6 v_c}{v_c + v_d} \tag{4.1}$$

where v_c is the contaminant volume and v_d is the diluent volume. Below 5000 ppm the v_c term in the denominator is usually insignificant and the equation simplifies to

$$C_{ppm} = \frac{10^6 v_c}{v_d} \tag{4.2}$$

Static Systems for Producing Gas Mixtures

If the concentration in percent is desired, then

$$C_\% = \frac{10^2 v_c}{v_c + v_d} \quad (4.3)$$

or

$$v_c = \frac{C_\% v_d}{(100 - C_\%)} \quad (4.4)$$

Table 4.2 and Examples 4.1 through 4.3 illustrate the use of these equations.

Example 4.1. What is the concentration in percent when 800 mL of propane is added to a 40.0-L container? Assume that no diluent air is lost when the sample is added.

| Contaminant volume | 0.800 L |
| Container volume | 40.0 L |

From Equation 4.3, the concentration in percent is

$$C = \frac{(0.800)(100)}{(40.0 + 0.800)} = 1.96\%$$

Example 4.2. What is the concentration in parts per million when 10 mL of chlorine gas is added to a 5-gal container?

Container volume	5 gal or 18,925 mL
Contaminant volume	10 mL
mL/gal	3785

From Equation 4.2, the concentration in parts per million is

$$C = \frac{(E+06)(10)}{(18,925)} = 528 \text{ ppm}$$

Example 4.3. What volume of carbon monoxide must be added to a plastic bag containing 10 L of air to produce a concentration of 5% carbon monoxide?

| Concentration of CO | 5% |
| Container volume | 10 L |

Table 4.2 Conditions for Producing Various Concentrations in Static Systems

Contaminant Vapor	Mol. wt. (g/mol)	Density (g/mL)	Vessel size (L)	Desired conc. (ppm)	Contaminant Required Standard Conditions Weight (mg)	Contaminant Required Standard Conditions Volume (μL)
Acetic acid	60.05	1.049	5.64	100	1.38	1.32
Acetone	58.08	0.791	5.64	2000	26.7	33.8
Benzene	78.11	0.879	5.64	50	0.899	1.02
Butyl acetate	116.16	0.882	5.64	500	13.4	15.2
Chlorobenzene	112.56	1.106	5.64	50	1.30	1.17
Chloroform	119.38	1.489	5.64	100	2.75	1.85
Cyclohexane	84.16	0.779	20.0	250	17.2	22.0
Decane	142.29	0.730	20.0	1000	116	159
Dibromoethane	187.87	2.055	20.0	25	3.83	1.87
Diethyl ether	74.12	0.708	20.0	2000	121	171
Hexane	86.18	0.659	20.0	500	35.2	53.4
Methanol	32.04	0.792	100	1000	131	165
Methyl ethyl ketone	72.11	0.805	100	500	147	183
Styrene	104.15	0.906	100	25	10.6	11.7
Trichloroethylene	131.39	1.466	100	200	107	73.2
Xylene	106.17	0.880	100	100	43.3	49.2

From Equation 4.4, the required carbon monoxide volume is

$$v(CO) = \frac{(5)(10)}{(100-5)} = 0.526 \text{ L}$$

If a liquid is added to a closed system, the resulting concentration, expressed in parts per million, can be calculated from the expression,

$$C = \frac{22.4 \times 10^6 \left(\frac{T}{273° K}\right)\left(\frac{760 \text{mm}}{P}\right) W}{VM} \quad (4.5)$$

where C is the concentration (ppm), T is the absolute temperature (°K), P is the system pressure (mmHg), W is the weight (g), V is the volume of the system (L), and M is the molecular weight (g/mol).

At normal room temperature and pressure (25°C and 760 mmHg), Equation 4.5 reduces to

$$C = \frac{24.5 \times 10^6 \, W}{VM} \quad (4.6)$$

Static Systems for Producing Gas Mixtures

Example 4.4. What is the concentration when 50 mg of acetone is added to a 20-L container at 25°C and 760 mmHg?

Weight	50 mg or 0.050 g
Chamber volume	20 L
Molecular weight	58.1 g/mol

From Equation 4.6, the concentration in parts per million is

$$C = \frac{(24.5\mathrm{E}+06)(0.050)}{(20)(58.1)} = 1054 \text{ ppm}$$

Example 4.5. What is the weight in grams of carbon tetrachloride needed to achieve a concentration of 500 ppm in a 10 ft³ chamber? The atmospheric pressure and temperature are 25°C and 745 mmHg.

Concentration	500 ppm
Chamber volume	10 ft³
Molecular weight	153.8 g/mol
Room temperature	25°C or 298°K
Room pressure	745 mmHg
L/ft³	28.3

From Equation 4.5, the weight in grams is

$$W = \frac{(500)(10)(28.3)(153.8)}{(22.4\mathrm{E}+06)\left(\dfrac{298}{273}\right)\left(\dfrac{760}{745}\right)} = 0.872 \text{ g}$$

More often, however, the volume of material needed is a more useful variable. If

$$W = \rho v_L \qquad (4.7)$$

where ρ is the density (g/mL) and v_L is the liquid volume (mL), then Equation 4.5 becomes

$$C = \frac{22.4 \times 10^6 \left(\dfrac{T}{273°K}\right)\left(\dfrac{760\,\mathrm{mm}}{P}\right)\rho v_L}{VM} \qquad (4.8)$$

Example 4.6. What is the concentration when 50 μL of benzene are injected into a 5-gal bottle? The temperature and pressure are 18°C and 14 lb/in.,² respectively.

Injected volume	50 μL or 0.050 mL
Chamber volume	5 gal
Molecular weight	78.1 g/mol
Benzene density	0.880 g/mL
Room temperature	18°C or 291°K
Room pressure	14 lb/ft²
L/gal	3.78

From Equation 4.8, the concentration in parts per million is

$$C = \frac{(22.4\text{E}+06)(0.88)(0.05)\left(\frac{291}{273}\right)\left(\frac{14.7}{14.0}\right)}{(78.1)(5)(3.78)} = 747 \text{ ppm}$$

Example 4.7. What volume of ethanol must be vaporized in a 30 ft³ chamber to produce a concentration of 1000 ppm at 20°C and 750 mmHg?

Concentration	1000 ppm
Chamber volume	30 ft³
Molecular weight	46.1 g/mol
Ethanol density	0.789 g/mL
Room temperature	20°C or 293°K
Room pressure	750 mmHg
L/ft³	28.3

From Equation 4.8, the volume in milliliters is

$$V_L = \frac{(1000)(46.1)(30)(28.3)\left(\frac{273}{293}\right)\left(\frac{750}{760}\right)}{(0.789)(22.4\text{E}+06)} = 2.04 \text{ mL}$$

4.1.1.6 Volume Dilution Calculations

After the standard gas mixture is generated, it is sampled or moved to a site where appropriate measurements are made. Removing the test material from the container usually requires the introduction of some type of dilution gas. It is useful, therefore, to be able to calculate what error might be introduced when known volumes of a gas sample are removed and replaced with a diluent gas. If

Static Systems for Producing Gas Mixtures

one assumes instantaneous and perfect mixing with the incoming dilution gas and representative sample removal at a constant rate, then the concentration is [49]

$$C = C_o e^{\frac{-V_w}{V}} \quad (4.9)$$

and taking logs yields

$$2.303 \log\left(\frac{C_o}{C}\right) = \frac{V_w}{V} \quad (4.10)$$

or

$$2.303\left(\log C_o - \log C\right) = \frac{V_w}{V} \quad (4.11)$$

where C is the resultant concentration, C_o is the original concentration, V_w is the volume of sample withdrawn, and V is the chamber volume.

The validity of Equation 4.10 has been checked experimentally. Stead and Taylor,[50] for example, sampled benzene, carbon monoxide, and mercury vapors at various flow rates in a 5-gal jug and found close agreement between actual and theoretical concentrations. Silver,[51] using a 20,000-L room and a flow rate of 3,600-L/min, found the actual concentration to be within ±3% of the expected theoretical values.

Example 4.8. How many chamber volumes are required to purge any given vessel to 1% of its original concentration? To 0.1% of the original concentration?

From Equation 4.10, the number of chamber volumes required are

$$\frac{V_w}{V} = 2.303 \log\left(\frac{100}{1}\right) = 4.61 \text{ chamber volumes}$$

$$\frac{V_w}{V} = 2.303 \log\left(\frac{100}{0.1}\right) = 6.91 \text{ chamber volumes}$$

Example 4.9. What is the theoretical concentration of methyl bromide left in a 20-L container after 10-L of the gas mixture have been continuously removed and replaced with diluent gas? The initial concentration is 500 ppm.

Concentration	500 ppm
Container volume	20 L
Volume removed	10 L

From Equation 4.11, the concentration is

$$(2.303 \log 500) - (2.303 \log C) = \frac{10}{20}$$

$$\log C = \log 500 - \left(\frac{10}{(20)(2.303)}\right)$$

$$\log C = 2.482 \quad C = 303 \text{ ppm}$$

If some part of the withdrawn sample is used and then returned as part of the makeup gas, the situation is somewhat more complicated, but such calculations have been made by Buchberg and Wilson.[52,53]

Often it is desirable to know how long a given size chamber will produce a mixture above a certain concentration. If

$$V_w = Q\tau \qquad (4.12)$$

where Q is the flow rate (volume/min) and τ is the time (min), is combined with Equation 4.11, then

$$\tau = 2.303 \left(\frac{V}{Q}\right)(\log C_o - \log C) \qquad (4.13)$$

Example 4.10 How long can a sample be withdrawn at 0.1 ft³/min before the concentration in a 2 m³ chamber is reduced by 5%?

Chamber volume	2 m³
Removal rate	0.1 ft³/min
Final concentration	95%
ft³/m³	35.3

From Equations 4.10 and 4.12, the time required is

Static Systems for Producing Gas Mixtures

$$V_w = (2)(35.3)\left[2.303 \log\left(\frac{100}{95}\right)\right] = 3.62 \text{ ft}^3$$

$$\tau = \frac{3.62}{0.1} = 36.2 \text{ min}$$

A variation of Equation 4.9 can be used to calculate how long it takes any given mixture to fill a test chamber.[51] If

$$C = C_o\left(1 - e^{\frac{Q\tau}{V}}\right) \qquad (4.14)$$

then

$$\tau = 2.303 \frac{V}{Q} \log\left[\frac{C_o}{(C_o - C)}\right] \qquad (4.15)$$

Here, τ is the time to achieve a concentration C.

Example 4.11. How long would it take to fill a 20-m³ chamber to a concentration of 5-ppm using a 20-ppm contaminant gas flowing at 50 L/min?

Chamber volume	20 m³
Flow rate	50 L/min
Initial concentration	20 ppm
Desired concentration	5 ppm
L/m³	1000

From Equation 4.13, the time is

$$\tau = 2.303\left[\frac{(20)(1000)}{50}\right] \log\left[\frac{20}{(20-5)}\right] = 115 \text{ min}$$

Example 4.12. What is the time required to purge any size chamber until it is 99.9% filled with test gas? Express the answer in terms of V and Q.

Initial concentration	100.0 ppm or %
Desired concentration	99.9 ppm or %

From Equation 4.13, the time is

$$\tau = 2.303\left(\frac{V}{Q}\right)\log\left[\frac{100}{(100.0-99.9)}\right] = 6.91\left(\frac{V}{Q}\right)$$

Exponential dilution can be used directly to produce concentrations as low as 0.1 ppb.[54] As before, a small amount of gas or liquid is introduced into a vessel of known volume that contains a stirring mechanism. The diluting gas passes through at a known constant rate. If one rearranges Equation 4.13 to

$$\log C = \log C_o - \frac{\tau}{\left[2.303\left(\frac{V}{Q}\right)\right]} \tag{4.16}$$

the initial concentration, C_o, then decays at a known rate. If log C is plotted as a function of the time, τ, and a straight line results, then the decay is truly exponential. A number of examples of this technique are reported in the literature.[55-66]

Sometimes it is desirable to sample from a rigid container without suffering makeup gas dilution or significant internal pressure decrease. This is accomplished by attaching a deflated plastic bag to the dilution gas inlet,[49,67-70] as shown in Figure 4.7. When the mixture is sampled, dilution air fills the plastic bag as the mixture is displaced. Care must be taken, however, in selecting a bag that does not adsorb appreciable quantities of the material under study.

4.1.1.7 Discussion

The greatest advantage of the rigid static system is its simplicity of design. All of its components consist of readily available laboratory equipment that are also relatively inexpensive and easy to operate. Accuracies of ±5% are routinely obtainable in the parts per million range if a few basic precautionary measures are undertaken.

The accuracy of the concentration produced from rigid containers generally depends on a knowledge of three parameters: the contaminant purity, the volume of the system, and the volume of the liquid or gas dispersed in the vessel. The other variables of molecular weight and gas density are available in the literature. Room temperature and pressure can be measured accurately with standard laboratory equipment.

Contaminant purity is usually specified by the manufacturer, but if doubt exists, liquids can be checked by distillation, refractive index, or density determinations. Gas and liquid purities can be checked by gas chromatography.

If the injected volume of an ideal pure gas is known to within ±2%, the chamber volume to within ±1%, and the gas purity to ±1%, the expected concentration in known to within ±3%. Even if great care is taken and all the proper correction factors are applied for nonideality, gas purities, and wall adsorption or reactions, accuracies of better than 1% are usually difficult to achieve in the parts per million range. Mixtures in the percent range, however, can be made more

Static Systems for Producing Gas Mixtures

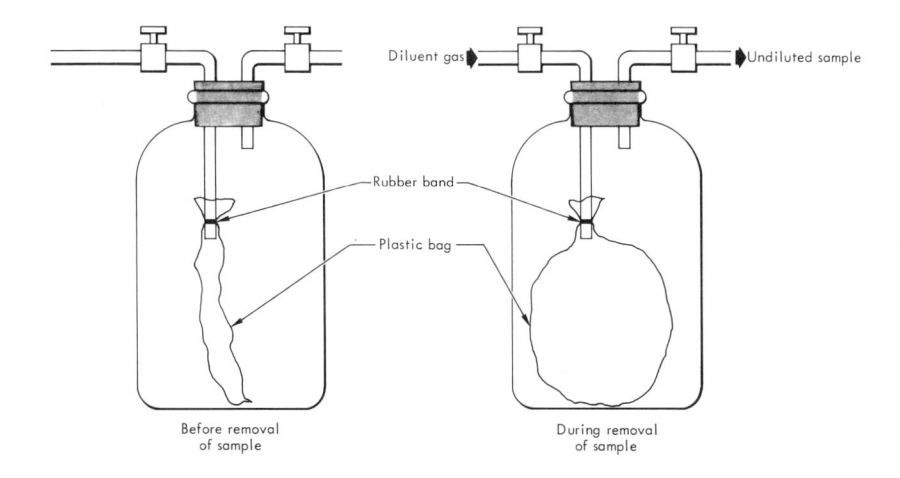

Figure 4.7 Setup for removing a test gas from a rigid static system without causing sample dilution.

accurately,[70] since the volumes of the contaminant gas can be measured more closely than 1%.

The disadvantages of a static system are numerous, but none is sufficiently critical to rule it out as a valid experimental technique. One of the main disadvantages is adsorption and reaction on the container walls.[72,73] Adsorption losses of 50% have been observed,[74] and particular care must be taken when low concentrations of strong oxidizing and reducing agents, and polymerizable materials are prepared. Materials such as ozone, nitrogen dioxide, hydrogen fluoride, nitrated compounds, and styrene monomer are a few examples of such compounds. Adsorption losses can be reduced somewhat by increasing the size of the chamber, which decreases the surface-to-volume ratio, but even in large chambers (156 m^3) the concentration of sulfur dioxide can decrease significantly after 2 hr.[75] Enlarging the chamber also makes it possible to sample larger volumes with less dilution, but then it takes correspondingly longer to flush the chamber between individually produced concentrations.

Static chambers are occasionally subject to implosion or explosion. Sometimes it is necessary to withdraw some of the chamber volume and work with a partial vacuum in a glass container if a large amount of contaminant is to be introduced.[68,76,77] Glass containers have been known to implode if there are weaknesses in the boundary wall. Vessel explosion can occur when working with concentrations above the lower explosive limit, and if a potentially dangerous situation might occur, the chamber can be surrounded with another nonbreakable vessel to prevent the spread of glass fragments. Van Sandt, for example, routinely encloses his 40-L bottles within steel containers with a wire mesh top.[77]

4.1.2 Rigid Chambers in Series

When laboratory space is at a premium, the use of relatively small containers to obtain large volumes of test gas at an almost constant concentration is still possible. This can be accomplished by connecting vessels in series, as shown in Figure 4.8. Here, each container is filled with the desired concentration of contaminant and connected in series with some type of nonreactive tubing, usually glass. As the sample is withdrawn from the last container, dilution air enters the first vessel and dilutes the gas or vapor sample. The diluted gases then move to the second flask and again dilute the test mixture, but to a lesser degree. By adding more containers, large volumes can be withdrawn without seriously decreasing the expected concentration. For example, 2.5 times the first vessel's volume can be withdrawn from a five-container system before the residual concentration is reduced to 90% of its initial value.[78]

Chamber materials, mixing devices, and evaporation methods can be any of the types discussed for single rigid containers. Gases and liquids are added with gastight syringes, microburets, or ultramicropipets, depending on the accuracy and concentration desired.

The concentrations emerging from each individual vessel are obtained from Equation 4.17. When the vessels are connected in series, the residual concentration in the final nth container, assuming perfect and instantaneous mixing, is calculated from the expression

$$C_n = C_o \left[1 + \frac{1}{1!}\left(\frac{V_w}{V}\right) + \frac{1}{2!}\left(\frac{V_w}{V}\right)^2 + \cdots + \frac{1}{(n-1)!}\left(\frac{V_w}{V}\right)^{n-1} \right] e^{\frac{-V_w}{V}} \quad (4.17)$$

where C_n is the concentration of the nth vessel after removal of V_w, C_o is the initial concentration of all vessels, V is the total vessel volume, V_w is the volume of material withdrawn from the system, and n is the number of containers.

Example 4.13. What is the theoretical concentration of methylene chloride left in a two- and four-bottle system (each bottle holds 20 L) after 40 L have been withdrawn? The initial concentration is 1000 ppm.

Initial concentration	1000 ppm
Bottle volume	20 L
Volume withdrawn	40 L

From Equation 4.17, the concentration left in a two-bottle system is

$$C_2 = (1000)\left[1 + \left(\frac{40}{20}\right)\right] e^{-\frac{40}{20}} = 406 \text{ ppm}$$

$$C_4 = (1000)\left[1 + \left(\frac{40}{20}\right) + \left(\frac{1}{2}\right)\left(\frac{40}{20}\right)^2 + \left(\frac{1}{6}\right)\left(\frac{40}{20}\right)^3\right] e^{-\frac{40}{20}} = 857 \text{ ppm}$$

Static Systems for Producing Gas Mixtures

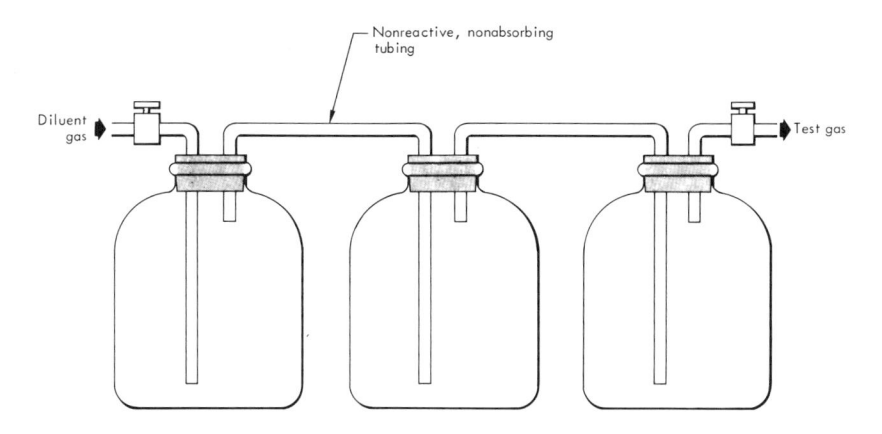

Figure 4.8 Sketch of rigid chambers connected in series.

Table 4.3 gives some of the dilution calculations up to a seven-bottle system. Note how the addition of each container greatly increases the sample volume produced before a given minimum concentration is reached. Equation 4.17 has been experimentally checked several times[22,24] and has been found to be valid after removal of up to 200% of the container volume.

Multiple-vessel arrangements generally have the same advantages and disadvantages as single-vessel systems. The multiple-container system will produce large volumes if it conforms to the dilution equation. This system is still relatively simple to operate and maintain in comparison with some of its dynamic counterparts.

There is an additional disadvantage besides the usual ones of wall adsorption and reaction, and time-consuming flushing between individually prepared concentrations. Series systems usually require vast areas of bench space and hence become unwieldy for the small laboratory. Five 40-L glass bottles with all the accompanying paraphernalia, for example, will generally require about 12 ft of laboratory table top.

4.1.3 Nonrigid Chambers

A valuable alternative to the single or series arrangement of rigid vessels is the nonrigid container, or more simply, the plastic bag. These containers allow the entire sample to be withdrawn without any troublesome volume dilution with replacement air. As the sample is removed from the bag, the boundaries of the bag conform to the remaining volume and the resulting internal pressure change is usually negligible. Plastic bags not only have wide appeal for use in the laboratory, but gas mixtures can be carried into the field and used for instrument calibration. Bags can also be filled with gas samples at remote sites and sent to the laboratory for a complete analysis.

The flexible chamber has several advantages over the rigid static system. The plastic containers are light, compact, and easily portable, and they are usually less expensive orders of magnitude. They also require no air dilution during sampling.

Table 4.3   Residual Concentrations Present in Vessels Connected in Series After Removal of a Given Volume

Volume Withdrawn Percent (V_w/V)100	Residual Concentration in the Designated Vessel — Percent						
	First Vessel	Second Vessel	Third Vessel	Fourth Vessel	Fifth Vessel	Sixth Vessel	Seventh Vessel
0	100.00	100.00	100.00	100.00	100.00	100.00	100.00
10	90.48	99.53	99.98	100.00	100.00	100.00	100.00
20	81.87	98.25	99.89	99.99	100.00	100.00	100.00
40	67.03	93.84	99.21	99.92	99.99	100.00	100.00
60	54.88	87.81	97.69	99.66	99.96	100.00	100.00
80	44.93	80.88	95.26	99.09	99.86	99.98	100.00
100	36.79	73.58	91.97	98.10	99.63	99.94	99.99
125	28.65	64.46	86.85	96.17	99.09	99.82	99.97
150	22.31	55.78	80.88	93.44	98.14	99.55	99.91
175	17.38	47.79	74.40	89.92	96.71	99.09	99.78
200	13.53	40.60	67.67	85.71	94.73	98.34	99.55
250	8.21	28.73	54.38	75.76	89.12	95.80	98.58
300	4.98	19.91	42.32	64.72	81.53	91.61	96.65
350	3.02	13.59	32.08	53.66	72.54	85.76	93.47
400	1.83	9.16	23.81	43.35	62.88	78.51	88.93
450	1.11	6.11	17.36	34.23	53.21	70.29	83.11
500	0.674	4.04	12.47	26.50	44.05	61.60	76.22
550	0.409	2.66	8.84	20.17	35.75	52.89	68.60
600	0.248	1.74	6.20	15.12	28.51	44.57	60.63
700	0.091	0.730	2.96	8.18	17.30	30.07	44.97
800	0.034	0.302	1.38	4.24	9.96	19.12	31.34
900	0.012	0.123	0.623	2.12	5.50	11.57	20.68
1000	0.005	0.050	0.277	1.03	2.93	6.71	13.01

However, plastic bags cannot be used indiscriminately. Diffusion through the walls and sample decomposition and adsorption on the container walls must be taken into account, especially when samples are to be stored for long periods of time.

The plastic bag consists of a sealed, flexible-wall container that can be inflated to full volume without stretching the bag boundaries. There is usually some type of inlet port or septum for gas addition. Stainless steel and polypropylene valves stop gas entry into the bag. Filling snouts with hose clamps are also available. Bags are constructed from a wide variety of materials, but usually polymers of some type are the most useful. Tedlar®, Kel-F®, Teflon®, and five-layer (polyethylene, polyamid, aluminum foil, polyvinyldechloride, and polyester) are the most popular. However, Mylar®, Saran Wrap®, Scotchpak®, sandwich bags,[79] and even football bladders[80] have been used. Some of the physical properties are summarized in Tables 4.4 and 4.5.

The materials used in bag construction are generally from 1 to 20 mils in thickness. This allows the wall boundaries to assume any shape as long as the bag is not filled to capacity and the bag pressure does not exceed atmospheric pressure. The polymer is either used alone or is laminated to some other material, usually

Static Systems for Producing Gas Mixtures

Table 4.4 Characteristics of Some Common Polymeric Films

Generic or Brand Name	Polymeric Material	Thickness (mils)	References
Aclar	Polyfluorocarbon	1–20	87
Cellophane	Cellulose	1.5	90
Kel-F	Polyfluorocarbon	5	90
Kynar	Polyfluorocarbon	1–20	87
Mylar	Polyester	2	15, 81, 91, 95
Mylar, aluminized	Polyester	2	81, 95
Polyethylene	Polyethylene	5	90
Polyethylene, 4 others	Five layers[a]	5.5	83
PVC	Polyvinyl chloride	4	90
Saran wrap, type 12	Polyvinylidine chloride	2	81
Saran wrap, type 5-517	Polyvinylidine chloride	2	90
Scotchpak, type 20A5	Polyester	2	81, 91
Tedlar	Polyfluorocarbon	1–20	82, 84, 85, 87
Teflon	Polyfluorocarbon	1–20	81, 85, 87, 90

[a] 75 μ polyethylene, 40 μ polyamid, 12 μ aluminum foil, 3–4 μ polyvinyldechloride, and 12 μ polyester.

aluminum. This aluminizing apparently seals the pores and makes the walls less permeable to the sample gases.[77] At any rate, the inner layer should be impermeable to the gases, and the outer layer should be impermeable to moisture.[81]

Flexible containers may be purchased commercially[82-87] or made in the laboratory. Large sheets of polymers can be purchased, cut into the desired size, and sealed with a hot iron. Mylar, for example, can be obtained in 100-ft rolls, and Mylar tape can be used to seal the plastic layers together with a household iron[88] or rotary heat sealers. Double sealing is sometimes needed to ensure that bags are indeed free from leaks.[15] Bag sizes of 1 to 200 L have been used.

4.1.3.1 Sample Introduction

Before any contaminant or diluent gas is added, the bag is rolled as tightly as possible to minimize the dead volume. The component gases are normally metered into the bag and stopped just before the bag is completely full. Figure 4.9 shows several methods of sample introduction. Gaseous contaminants are normally added either slowly into the filling air stream or directly into the bag through a rubber patch or septum after the bag is filled. Gas-tight, Teflon-tipped syringes are used for either gases or liquids. If liquids are to be injected, they should be vaporized, if possible, by the incoming air outside the bag, rather than injected directly onto the container walls. The bag opening can be closed with a stopcock, pinch clamp, or screw-type valve, and mixing is accomplished by kneading the bag for several minutes.

4.1.3.2 Sample Decay

The initial concentration in a plastic bag will, with time, slowly decay toward a zero concentration. Some substances, such as nitrogen dioxide and ozone, will decay quickly, while other materials, such as sulfur dioxide and hydrocarbons,[89,127] will exhibit less tendency to decrease in concentration. The decay of various

Table 4.5 Properties of Some Plastic Films[a]

Film Type	Chemical Resistance[b] to					Physical Properties		
	Strong Acids	Strong Alkalis	Ketone and Esters	Chlorohydro-carbons	Hydrocarbons	Sealing Temp. (°F)	Water Absorption (g/in.²/hr)	Water Diffusion (g/in.²/hr)
Cellophane (coated)	P	P	P	E	F	200–350	High	0.002–0.01
Cellophane (plain)	P	P	P	E	E	None	High	High
Cellulose diacetate	P	P	P	P	G	400–500	Low	0.1–0.4
Cellulose nitrate	G	F	P	F	F	None	Low	—
Cellulose triacetate	P	P	P	P	G	None	Low	0.1
Ethyl cellulose	P	E	P	P	F	—	Low	0.1–0.5
Polyester[c]	E	E	E	E	E	490	Very low	0.018
Polyethylene	E	E	G	F	F	250–400	Very low	0.012–0.014
Polyvinyl alcohol	P	P	E	E	E	300–400	0.30	0.1
Polyvinylidene chloride	E	G	F	F	E	285	Very low	0.002
Styrene	G	E	P	G	G	250–300	0.0004	0.062
Teflon	E	E	E	E	E	None	None	None
Vinyl chloride	G	E	P	F	G	200–400	Very low	0.01
Vinyl chloride, GRS	G	G	P	E	G	270–350	Very low	0.07

[a] Data taken from Reference 156.
[b] E = excellent; G= good; F = fair; P = poor.
[c] Includes Mylar, Videne, and Scotchpak.

Static Systems for Producing Gas Mixtures

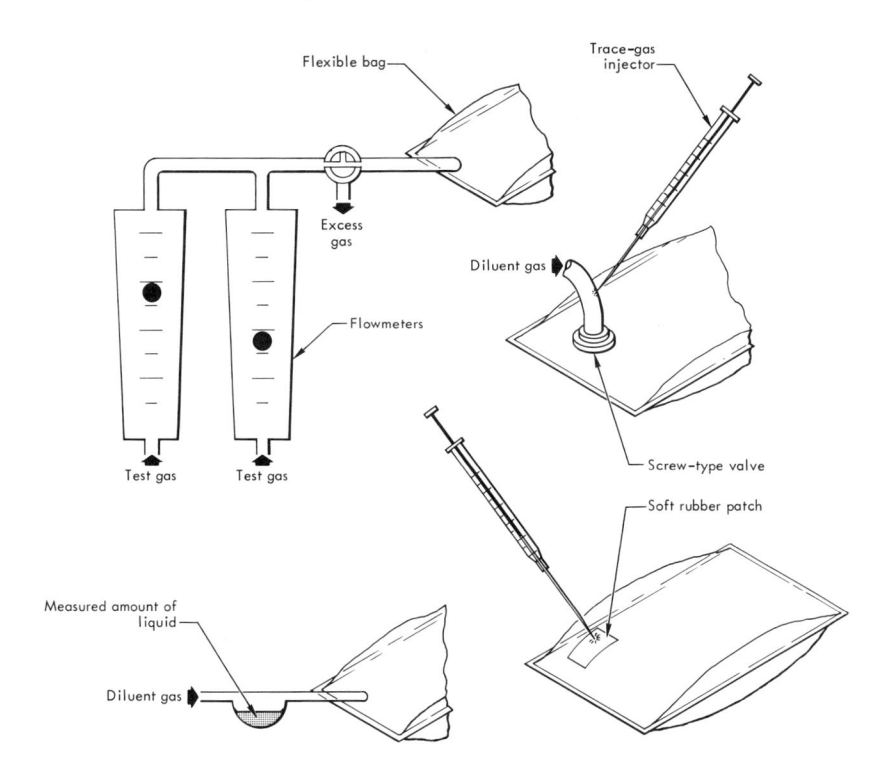

Figure 4.9 Four methods of introducing a sample into a nonrigid chamber.

materials in several types of flexible containers is shown in Table 4.6,[81] and typical data for sulfur dioxide are plotted in Figure 4.10. Sample losses are usually influenced by the bag material, the nature of the contaminant, the relative humidity, and the transparency to radiation.[90] Examples of the effect of humidity are shown in Figure 4.11.[81] Note that higher humidity accelerates the decay process of a typical concentration. If there is doubt and the mixtures are to be stored for long periods of time, they should be checked before use to determine to what extent they have deteriorated.

Sample decay can be lessened if the container is first preconditioned to the test substances. Preconditioning requires flushing out the bag several times with the test material. Some preconditioning cycles recommended require at least six refills, with at least one of them remaining in the bag overnight.[91] Although preconditioning is usually helpful, nitrogen dioxide, for example, decays at the same rate, whether or not the preconditioning cycles have been used.

4.2 PRESSURIZED SYSTEMS

Pressurized systems are frequently used in the laboratory to produce large volumes of gas mixtures. Technically speaking, a pressurized system is a dynamic

Table 4.6 Evaluation of Some Polymeric Films for Storing Various Gases[a]

Film Type	Thick. (mils)	Size (in.)	Gas or Vapor	Concentration Initial (ppm)	Concentration Final (ppm)	Storage Time (hrs)
Mylar	2	18 × 50	Acetone	60 (39)[b]	62 (20)	18 (65)
			Benzene	76 (31)	81 (26)	18 (65)
			Butyraldehyde	55 (66)	50 (69)	24 (17)
			Ethylene	64	63	19
			Nitrogen dioxide	52 (78)	41 (13)	18 (65)
			Sulfur dioxide	69	68	18
Aluminized mylar	1	18 × 52	Butyraldehyde	63 (79)	69 (77)	18 (24)
			2-Methyl pentane	71 (72)	73 (79)	22 (19)
			Nitrogen dioxide	71 (81)	47 (52)	22 (19)
			Sulfur dioxide	73 (70)	78 (77)	22 (19)
Saran wrap	2	36 × 36	Benzene	72 (79)	76 (79)	19 (20)
			Butyraldehyde	49	51	17
			2-Methyl pentane	65	65	17
			Nitrogen dioxide	64 (66)	39 (37)	19 (20)
Scotchpak	2	36 × 32	Acetone	56	78	17
			Benzene	70	74	17
			Butyraldehyde	49	50	23
			2-Methyl pentane	60	4	23
			Nitrogen dioxide	60 (66)	30 (0)	17 (23)
			Sulfur dioxide	65	44	23
Aluminized scotchpak	2	36 × 34	Acetone	54	57	17
			Benzene	63	68	17
			Butyraldehyde	50	45	17
			2-Methyl pentane	66	68	17
			Nitrogen dioxide	51 (76)	0 (8)	17 (18)
			Sulfur dioxide	83 (77)	73 (55)	17 (18)

[a] Data taken from Baker and Doerr[81] with permission.
[b] The items in parentheses correspond to the second experiment.

method, since the usual procedure is to mix moving gas streams. However, the process of making and storing mixtures of gases in an appropriate cylinder is generally termed a "static method" of calibrated gas mixture production. Pressurized systems are most easily obtained from commerical manufacturers at pressures of about 2500 psig. Higher pressures of 3500 to 6000 can be obtained on request for certain gases.[97] Pressurized mixtures of gases and pure gases are available commercially in a wide variety of cylinder sizes (see Appendix P). They can be produced in sizes from large steel cylinders with volumes up to 220 ft³ to small aerosol cans, which are good for volumes up to several liters.[98] They can, however be produced in the laboratory if certain basic equipment is available. The three main types of pressurized systems are gravimetric, partial pressure, and volumetric. The first two types are the most commonly used commercial tech-

Static Systems for Producing Gas Mixtures

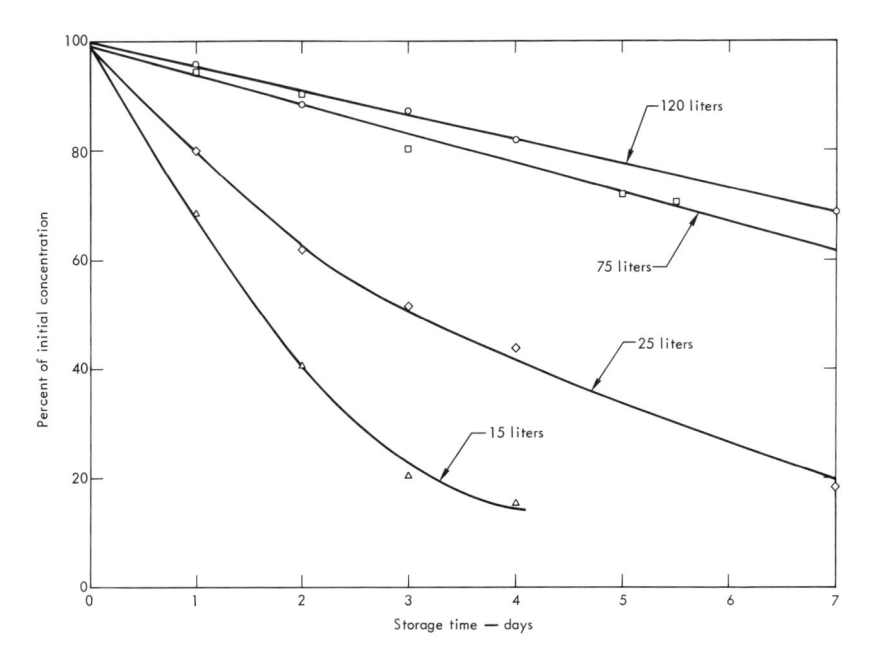

Figure 4.10 Concentration of SO_2 vs storage time in Mylar bags having capacities of 15, 25, 75, and 120 L (initial SO_2 concentration was 1 ppm). Data taken from Conner and Nader[92] with permission.

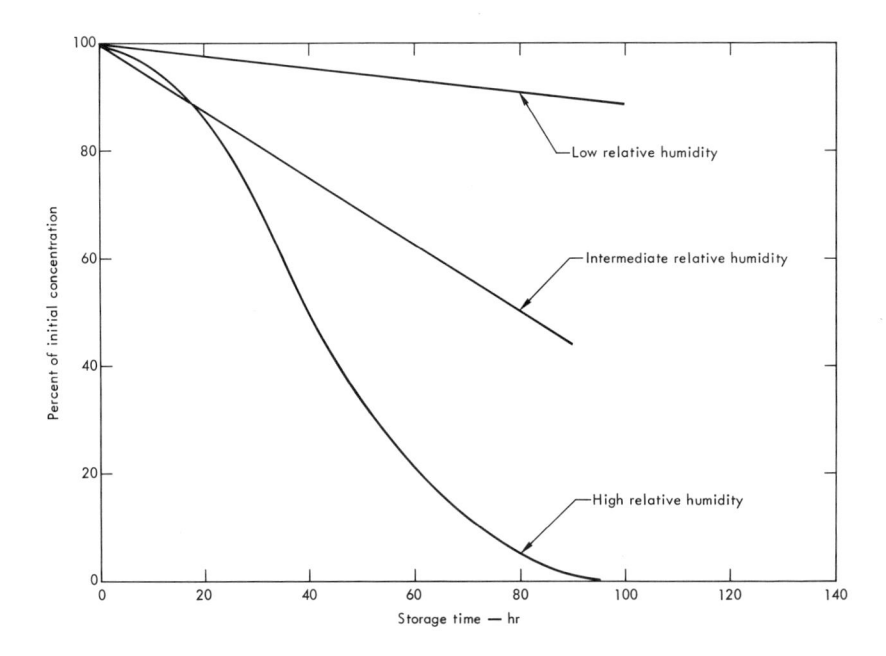

Figure 4.11 Concentration of NO_2 vs storage time in a 1-mil thick aluminized Mylar bag at three different relative humidities (initial NO_2 concentration was 70 ppm). Data taken from Baker and Doerr[81] with permission.

niques. However, all three types can be used efficiently to produce large volumes of standard gas mixtures from the parts per million to the percent range.

4.2.1 Gravimetric Methods

Gravimetric methods are considered to be the most accurate technique of producing primary analytical standards.[99] Representative industrial accuracies are shown in Tables 4.7 and 4.8.[100] Such standards are prepared in cylinders made of steel, aluminum, and stainless steel, and lined with a variety of proprietary substances,[101,102] or preconditioned with nitric oxide or silane.[102] The cylinder or vessel to be filled is weighed before and after the addition of each of the components on a suitable high-load balance. Most commercial manufacturers use specially designed and engineered high-load balances for this purpose. Cylinders such as the lecture bottle and the 7-kg cylinder can be weighed on commercially available balances. Smaller, specially made containers have also been designed for small-volume mixtures.[103] High-capacity top-loading balances, for example, can weigh containers from 5 to 11 kg to within ±50 mg. Although the tared weights of cylinders can be recorded to within ±0.5 mg for gas mixtures in the percent range, a lack of knowledge concerning sample purity, adsorption, and side reactions make this method accurate to between ±0.1 to 0.01%, depending on the magnitude of the concentrations encountered.

Almost any combination of gases can be mixed (even to levels of 1 ppb[104]), but there are some limitations. The most common problems occur when the components have insufficient partial pressure, are flammable, react with each other or the cylinder, or suffer a concentration decrease with time. Insufficient partial pressure occurs when the components liquify at or near room temperature. These mixtures are controlled by limiting either the maximum concentration or the total cylinder pressure. Care must be taken not to let the cylinder storage temperature drop below 50°F or component condensation may result. Flammable mixtures are limited by the lower explosive limit. Mixtures above 50% LEL are normally not available. Component reaction is also a problem with a number of gas combinations. For example, hydrogen sulfide can react with carbon monoxide, sulfur dioxide, air, carbonyl sulfide, or even the cylinder walls.[100] Gas mixtures, especially those that contain low concentrations of reactive gas, may undergo a concentration decrease with time, even after extensive cylinder treatment. Examples of this phenomena are shown in Tables 4.9 and 4.10.

Once the gases are added to the cylinders, they must be made homogeneous before use. The simple act of adding gases to the cylinders does not ensure a totally mixed cylinder. For example, mixtures having a component in the parts per million range in a single background gas will be 80 or 90% homogeneous. However, a mixture containing 5 to 25% of one component in another may be only 25 to 50% homogeneous.[100] Mixing can be accomplished either by rolling or thermal convection. Rolling, the most common commercial technique, is accomplished by placing the cylinders on their sides on rollers. The cylinders are then spun for about an hour. The thermal technique is only employed in some special applications. This involves heating the cylinder bottom with a hot-water bath to a temperature of no greater than 125°F.

Static Systems for Producing Gas Mixtures

Table 4.7 Gas Mixture Specifications

Category	Range	Preparation Tolerance (±)	Certification Accuracy
Primary standard (gravimetric method)	5–50%	1% of component	±0.02% absolute or 1% of the component, whichever is smaller
	1–5%	2% of component	
	<1%	5% of component	
Certified standard (partial pressure or volumetric method)	10–50%	5% of component	2% of component
	50 ppm to 10%	10% of component	2% of component
	10–50 ppm	20% of component	5% of component
	3–10 ppm	2 ppm	5% of component
Unanalyzed mixture (partial pressure or volumetric method)	10–50%	5% of component	Not analyzed
	50 ppm to 10%	10% of component	Not analyzed
	10–50 ppm	20% of component	Not analyzed
	3–10 ppm	2 ppm	Not analyzed

Data taken from Scornavacca[100] with permission.

Table 4.8 Gas Mixture Specifications for a Five-Gas Blend

Category	Mixture Requested	Preparation Range	Certification Accuracy
Primary standard	20 ppm methane	19–21 ppm	±0.2 ppm
	1% carbon monoxide	0.98–1.02%	±0.01%
	5% carbon dioxide	4.9–5.1%	±0.02%
	40% nitrogen	39.6–40.4%	±0.02%
	Balance helium		
Certified standard	20 ppm methane	16–24 ppm	±1 ppm
	1% carbon monoxide	0.9–1.1%	±0.02%
	5% carbon dioxide	4.5–5.5%	±0.1%
	40% nitrogen	38.0–42.0%	±0.8%
	Balance helium		

[a] Data taken from Scornavacca[100] with permission.

There has been some concern that, because of the differences in gas density, the gas mixtures might stratify on standing. Tests have been conducted at Matheson Gas Company that indicate that once the gases are mixed, they remain mixed and do not separate.[100] Stratification can occur, however, if a component condenses out due to extreme cooling or overpressurization.

Once the gases are mixed, their final composition is determined by either the gravimetric weight or analysis using NBS Standard Reference Materials. The gases are then shipped or stored until used. Once the gases are shipped, they must be stored out of the sunlight at about 70°F. The storage temperature should not be allowed to go above 125°F or below 40°F. The cylinders should always be chained or belted securely to the wall or bench top to prevent falling. This is especially important in earthquake prone areas such as California.

Although gravimetric calculations would seem to be trivial, the best results are obtained by correcting for any buoyancy changes between measurements. Anything that would affect the air density around the cylinders and balance weights, such as temperature changes, relative humidity, and barometric pressure, must be corrected. A detailed account of these calculations is given by Miller.[105] Sample corrections for gas purity in a mixture containing hydrogen, methane, nitrogen, and helium are shown in Table 4.11. Notice that the total weight of any component is the amount added as the pure component plus any that might be present as an impurity in the other components.

Performance audits comparing industrial gas accuracies are available.[106] The accuracy of a good system should not fall below ±0.1% if the component purity is known and if negligible gas interaction or absorption is present. Values of ±0.05%[99,107] are common, and ±0.02% have been routinely achieved[107] with components in the percent range. Standards in the parts per million range, however, cannot approach this accuracy unless a predilution has been made. Most of the errors inherent in pressurized measurements, such as temperature equilibrium and deviation from ideality, can be neglected. The main disadvantage of the gravimetric technique to the casual gas blender is the expense of the pressurizing equipment and a high-load, high-precision balance.

4.2.2 Partial Pressure Methods

Test gases can be admitted to steel cylinders in known amounts if the partial pressure of each component is precisely known. Gas cylinders can be pressurized safely to 2500 psig and above on a routine basis commercially, but it is not usually advisable or practical for most laboratories to work at these elevated pressures. Nevertheless, gas mixtures can be made using pressures up to 12 atm with little specialized equipment.[110,111] Partial pressure or manometric techniques have the advantage that fairly large volumes of complex gas mixtures can be generated and stored for future use. Several cylinders can be filled at the same time for relatively little cost.

A representative system is shown in Figure 4.12. The cylinders to be filled are connected to a manifold with an appropriate system of gauges over the pressure range of interest. The system should include a gauge in the vacuum region, since the manifold is usually evacuated to less than 100 µm before each component gas is admitted to the cylinder. In the psia range, if large (16-in.) Bourdon or aneroid gauges are used, an accuracy of 1 part in 500 is claimed and is usually proven to be even better.[112] The supply gases are normally connected to a separate manifold, and the two manifolds are connected with a valve.

A typical system is filled using the following procedure. The system of manifolds and cylinders to be filled is evacuated and component 1 flows into the empty cylinders. The desired pressure is reached and recorded. This must be done slowly to avoid errors from excessive temperature changes.[112] All valves are shut off and the manifold is again evacuated. Component 2 is admitted to the manifold at a pressure slightly in excess of component 1 to prevent backflow of the filled component. Component 2 is then admitted to the test cylinder to the desired pressure. Additional components are admitted by repeating this same technique

Static Systems for Producing Gas Mixtures

Table 4.9 Gas Stability in Various Cylinders

Gas	Cylinder Type	Concentration (ppm) Initial	2 months	2 years
Nitrogen dioxide	Wax lined	200	135	<1
Nitrogen dioxide	Steel (Cr, Mo)	200	179	151
Nitrogen dioxide	Treated aluminum	200	197	200
Sulfur dioxide	Wax lined	160	148	120
Sulfur dioxide	Steel (Cr, Mo)	160	148	132
Sulfur dioxide	Treated aluminum	160	159	157
Carbon monoxide	Steel (Cr, Mo)	25	21	15
Carbon monoxide	Treated aluminum	25	25	25

Data taken from Wechter[102] with permission.

Table 4.10 Stability of Nitrous Oxide in Air in Steel Cylinders After Specified Number of Days

Period (days)	Concentration after specified number of days (ppm)						
	1	2	3	4	5	6	7
Initial	5.00	20.0	50.0	100	250	500	1000
1	4.65	19.5	49.5	100	253	518	1030
3	4.55	19.2	49.2	100	253	510	1050
4	4.40	19.3	49.2	100	252	507	1010
7	4.30	19.2	49.2	100	252	510	1040
11	4.40	19.3	49.2	100	252	508	1030
18	4.20	19.2	49.2	99	252	515	1040
24	4.10	19.0	49.0	99.5	252	512	1040
29	3.95	19.1	49.2	100	251	513	1040

Data taken from Lee and Paine[109] with permission.

until all have been added. Low parts per million concentrations can be prepared by evacuating the system and introducing the trace contaminant through the system. This material is then swept into a storage cylinder and pressurized with the appropriated dilution gas.[113]

Materials that are normally liquids at room temperature can also be used to fill the cylinder, since they exert a certain vapor pressure, depending on the nature and volatility of the solvent. The vapor pressure of these liquids, however, must be greater than their partial pressure in the gas mixture at the lowest working or storage temperature at which the sample is used. In other words, if the pressure of the system becomes too high (or the temperature too low), the vapors will condense out and the usable gas mixture will have a variable concentration.

As with the gravimetric systems, the time required for mixing must be considered. The time depends on the vessel size and geometry, injection turbulence, and the interdiffusion coefficients of the gases.[114] Cylinders are mixed by rolling, heating, or using a steel shot to aid in the mixing process.[115]

Table 4.11 Corrections for the Purity of Component Gases[a]

Components	Mass Spec. Analysis Mole (%)	Wt (%)	Weight (g)	Weight Corr. for Purity (g)	Actual Weight (g)
Hydrogen	99.8	97.29	0.3349	0.3258	0.3258
Nitrogen	0.2	2.71		0.0091	
Methane	98.93	97.83		11.9763	
Nitrogen	0.71	1.22		0.1492	
Carbon dioxide	0.21	0.57	12.2294	0.0697	11.9763
Ethane	0.15	0.28		0.0342	
Nitrogen	100.00	100.00	176.4653	176.4653	176.6236[b]
Helium	100.00	100.00	33.1762	33.1762	33.1762
Ethane[c]					0.0342
Carbon dioxide[c]					0.0697
Total			222.2058	222.2058	222.2058

[a] Data taken from Miller, et al.[105] with permission.
[b] The actual weight of nitrogen in the mixture is 176.4653 g plus 0.0091 g from the hydrogen plus 0.1492 g from the methane.
[c] These components were initially present in the methane.

Calculation of the gas mixture concentration appears to be straightforward if Dalton's Law is assumed:

$$P = p_a + p_b + \cdots + p_n \tag{4.18}$$

where P is the total pressure and $p_{a,b,\ldots,n}$ are the contributing partial pressures of the component gases. The concentration in percent by volume of each component at one temperature is

$$C_\% = \frac{10^2 p_n}{P} \tag{4.19}$$

and in parts per million by volume would be

$$C_{ppm} = \frac{10^6 p_n}{P} \tag{4.20}$$

Static Systems for Producing Gas Mixtures

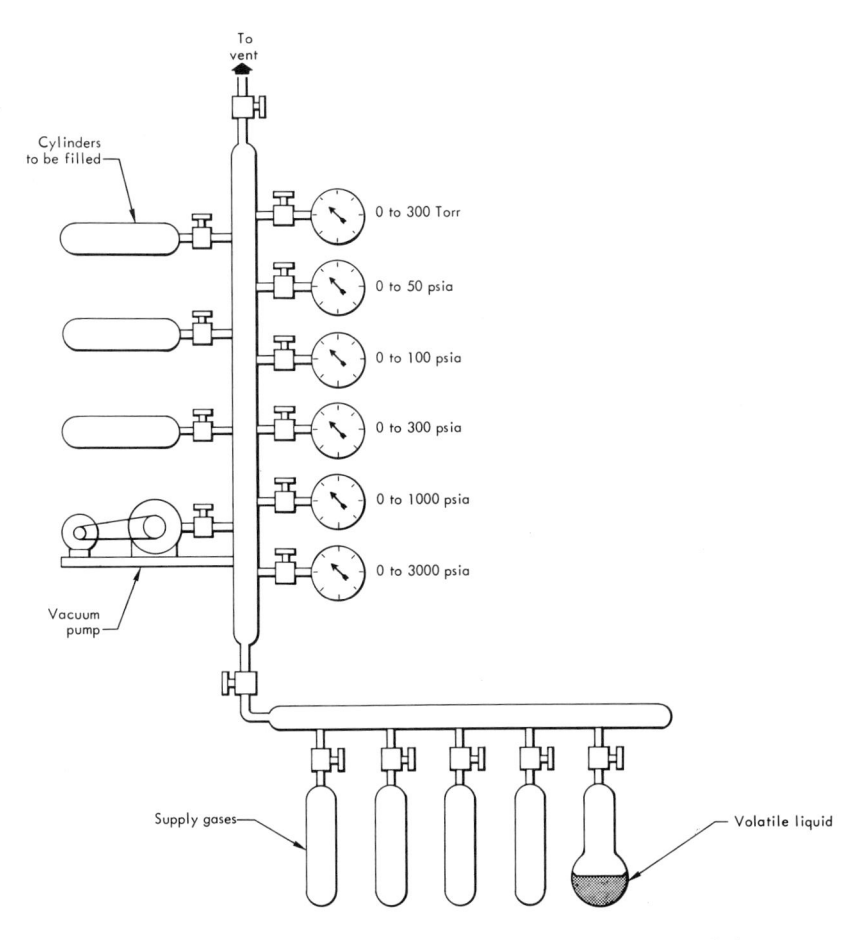

Figure 4.12 Sketch of a partial pressure system for filling gas cylinders.

Example 4.14. What partial pressure of carbon monoxide is required to produce a 20% mixture by volume of carbon monoxide in air?

| Desired CO concentration | 20% |
| Final cylinder pressure | 2500 psia |

From Equation 4.19, the required partial pressure of carbon monoxide is

$$p_{(CO)} = \frac{(20)(2500)}{100} = 500 \text{ psia}$$

Example 4.15. Calculate the percentage concentration of a nitrogen, oxygen, and helium mixture when an evacuated vessel is first filled to 200 mmHg with nitrogen, to 740 mmHg with oxygen, and then to 5 atm with helium. Assume all pressures are recorded at the same temperature and there is no compressibility correction.

Nitrogen partial pressure	200 mmHg
Oxygen and nitrogen partial pressure	740 mmHg
Total system pressure	5 atm or 3800 mmHg
mmHg/atm	760

$P_{(N_2)} = 200$ mmHg
$P_{(O_2)} = 740 - 200 = 540$ mmHg
$P_{(He)} = 3800 - 740 = 3060$ mmHg

From Equation 4.19, the concentrations are

$$C_{(N_2)} = \frac{(100)(200)}{3800} = 5.26\%$$

$$C_{(O_2)} = \frac{(100)(540)}{3800} = 14.2\%$$

$$C_{(He)} = \frac{(100)(3060)}{3800} = 80.5\%$$

At room temperature and pressure, most gases conform to the perfect gas law. However, at elevated pressures, the deviations are pronounced and corrections of ±10 to 20% are not uncommon (as shown in Figure 1.1). This can sometimes be corrected by defining a new quantity, κ, the compressibility, which is given by

$$\kappa = \frac{PV}{RT} \qquad (4.21)$$

where P, V, and T are measured experimentally to yield κ, a correction factor for nonideality. Thus corrected, the concentration on the nth component can be expressed by

$$C_n = \frac{\dfrac{10^2 p_n}{\kappa_n}}{\dfrac{p_a}{\kappa'_a} + \dfrac{p_b}{\kappa'_b} + \cdots + \dfrac{p_n}{\kappa'_n}} \qquad (4.22)$$

where κ_n is the compressibility of the pure components at the filling pressure and

Static Systems for Producing Gas Mixtures

$\kappa'_{a,b,...,n}$ are the compressibilities of the gas mixtures at the final pressure. It must be remembered that as each component is added to the system under pressure, any temperature rise must be accounted for. Either an appropriate calculation correction must be made or the system must be allowed to return to the initial ambient temperature before a pressure reading is made and the next component is added. In general, it is best to add the most compressible gas first, unless it happens to be the major component of the mixture. If the compressible gas is admitted last, it is in the most compressed state and usually introduces the greatest error.[112]

Example 4.16. Calculate the concentrations of a butane and propane mixture when the tank to be filled is evacuated and filled with butane to 10 psia (the compressibilities are 0.97 at 10 psia and 0.88 at 3 atm), and propane is added until a pressure of 3 atm is reached (the compressibility is 0.95 at 3 atm). Assume that all pressures are recorded at the same ambient temperature.

Butane partial pressure	10 psia
Total system pressure	3 atm
Butane compres. at 10 psia	0.97
Butane compres. at 3 atm	0.88
Propane compres. at 3 atm	0.95
psia/atm	14.7

At 10 psia

$$P_{(C_4)} = \frac{(10)}{(14.7)(0.97)} = 0.701 \text{ atm}$$

At 3 atm

$$P_{(C_4)} = \frac{(10)}{(14.7)(0.88)} = 0.773 \text{ atm}$$

$$P_{(C_3)} = \left(\frac{3}{0.95}\right) - 0.773 = 2.38 \text{ atm}$$

From Equation 4.19, the concentrations are

$$C_{(C_4)} = \frac{(100)(0.701)}{(0.773 + 2.38)} = 22.2\%$$

$$C_{(C_3)} = \frac{(100)(2.38)}{(0.773 + 2.38)} = 75.5\%$$

The 2.3% is unaccounted for because of compressibility errors.

Compressibilities for the pure gases can be found in most chemical engineering handbooks. However, data for gas mixtures are not generally available for all possible combinations of gases, temperatures, and pressures, so the value used is usually the one for the pure component.

Partial pressure techniques are used mostly in the laboratory to produce preliminary analytical standards. The routine accuracies are usually ±5% at reduced pressures where compressibilities approach unity.[116] These methods can be used to produce mixtures in the percent range, but gas concentrations as low as 10 ppm have been achieved only after making a preblend.[111] Accuracies of ±1% are claimed for blends in the 180 to 380 ppm range for small containers,[117] and such techniques have been used successfully in producing standards for mass spectrometric analysis.[112,117]

There are several limitations that must be understood before these systems can be employed. Liquefaction of one or more of the components can cause large deviations from the predicted concentrations. Deviations can also occur if compressibility data for the condition of interest are not available for the gaseous mixtures, if the heat created by the compression is not accounted for in the calculation, or if heat is not allowed to dissipate before the addition of successive components. Errors also occur from the effects of polymerization of certain mixtures (e.g., 0.5% hydrogen cyanide) and decomposition (e.g., nickel carbonyl) unless proper stabilizing agents are added.[118]

Probably the greatest disadvantage of partial pressure techniques is the inherent danger of working with a pressurized system that may contain reactive mixtures. Pressurizing with pure oxygen when hydrocarbons are present can be hazardous, and certain pure gases, such as acetylene and monovinyl acetylene, detonate on pressurizing. Other materials that react, even though one component is present in parts per million quantities, include oxygen and hydrogen, chlorine and hydrogen, nitrogen dioxide and nitric acid, and unsaturated hydrocarbons with sulfur-bearing compounds.[116]

4.2.3 Volumetric Methods

The volumetric method consists of metering known flows or volumes of gases and compressing them into a pressurized vessel. These methods are able to produce concentrations from the medium parts per million range to the percent range with accuracies in the neighborhood of ±1 to 10% depending on the technique chosen.

Figure 4.13 shows a typical volumetric system. Pure component gases from pressurized cylinders are metered through a flow meter and compressed into another cylinder of suitable strength. The pure starting materials may have been previously diluted and may even be contained in vessels at atmospheric pressure or slightly above. Plastic bags and spirometers have been known to house the starting materials. If small volumes are required, then high-pressure, gas-tight syringes can be employed. Introduction of larger gas volumes can be done with calibrated gas volumes,[119-122] much like the system shown in Figure 4.5. A chamber of known size is filled with the gas of interest and is subsequently swept into the source cylinder. The flow meter can be almost any of the devices

Static Systems for Producing Gas Mixtures

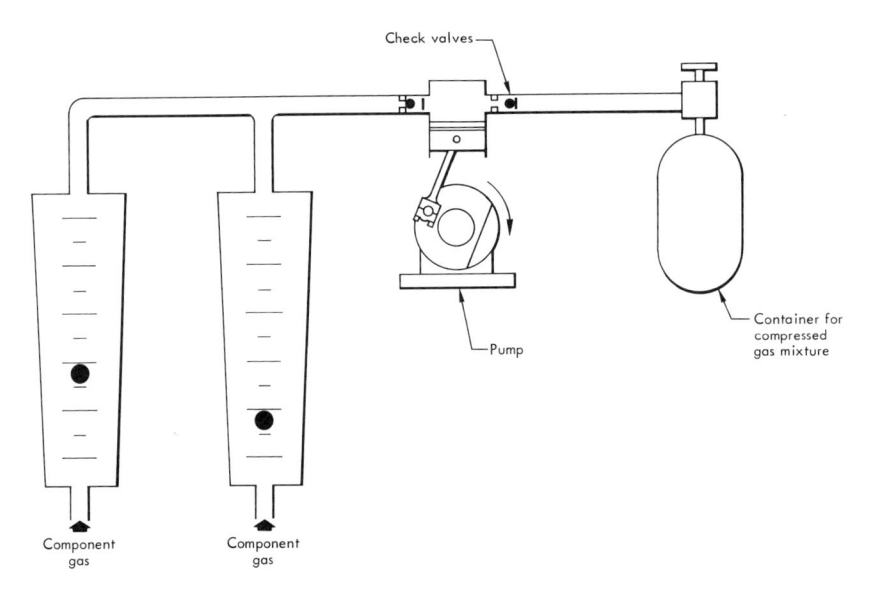

Figure 4.13 Sketch of a volumetric system for filling gas cylinders.

discussed in Chapter 3. Mass flow meters, rotameters, and wet and dry gas meters are the ones most frequently used. The compressor chosen is usually the diaphragm type in which the test gases do not come in contact with oil mists or other contaminants. Gases are usually metered and compressed into the storage vessel one at a time. This often leads to a layering effect of the gases in the cylinder and incomplete mixing. Methods of mixing are discussed in Section 4.2.1.

The concentration of the individual components is proportional to the ratio of the component volumes to the total volume and can be computed from Equations 4.2 and 4.3 in the parts per million and percent range, respectively.

Volumetric methods enjoy several advantages over most other static calibration methods. Low parts per million mixtures of nonreactive, nonadsorbing materials can be generated and easily stored for future use with best accuracies at ±2%. Mixtures in the percent range are also conveniently stored, but explosive mixtures near or above the lower explosive limit must be scrupulously avoided.

4.3 PARTIALLY EVACUATED SYSTEMS

Gaseous mixtures are rarely useful if they are made in a closed container at a pressure of less than 1 atm. Mixtures must usually be transported to another location where the test, standardization, or evaluation takes place. There are, however, several occasions when it is desirable to produce a standard gas mixture at atmospheric pressure or below in a closed vessel for calibration work. A classic example is the production of analytical standards for gas phase infrared and ultraviolet absorption measurements.[123,124] This is illustrated in Figure 4.14.

The major components of the system are an air purifier, an optical cell, a vacuum pump, and a pressure measuring system. To operate the system, stopcock

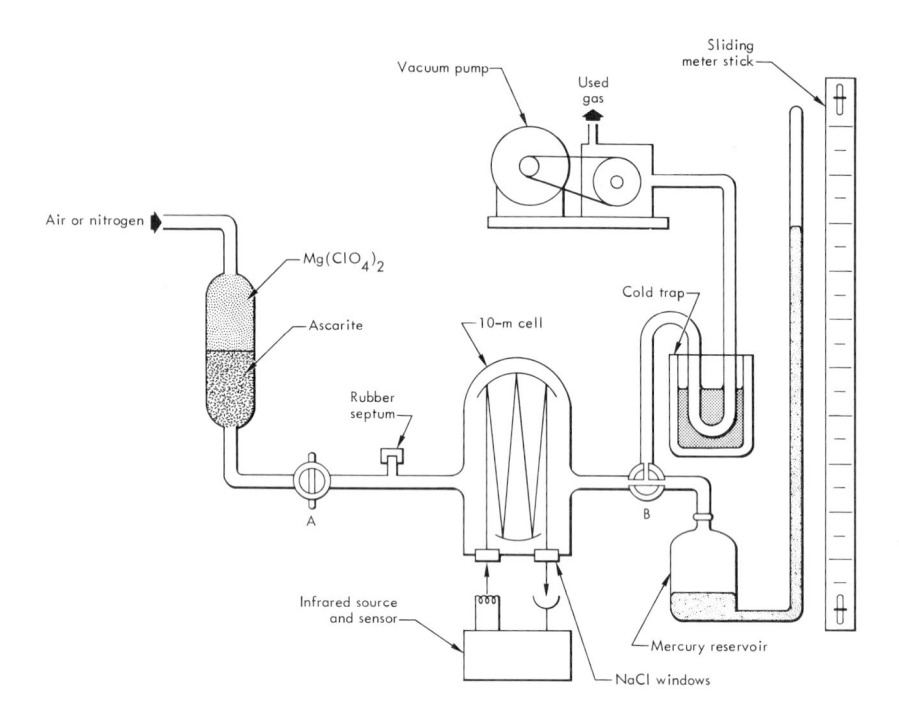

Figure 4.14 Sketch of a partially evacuated system for producing analytical standards for gas-phase infrared absorption measurements.

A is closed and stopcock B is set in the position shown. The system is evacuated to less than 1 mmHg. Stopcock B is then turned 180° to eliminate the vacuum pump from the system. An appropriate amount of gas or liquid is injected into the cell through a rubber septum as the diluent gas is slowly bled into the cell by just cracking stopcock A. The reduced pressure and flowing gas is normally sufficient to vaporize and mix the component gases. The diluent gases (usually air or nitrogen) must be free from water vapor to prevent fogging of the sodium chloride windows. Carbon dioxide should be removed with Ascarite®, for it could interfere significantly with some absorption measurements.

Concentrations are adjusted by pumping the cell down to a known pressure and diluting with more gas. Usually, the most concentrated standard is measured first and progressively more dilute standards are made by adjusting the pressure ratios.

The accuracy of the system just described is ±1 to 5% and depends on the concentration desired and the magnitude of the pressure dilution made. If a 10- to 100-fold dilution is made, the error involved will be larger than if a 2- or 3-fold dilution is made. Materials that are readily adsorbed, reacted, or polymerized should be avoided, for they can adversely affect the optics of a multiple-pass absorption cell.

DeGrazio and Auge[125] have developed a somewhat similar method for producing mixtures of C_1 to C_4 hydrocarbons in helium. The system is similarly evacuated, and trace gases are introduced until the required pressure is attained and

Static Systems for Producing Gas Mixtures

noted with a 0 to 1000 mmHg precision pressure gauge. In this manner, a somewhat better accuracy of ±4% is obtained in the percent range, but a decrease to ±15% in the 100 ppm range is noted.

The starting concentrations of the mixtures can be computed from Equations 4.2 and 4.3 for gases, and from Equations 4.5, 4.6, and 4.8 for liquids.

When injecting liquids or gases into a nearly evacuated system, the dead volume of the syringe needle must be taken into account, especially when generating parts per million concentrations in small-volume systems. Syringe dead volume is not always constant and depends a great deal on the solvent employed, the syringe type, and the needle configuration, as well as on the prevailing experimental conditions. King and Dupre, for example, show the dead volume for methanol (0.30 mL) to be more than twice that of water (0.13 mL) when using certain 5-µL syringes.[126] Experimentally, dead volume for each syringe can be determined by making absorption measurements at various apparent syringe volumes, plotting them, and extrapolating them to zero absorbance. Figure 4.15 shows the results of three sets of such measurements carried out on a typical 10-µL syringe with a 2-in. needle. Notice that the dead volumes are 0.45, 0.60, and 0.80 µL, an appreciably large error, even when the syringe is used at full capacity.

Concentration changes are made by evacuating the system. The resulting concentration is then computed from

$$C = \frac{C_o P}{P_o} \quad (4.23)$$

where C_o is the initial concentration in either parts per million or percent, P is the new reduced pressure, and p_o is the starting pressure.

Example 4.17. What volume of methane is needed to produce a concentration of 750 ppm in air in an 8.05-L cell at a test pressure of 1 atm? What pressure reduction is required to achieve a final concentration of 450 ppm if the pressure is returned to 1 atm?

Initial concentration	750 ppm
Final concentration	450 ppm
Starting pressure	760 mmHg
Cell volume	8.05 L
mL/L	1000

From Equation 4.2, the volume required is

$$v_c = (750)(8.05)(E-06)(1000) = 6.04 \text{ mL}$$

From Equation 4.23, the pressure required is

$$P = \frac{(450)(760)}{750} = 456 \text{ mmHg}$$

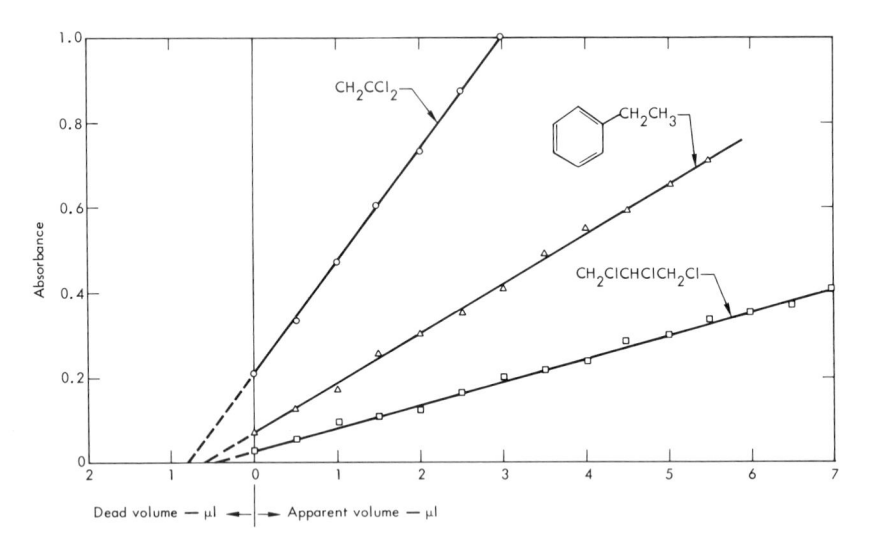

Figure 4.15 Graph for determining the dead volume of a syringe. The actual syringe volume injected is the sum of the apparent and dead volumes.

Example 4.18. What is the apparent volume of benzene required to produce a concentration of 200 ppm in a 7.35-L absorption cell at 700 mmHg and 25°C? The syringe dead volume is 1.2 µL.

Desired concentration	200 ppm
Molecular weight	78.1 g/mol
Density	0.880 g/ml
Cell volume	7.35 L
Standard temperature	273°K
Actual temperature	25°C or 298°K
Standard pressure	760 mmHg
Actual pressure	700 mmHg
Molar volume at std. cond.	22.4 L/mol
Syringe dead volume	1.2 µL

From Equation 4.8, the total volume of benzene required is

$$v_L = \frac{(200)(78.1)(7.35)\left(\dfrac{273}{298}\right)\left(\dfrac{700}{760}\right)}{(22.4)(E+06)(0.880)}$$

$$v_L = 0.00491 \text{ mL or } 4.91 \text{ µL}$$

The apparent syringe volume is then

$$v_a = 4.91 - 1.2 = 3.7 \text{ µL}$$

REFERENCES

1. Axelrod, H. D. and J. P. Lodge. "Sampling and Calibration of Gaseous Pollutants," in *Air Pollution,* 3rd Ed, Vol. III. A. C. Stern, Ed., San Francisco: Academic Press, 1976, chap. 4.
2. Namiesnik, J. *J. Chromatogr.* 300:79 (1984).
3. Smith, A. F. "Standard Atmospheres," *Detection and Measurement of Hazardous Gases,* C. F. Cullis and J. G. Firth, Eds. Exeter, NH:Heinemann Education Books, 1981, chap. 7.
4. Fox, E. A. and V. E. Gex. *J. Air Pollut. Control Assoc.* 7:60 (1957).
5. Benforado, D. M., W. J. Rotella, and D. L. Horton. *J. Air Pollut.* 19:101 (1969).
6. Cotabish, H. N., P. W. McConnaughey, and H. C. Messer. *Am. Ind. Hyg. Assoc. J.* 22:393 (1961).
7. Williams, H. P. *J. Gas Chromotog.* 6:468 (1968).
8. Willson, K. W. and W. Buchberg. *Ind. Eng. Chem.* 50:1705 (1958).
9. Altshuller, A. P. and C. A. Clemons. *Anal. Chem.* 34:466 (1962).
10. Altshuller, A. P., A. F. Wartberg, I. R. Cohen, and S. F. Sleva. *Int. J. Air Water Pollut.* 6:75 (1962).
11. DeGrazio, R. P. *J. Gas Chromotog.* 6:468 (1968).
12. Russell, J. W. and L. A. Shadoff. *J. Chromatogr.* 134:375 (1977).
13. Tanaka, T. *J. Chromatogr.* 153:7 (1978).
14. Morris, C., R. Berkley, and J. Bumgarner. *Analyt. Lett.* 16:1585 (1983).
15. Samimi, R. *Am. Ind. Hyg. Assoc. J.* 44:40 (1983).
16. Thuman, W. C. "Development of Technology for Production, Sampling and Assay of Simulated Atmospheres in Closed Chambers," unnumbered report, (Palo Alto, CA:Stanford Research Institute, 1958).
17. Brief, R. S. and F. W. Church. *Am. Ind. Hyg. Assoc. J.* 21:239 (1960).
18. Campbell, E. E., M. F. Milligan, and H. M. Miller. *Am. Ind. Hyg Assoc. J.* 20:138 (1959).
19. Leonardos, G., D. Kendall, and N. Barnard. *J. Air Pollut. Control Assoc.* 19:91 (1969).
20. Baretta, E. D., R. D. Stewart, and J. E. Mutchler. *Am. Ind. Hyg. Assoc. J.* 30:537 (1969).
21. Stead, F. M. and G. J. Taylor. *J. Ind. Hyg. Tox.* 29:408 (1947).
22. Van Sandt, W., V. Santomassimo, and R. D. Taylor. "Determination of Dilution of Vapors in Calibration Bottles by Infrared Analysis," in *Hazards Control Quarterly Report,* No. UCRL-7450, (Livermore, CA: Lawrence Livermore National Laboratory 1963).
23. Apol, A. G., W. A. Cook, and E. F. Lawrence. *Am. Ind. Hyg. Assoc. J.* 27:149 (1966).
24. Morley, M. J. and B. D. Tebbens. *Am. Ind. Hyg. Assoc. J.* 14:303 (1953).
25. Harsch, D. E. *Atmos. Environ.* 14:1105 (1980).
26. McCaldin, R. O. and E. R. Hendrickson. *Am. Ind. Hyg. Assoc. J.* 20:509 (1959).
27. Paule, R. C. *Anal. Chem.* 44:1537 (1972).
28. Conner, S. W. and A. V. Jenson. *Anal. Chem.* 36:799 (1964).
29. Sinterland, A. N. *Am. Ind. Hyg. Assoc. J.* 14:113 (1953).
30. Lodge, J. P., J. B. Pate, and H. A. Huitt. *Chemist Analyst* 52:53 (1963).
31. *Matheson Gas Data Book,* 6th ed. (East Rutherford, NJ: Matheson Company Inc., 1980)

32. Scientific Gas Products Inc., General Catalog No. 91 (1982).
33. Feldstein, J. and J. D. Coons, H. C. Johnson, and J. E. Yocom. *Am. Ind. Hyg Assoc. J.* 20:374 (1959).
34. Roccanova, B. "Present State of the Art of the Preparation of Gaseous Standards," paper presented at the Conference on Analytical Chemistry and Spectroscopy, Pittsburgh, PA, 1968.
35. Cole-Parmer Instrument Company, Catalog 1991–1992, Chicago, IL.
36. Crawford, R. W. "An Automatic, Variable-Volume, Gas Dispenser, report No. UCID-15138, revision 1, (Livermore, CA: Lawrence Livermore National Laboratory, 1965).
37. Silverman, L. "Experimental Test Methods," in *Air Pollution Handbook,* P. L. Magill, F. R. Holden, and C. Ackley, Eds. (New York: McGraw-Hill Book Company, Inc., 1956).
38. Vango, S. P. *Chemist-Analyst* 52:53 (1963).
39. Urone, P., J. B. Evans, and C. M. Noyes. *Anal. Chem.* 37:1104 (1965).
40. Russell, S. *Am. Ind. Hyg. Assoc. J.* 25:359 (1964).
41. Mullet, C. W. *Gas* 37:59 (1961).
42. Langer, A. *Rev. Sci. Instr.* 18:101 (1947).
43. Roberts, R. M. and J. J. Madison. *Anal. Chem.* 29:1555 (1957).
44. Hanson, D. N. and A. Maimoni. *Anal. Chem.* 31:158 (1959).
45. Back, R. A., N. J. Friswell, J. C. Boden, and J. M. Parsons. *J. Chromatog. Sci.* 7:708 (1969).
46. Stewart, R. D. *Am. Ind. Hyg. Assoc. J.* 22:252 (1961).
47. Brown, J. R. and E. Mastromatteo. *Am. Ind. Hyg. Assoc. J.* 25:560 (1964).
48. Brief, R. S., R. S. Ajemian, and R. G. Confer. *Am. Ind. Hyg. Assoc. J.* 28:21 (1967).
49. Powell, C. H. and A. D. Hosey, Eds. "The Industrial Environment — Its Evaluation and Control" (Washington, D. C.: U. S. Government Printing Office, 1965).
50. Stead, F. M. and G. J. Taylor. *J. Ind. Hyg. Tox.* 29:408 (1947).
51. Silver, S. D. *J. Lab. Clin. Med.* 31:1153 (1946).
52. Buchberg, H. and K. W. Wilson. *J. Air Pollut. Control Assoc.* 8:285 (1959).
53. Buchberg, H, K. W. Wilson, and R. P. Lipkis. Preliminary Design Study — Air Resource Test Facilities, Report No. 56-59, (Los Angeles, CA: University of California 1956).
54. De Souza, T. L. C. and S. P. Bhatia. *Anal. Chem.* 48:2234 (1976).
55. Ritter, J. J. and N. K. Adams. *Anal. Chem.* 48:612 (1976)
56. Freeland, L. T. *Am. Ind. Hyg. Assoc. J.* 38:172 (1977).
57. Sedlak, J. M. and K. F. Burton. *Anal Chem.* 48:2020 (1976).
58. Pellizzari, E. D., J. E. Bunch, B. H. Carpenter, and E. Sawicki. *Environ. Sci Technol.* 9:552 (1975).
59. Turk, A. "Basic Principles of Sensory Evaluation," ASTM Spec. Tech. Publ. No. 433, 79 (1969).
60. Braun, W., N. C. Peterson, A. M. Bass, and M. J. Kurylo. *J. Chromatogr.* 55:237 (1971).
61. Fontijn, A., A Sabedel, and R. J. Ronco. *Anal. Chem.* 42:575 (1970).
62. Nozoye, H. *Anal. Chem.* 50:1727 (1978).
63. Spangler, G. E. and P. A. Laweless. *Anal. Chem.* 50:884 (1978).
64. Bruner, F., C. Canulli, and M. Possanzini. *Anal. Chem.* 45:1790 (1973).
65. R. S. Braman. *Chromatographic Analysis of the Environment,* in R. L. Grob, Ed. (New York: Marcel Dekker 1975), p. 82.

66. Fowlis, I. A. and R. P. W. Scott. *J. Chromatogr.* 11:1 (1963).
67. Gill, W. E. *Am. Ind. Hyg. Assoc. J.* 21:87 (1960).
68. Caplan, P. E. "Calibration of Air Sampling Instruments," in *Air Sampling Instruments for Evaluation of Atmospheric Contaminants* (Cincinnati, OH: American Conference of Governmental Industrial Hygienists, 1966).
69. Niedermayer, A. O. *Anal. Chem.* 42:310 (1970).
70. Katz, M., Ed. *Methods of Air Sampling and Analysis,* 2nd ed. (Washington D. C.: American Public Health Association, 1977), p.17.
71. Hanson, D. N. and A. Maimoni. *Anal. Chem.* 31:158 (1959).
72. Fultyn, R. V. *Am Ind. Hyg. Assoc. J.* 22:49 (1961).
73. Groth, R. H. and T. B. Doyle. *J. Gas Chromatog.* 6:138 (1968).
74. Saltzman, B. E. *Anal. Chem.* 37:1261 (1965).
75. Wohlers, H. C., N. M. Trieff, H. Hewsteir, and W. Stevens. *Atmos. Environ.* 1:121 (1967).
76. Sherberger, R. F., G. P. Happ, F. A. Miller, and D. W. Fassett. *Am. Ind. Hyg. Assoc. J.* 19:464 (1958).
77. Van Sandt, W., V. Santomassimo, and R. D. Taylor. "Determination of Dilution of Vapors by Infrared Analysis," in *Hazards Control Quarterly Report,* UCRL-7450, (Livermore, CA: Lawrence Livermore National Laboratory 1963).
78. Lodge, J. P. "Production of Controlled Test Atmospheres," in *Air Pollution,* A. C. Stern, Ed. (New York: Academic Press, 1962).
79. Williams, H. P., C. V. Overfield, and J. D. Winefordner. *J. Gas Chromatog.* 5:511 (1967).
80. Lacy, J. and K. G. Woolmington. *Analyst* 86:547 (1961).
81. Baker, R. A. and R. C. Doerr. *Int. J. Air Water Pollut.* 2:142 (1959).
82. SKC Product Bulletin, SKC Sample Bags, Fullerton, CA.
83. Calibrated Instrument Tech Bulletin A-5, Ardsley, NY.
84. Pollution Measurement Corporation, Bulletin GB-74, Chicago, IL.
85. Carborundum Product Bulletin, Avondale, PA.
86. Cole-Parmer Instrument Company Catalog 1991–1992, Chicago, IL, p. 16.
87. Fluorodynamics Inc. Product Bulletin, Newark, DE.
88. Vanderkolk, A. L. and D. E. Van Farowe. *Am. Ind Hyg. Assoc. J.* 26:321 (1965).
89. Clemons, C. A. and A. D. Altshuller. *J. Air Pollut. Control Assoc.* 14:407 (1964).
90. Wilson, K. W. and H. Buchberg. *Ind. Eng. Chem.* 50:1705 (1958).
91. Pate, J. B., J. P. Lodge, and M. P. Neary. *Anal. Chem. Acta* 28:341 (1963).
92. Conner, W. D. and J. S. Nader. *Am. Ind. Hyg. Assoc. J.* 25:291 (1964)
93. Welch, A. F. and J. P. Terry. *Am. Ind. Hyg. Assoc. J.* 21:316 (1960).
94. Altshuller, A. P., A. F. Wartburg, I. R. Cohen, and S. F. Sleva. *Int. J. Air Water Pollut.* 6:75 (1962).
95. Cederlolf, R., M. Edfors, L. Friberg, and T. Lindvall. *J. Air Pollut. Control Assoc.* 16:92 (1966).
96. Simonds, H. R. and J. M. Church. *A Concise Guide to Plastics,* 2nd ed. (New York: Reinhold Publishing Company, 1963).
97. Bulletin No. 69022, Precision Gas Products, Inc., Linden, NJ.
98. Bulletin No. 7167, Scott Research Laboratories, San Bernardino, CA.
99. Collins, W. T. "A Gravimetric Standard for Primary Gas Flow Measurements," Report No. K-L-6181, Oak Ridge National Laboratory, Oak Ridge, TN, 1967.
100. Scornavacca, F. "Gas Mixtures — Facts and Fables," Matheson Gas Products Bulletin, 1975.

101. Kebbekus, E. and F. Scornavacca. *Am. Lab.* 9:51 (1977).
102. Wechter, S. G. "Preparation of Stable Pollution Gas Standards Using Treated Aluminum Cylinders," presented at the ASTM Calibration Symposium, Boulder, CO, August 5, 1975.
103. Gurvich, V. S. *Izv. Sibirsk. Otd. Akad. Nauk SSSR* 165 (1968).
104. Rhoderick, G. C. and W. L. Zielinski. *Anal. Chem.* 60:2454 (1988).
105. Miller, J. E., A. J. Carroll, and D. E. Emerson. "Preparation of Primary Standard Gas Mixtures for Analytical Instruments," Report No. 6674, (Washington, D.C.: U.S. Bureau of Mines, 1965).
106. Von Lehmden, D. J., E. L. Tew, and C. E. Decker. *J. Air Pollut. Cont. Assoc.* 37:384 (1987).
107. Roccanova, B. "Present State of the Art of the Preparation of Gaseous Standards," paper presented at the Conference on Analytical Chemistry and Spectroscopy, Pittsburgh, PA (1968).
108. Samini, B. S. *Am. Ind. Hyg. Assoc. J.* 44:40 (1983).
109. Lee, W. G. and J. A. Paine. Calibration in Air Monitoring, ASTM Spec. Pub. 598, (1975) p. 210.
110. Shannon, D. W. "Gas Chromatograph," in Quarterly Progress Report, Metallurgy Research Operation, April, May, June, 1964, Report No. HW-82651. General Electric Company, Richland, WA (1964).
111. Ruby, E. D. "An Apparatus for the Preparation of Standard Gas Mixtures Containing Trace Level Components," Report No. RFP-358, (Rocky Flats, CO: Dow Chemical Company, 1964).
112. Baker, W. J. and T. L. Zinn. *Perkin Elmer Instrument News* 11:1 (1960).
113. Zdrojewski, A. and J. L. Monkman. *Am. Ind. Hyg. Assoc. J.* 30:650 (1969).
114. Batt, L. and F. R. Cruickshank. *J. Chem. Soc. (London)* A2:261 (1967).
115. Silverman, L. "Experimntal Test Methods," in *Air Pollution Handbook,* P. L Magill, F. R. Holden, and C. Ackley, Eds. (New York: McGraw-Hill Book Company, Inc., 1956).
116. Roccanova, B. "Present State of the Art of the Preparation of Gaseous Standards," paper presented at the Conference on Analytical Chemistry and Spectroscopy, Pittsburgh, PA (1968).
117. Hughes, E. E. and W. D. Dorko. *Anal. Chem.* 40:750 (1968).
118. Cotabish, H. N., P. W. McConnaughey, and H. C. Messer. *Am. Ind Hyg. Assoc. J.* 22:393 (1968).
119. Roberts, R. M. and J. J. Madison. *Anal. Chem.* 29:1555 (1957).
120. Opler, A. and E. S. Smith. *Anal Chem.* 25:686 (1953).
121. French patent 1,087,140.
122. Fehringer, D. J. "Method for Preparing Low Concentration Standard Gas Samples," Report No. RFP-1626, (Rocky Flats, CO: Dow Chemical Company, 1971).
123. Feldstein, J. J., J. D. Coons, H. C. Johnson, and J. E. Yocom. *Am. Ind. Hyg. Assoc. J.* 20:374 (1959).
124. Altshuller, A. P. and A. F. Wartburg. *Appl. Spectr.* 15: 67 (1961).
125. DeGrazio, R. P. and R. G. Auge. "Evaluation of a Gas Mixing System," Report No. RFP-1143,(Rocky Flats, CO: Dow Chemical Company, 1968).
126. King, W. H. and G. D. Dupre. *Anal. Chem.* 41:1936 (1969).
127. Posner, J. C. and J. Woodfin. *Appl. Ind. Hyg.* 1:163 (1986).

CHAPTER 5

DYNAMIC SYSTEMS FOR PRODUCING GAS MIXTURES

The dynamic method of generating gas and vapor mixtures requires the uninterrupted blending of the component parts for some specified period of time. This technique enjoys many advantages over static methods and is especially useful in producing reactive gas mixtures. If the mixture is prone to decomposition or chemical reaction, the undesirable reaction products can be swept away and continually replaced by the relative pure and unreacted test gas mixture. Large volumes of concentrations from 50% down to the parts per billion range can be controlled and altered easily with convenient, compact equipment. Wall adsorption, a problem with most static systems, usually becomes negligible, since an equilibrium is established after operating for a long enough time period. However, these general advantages are sometimes offset by the cost and complexity of many dynamic methods, especially those required for producing parts per million and parts per billion mixtures from pure starting materials.

Dynamic methods have a much wider range of applicability than the static or closed systems. In any operation where unwanted waste gases must be swept away, they are indispensible. For example, in toxicological, inhalation, and odor investigations where oxygen and the gas of interest are consumed and carbon dioxide and water vapor are produced, a test mixture can be continuously supplied to maintain the desired component concentration. Other areas of applicability include gas-phase catalytic and kinetic studies, adsorption and absorption measurements, gas irradiation experiments, and analytical standards. Dynamic methods are especially useful for continuously producing standards for gas-phase infrared spectrophotometry and direct-reading gas-measuring instrumentation.

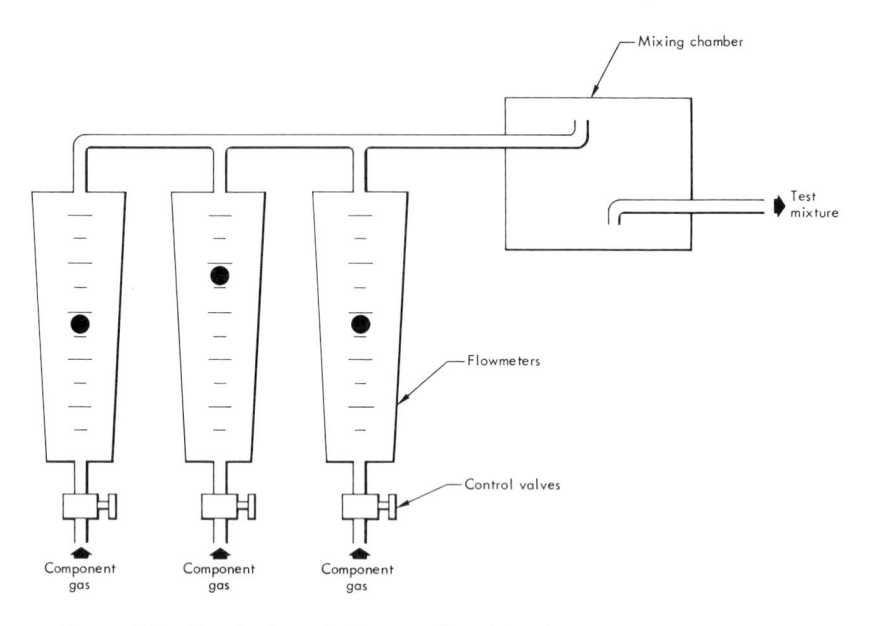

Figure 5.1 Sketch of a typical system for mixing three-component gas streams.

5.1 GAS STREAM MIXING

5.1.1 Single Dilution

The most widely used and successful method of mixing two or more gases is to dilute the gases with one another after measuring their flow rates.[1-9] Figure 5.1 shows a three-component test mixture being metered through calibrated flow meters (see Chapter 3). If the flow meters are operating smoothly, no mixing chamber is necessary. The turbulence of the entering components is normally sufficient to promote homogeneous mixing at the end of a normal length of tubing before exiting.[10]

Gases can either be metered or pumped as pure components,[11-22] or be previously diluted[23-30] to facilitate high dilution. Dilutions of 1,000:1 can be routinely achieved with commercial flow meters, and dilutions of 10,000:1 can be accomplished if a high-volume flow meter is chosen. The extreme low end of the flow meter range should not be used as accuracy as will be sacrificed. Nominal flow meter ranges are shown in Table 5.1. Any of the types shown may be used with any other in any given system. If small-volume dispensers (i.e., syringe injectors) are to be diluted with relatively high diluent gas flows, care must be taken to keep the inlets as small and short as possible. With too large an inlet, excessively long time periods are required to reach equilibrium.[23] Figure 5.2 shows an example of such an inlet. The 0.5-mm i.d. nozzle passes flows up to 0.3 L/min with only a fraction of an inch pressure drop.[23]

In addition to those flow meters shown in Table 5.1, mixing pumps with combined gases pumped through pistons at adjustable speeds are available. The

Dynamic Systems for Producing Gas Mixtures

Table 5.1 Ranges of Various Flow Rate Measuring Devices

Flow Device	Low	High
Asbestos plug flow meter	0.01 mL/min	100 mL/min
Controlled leak	0.001 µL/min	10 L/min
Critical orifice	10 mL/min	100 L/min
Dry gas meter	5 L/min	5,000 L/min
Heated-wire anemometer	1 mL/min	5,000 L/min
Mass flow controller	0.5 mL/min	1,000 L/min
Mass flow meter	1 mL/min	5,000 L/min
Orifice meter	5 mL/min	50 L/min
Pitot tube	600 ft/min	>10,000 ft/min
Rotameter	2 mL/min	300 ft^3/min
Rotating stopcock	0.01 mL/min	5 mL/min
Syringe injection	0.01 mL/min	100 mL/min
Wet test meter	1 L/min	80 L/min

(Approximate Useful Range)

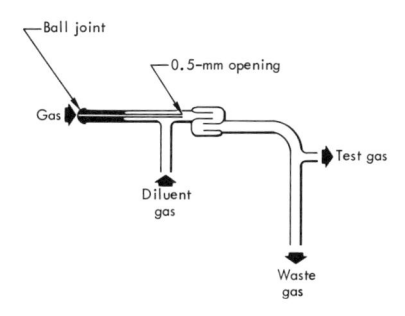

Figure 5.2 Sketch of a system with a low-flow inlet for making single dilutions.

methods of choice for dilution are usually rotameters, mass flow meters, or mass flow controllers.[31] However, orifice meters[11] and rotating stopcocks have been used.[32-34]

Equipment for any dynamic system is usually made with as inert a material as possible and practical. Glass and Teflon are normally used, and long sections of rubber or polymeric tubing should be avoided, especially when corrosive and reactive mixtures are involved. For example, a 1-in. section of Tygon tubing has been known to reduce a 0.1 mL/min flow of chlorine gas by 50%.[23]

The concentration calculation for a single dilution system is

$$C_\% = \frac{10^2 q_a}{q_a + q_b + \cdots + q_n} \tag{5.1}$$

where $C_\%$ is the concentration (percent), and $q_{a,b,n}$ are the flow rates of the individual pure components (volume/unit time). If component a is not pure, then Equation 5.1 becomes

$$C_{\%} = \frac{10^2 X q_a}{q_a + q_b + \cdots + q_n} \qquad (5.2)$$

where X is the mole fraction or decimal percent purity by volume.

For a two-component gas mixture, Equation 5.2 simplifies to

$$C_{\%} = \frac{10^2 X q_a}{q_a + q_D} \qquad (5.3)$$

where q_D is the dilution gas.

If the flow rate of component a is to be calculated, then Equation 5.3 for a pure gas rearranges to

$$q_a = \frac{C_{\%} q_D}{(100 - C_{\%})} \qquad (5.4)$$

The concentration in parts per million can be calculated from

$$C_{ppm} = \frac{10^6 q_a}{q_D} \qquad (5.5)$$

or

$$C_{ppm} = \frac{10^6 X q_a}{q_D} \qquad (5.6)$$

if the component gas is not pure.

Example 5.1. What is the flow rate of helium required to produce a concentration of 500 ppm in 100 L/min of air? Express the answer in mL/min.

Desired concentration	500 ppm
Air flow rate	100 L/min
mL/L	1000

From Equation 5.5, the required helium flow is

Dynamic Systems for Producing Gas Mixtures

$$q_{(He)} = \frac{(500)(100)(1000)}{(E+06)} = 50.0 \text{ mL/min}$$

Example 5.2. What flow rate of 1.0% hydrogen cyanide is required to produce a concentration of 1.5 ppm hydrogen cyanide in 20 L/min air?

Desired concentration	1.5 ppm
Air flow rate	20 L/min
HCN concentration	1.0% or 0.010 decimal percent
mL/L	1000

From Equation 5.5, the required flow of 1.0% hydrogen cyanide is

$$q_{(HCN)} = \frac{(20)(1000)(1.5)}{(0.010)(E+06)} = 3.0 \text{ mL/min}$$

Example 5.3 What flow rate of carbon monoxide is required to produce a 20% concentration in a stream of helium flowing at 10 ft³/min? Express the answer in L/min.

Desired concentration	20%
Helium flow rate	10 ft³/min
L/ft³	28.3

From Equation 5.4, the required carbon monoxide flow is

$$q_{(CO)} = \frac{(20)(10)(28.3)}{(100-20)} = 70.8 \text{ L/min}$$

If a cylinder of diluted gas is to be further diluted, the final concentration is

$$C_F = \frac{C_I q_a}{q_a + q_D} \quad (5.7)$$

where C_F is the final concentration (ppm or percent) and C_I is the initial concentration (ppm or percent) present in the cylinder. If the flow rate of component a is desired, then Equation 5.7 rearranges to

$$q_a = \frac{C_F q_D}{(C_I - C_F)} \quad (5.8)$$

Example 5.4. What is the concentration produced if 500 mL/min of 2000 ppm hydrogen fluoride is diluted with 2.00 ft³/min of air?

Flow rate of HF	500 mL/min or 0.500 L/min
HF concentration	2000 ppm
Air flow rate	2.00 ft³/min or 56.6 L/min
L/ft³	28.3
mL/L	1000

Using Equation 5.7, the final hydrogen fluoride concentration is

$$C_{(HF)} = \frac{(2000)(0.500)}{(0.500 + 56.6)} = 17.5 \text{ ppm}$$

Example 5.5. A cylinder contains 0.500% hydrogen chloride in nitrogen. What cylinder flow rate is required to produce a 600-ppm hydrogen chloride in 50 L/min nitrogen?

Initial HCl conc.	0.500% or 5000 ppm
Final HCl conc.	600 ppm
Nitrogen flow rate	50 L/min
ppm/%	10,000

Using Equation 5.8, the cylinder flow rate of hydrogen chloride is

$$q_{(HCl)} = \frac{(600)(50)}{(5000 - 600)} = 6.82 \text{ L / min}$$

5.1.2 Double Dilution

It is often desirable to generate a low parts per million concentration while avoiding high dilution ratios and the use of extremely low-flow metering devices. This can be accomplished by using the double- or multiple-dilution technique shown in Figure 5.3.[23,35-93] Here, the contaminant and diluent gases are combined in the first stage and the majority is allowed to escape using two control valves. The remaining mixture is again metered and combined with a second-stage dilution to form the final concentration. Using this method, the initial contaminant gas can be diluted by four to nine orders of magnitude.

Calculations for double or any multiple dilution are based on Equation 5.2 and can be expressed for an n-stage dilution by

$$C_{\%} = 10^2 \left(\frac{q_{a_1}}{q_{a_1} + q_{D_1}} \right) \left(\frac{q_{a_2}}{q_{a_2} + q_{D_2}} \right) \cdots \left(\frac{q_{a_n}}{q_{a_n} + q_{D_n}} \right) \quad (5.9)$$

where C is the concentration (percent), $q_{a_{1,2,\ldots,n}}$ is the flow rate of contaminant a

Dynamic Systems for Producing Gas Mixtures

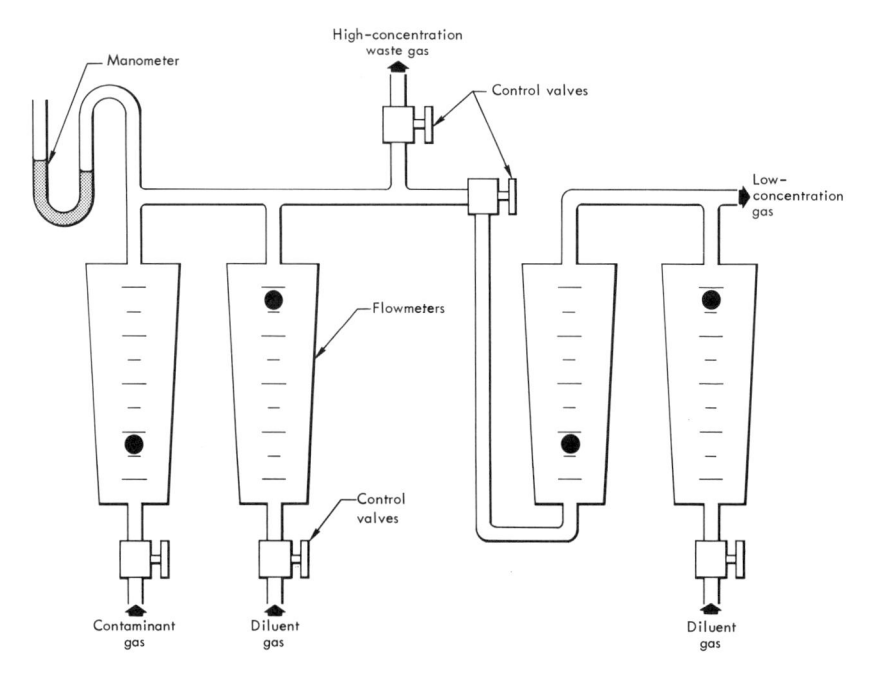

Figure 5.3 Sketch of a system for making double dilutions.

at stage 1, 2,...,n (L/min), and $q_{D1,2,...,n}$ is the flow rate of diluent gas at stage 1, 2,...,n (L/min).

Example 5.6. Nitric oxide (97% pure by volume) flows at 20 mL/min and is diluted with nitrogen flowing at a rate of 25 L/min. One percent of the resultant concentration is again diluted with 30 L/min of nitrogen. Calculate the resultant nitric oxide concentration.

Nitric oxide flow rate	20 mL/min
Primary nitrogen flow rate	25 L/min
Secondary nitrogen flow rate	30 L/min
Amount of first stage used	1.0% or 0.010 decimal percent
Nitric oxide purity	97% or 0.97 decimal percent
mL/L	1000

From Equation 5.6, the resultant nitric oxide concentration in the first stage is

$$C_{(NO)} = \frac{(0.97)(20)(E+06)}{(25)(1000)} = 776 \text{ ppm}$$

From Equation 5.7, the final nitric oxide concentration is

$$C_{(NO)} = \frac{(776)(0.010)(25)}{[30 + (0.010)(25)]} = 6.41 \text{ ppm}$$

Although the multiple dilution technique would seem to be simple enough, it actually becomes difficult after the initial dilution. Rotameters have been observed to oscillate wildly,[44] and a variety of resonances and instabilities make them extremely hard to control.[45] Moreover, as each stage is added, pressure buildups are inevitable from the control valve adjustments, and corrections become increasingly more difficult. Some of these instabilities can be side-stepped using mass flow controllers. They exhibit a steady flow control, and no pressure or temperature corrections are required. Accuracies, however, are ±5 to 10% after a typical two-stage dilution, although 2% has been reported by Schnelle.[36] A major disadvantage of the serial dilution method is the tremendous amount of contaminant gas that is wasted after each dilution stage. Fully 99% of the test gas in normally bled off and is not used in making up the test atmosphere.

5.2 INJECTION METHODS

Gases and liquids can be added to moving gas streams by a wide variety of mechanical dosers, injectors, and syringe peristaltic and high-pressure liquid chromatography pumps. The early attempts at producing standard test atmospheres containing one or more components used motor-driven syringes, belt drives, and geared mechanisms. These methods, when merged with modern speed control techniques, can generate gas mixtures of high accuracy and precision over extended time intervals. The concentrations generated can extend from the percent range to fractions of a part per million. Such methods are particularly useful in preparing standards for direct-reading instruments, detector tube and dosimeter evaluation, and analytical calibration curves for laboratory gas analyzers. These methods find considerable use in adsorption and absorption measurements, as well as in studying relatively slow chemical reactions by feeding in reactants at known controlled rates.

An idealized feed system is shown in Figure 5.4. The diluent gas supply (usually air) is pressure regulated, and an appropriate flow rate is chosen. Gas or liquid is forced into the injection port by some dispensing device, such as a pump, a motor-driven syringe, a piston, or even a gravity feed mechanism. If a liquid is involved, some method must be employed to vaporize and dispense it evenly into the moving diluent gas stream. A cooling unit is sometimes needed to remove or exchange unwanted heat if a heater is used to assist in the vaporization operation. A mixing chamber is required if the contaminant gases or vapors are added unevenly or if surges exist in the diluent gas flow or the vaporization process. A control unit is provided to alter feeding rates and to provide thermostatic control, if desired, over the evaporator.

5.2.1 Liquid Reservoirs

Motor-driven feeding devices have been used extensively to introduce gases, liquids, vapors, and even particulate matter into moving gas streams. The material to be injected is contained in some type of inert reservoir, the volume of which diminishes even when acted upon by the mechanical dispensing apparatus. Syringes have long been employed for this purpose,[46-48,65-72] and several types

Dynamic Systems for Producing Gas Mixtures

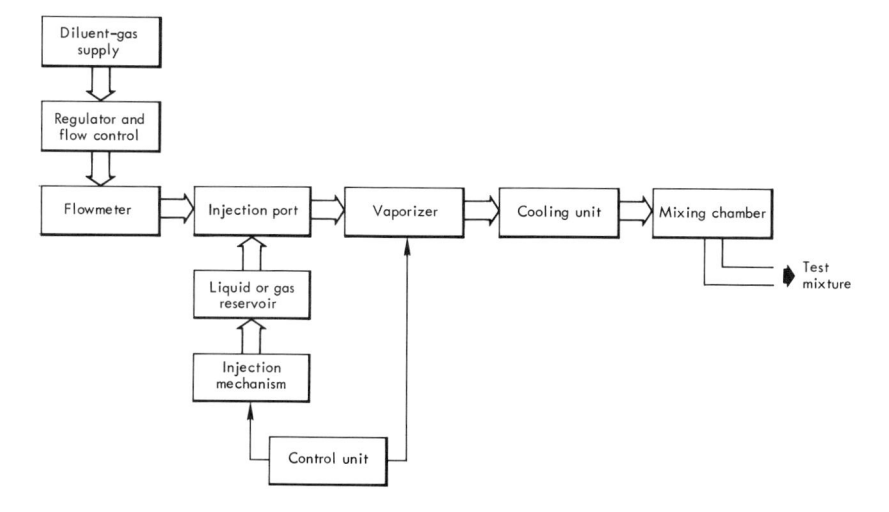

Figure 5.4 Block diagram of the major components of a dynamic injection system.

currently used and available are discussed in Chapter 4. Gas-tight syringes with a Teflon-tipped plunger are particularly useful for injecting most liquids or gases, but syringes made entirely of Teflon must be used in special instances, such as when working with hydrogen fluoride. When making low-rate injections, syringes made entirely of glass usually lose some material, particularly if it is a liquid, between the glass interfaces, and they are subject to large delivery errors, especially in the microliter per minute injection range. Syringes are available with 0.05- to 50-mL capacities and will deliver from 0.5 to 500 µL/min when dispensed at moderate or slow speeds.

Syringes can be used to make injections one at a time (Figure 5.5A),[49-56] two or more at a time (Figure 5.5B),[18,40,48,57-61] or alternately where one injects while the other fills itself from the reservoir (Figure 5.5C).[62] During long running times, some air may accumulate in the syringe, but this can be kept as low as 0.1 mL/day if tubing connections are short and system integrity is maintained.[62]

Syringe delivery rates will, of course, vary with syringe capacity and the rate of plunger advance. The amount of liquid dispensed for various syringe capacity and advance rates is shown in Table 5.2. Commercial syringe dispensers will deliver less than 1 mL/min to 720 µL/min if two 50-mL capacity syringes are used.[63]

The syringe can be mounted in either the upright or horizontal position. It must be securely clamped into position so that the driving arm of the pump makes firm contact with the plunger. Syringe mantles or removable glass jackets can be added to maintain constant temperatures or to circulate heating or cooling media to ensure that the material being injected is maintained at the proper phase.

Other reservoirs, such as ultramicroburets, have been used,[64] but some difficulties have been encountered during filling because of the general fragility of the glassware and the presence of the mercury interface.

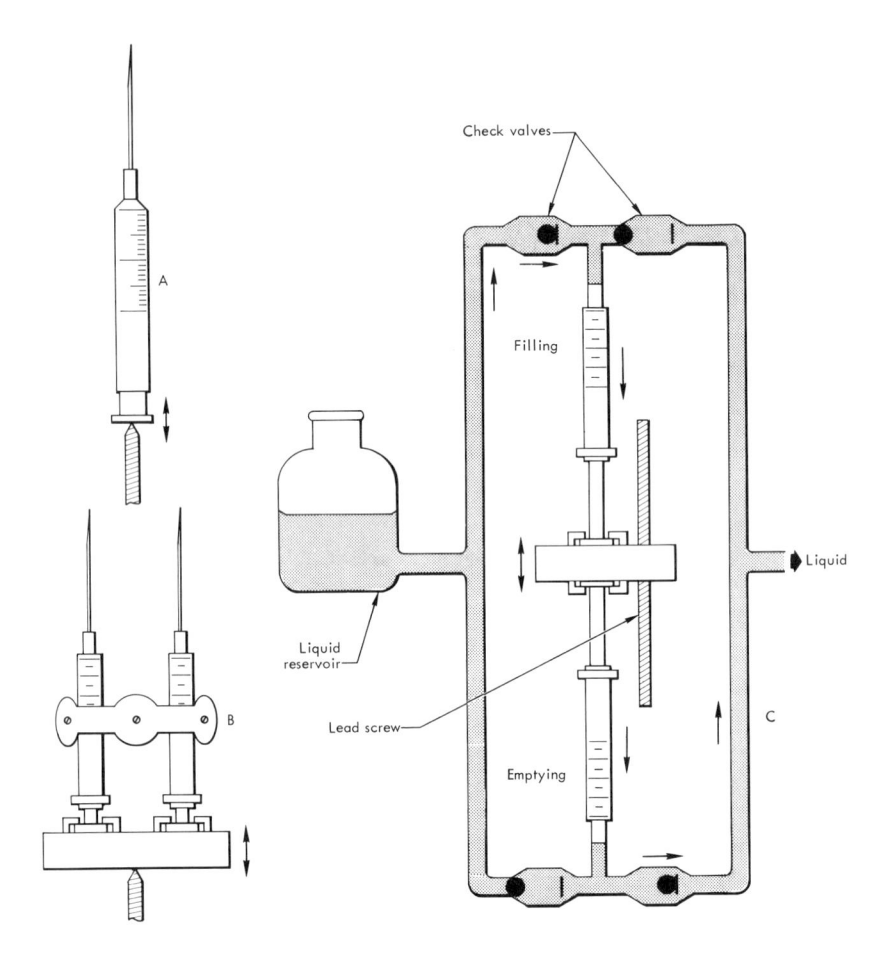

Figure 5.5 Sketch of a single-syringe injector (A) and two kinds of multiple syringe injectors (B and C). The check valves in C permit the syringes to alternately fill and empty as the lead screw moves up and down.

5.2.2 Syringe Drive Systems

Syringes, microburets, or any other contaminant reservoir can be emptied smoothly and evenly with a variety of techniques, but some type of electromechanical system with a motor is preferred. The major commercially available device usually chosen is the syringe pump.

Syringe injectors or pumps are summarized in Table 5.3. These pumps are available with discrete or completely variable speeds. The advance rate can be set with external dials or programmed via a key pad, such as in the Harvard unit (see Table 5.4 for the pertinent syringe dimensions for programming.) The pumps will accept syringe capacities from 1 µL to 140 mL and dispense the liquid contaminant from 0.01 µL/min to 140 mL/min. Gaseous contaminant can be injected using only gas-tight syringes. These pumps can accommodate as many as three

Dynamic Systems for Producing Gas Mixtures

Table 5.2 Syringe Injection Characteristics

Syringe Size Capacity (mL)	0.05	0.1	0.25	0.50	1.0	2.5	5.0	10	20	50	100
Length (mm)	60	60	60	60	60	60	60	61.8	66.4	80.8	102.5
µL/mm of Length	0.833	1.67	4.17	8.33	16.7	41.7	83.3	162	301	619	976

Syringe Advance Rate (mm/min)					Dispensed Volume (µL/min)						
0.010	0.0083	0.167	0.0417	0.0833	0.167	0.417	0.833	1.62	3.01	6.19	9.76
0.020	0.0167	0.0333	0.0833	0.167	0.333	0.833	1.67	3.24	6.02	12.4	19.5
0.050	0.0417	0.0833	0.208	0.417	0.833	2.08	4.17	8.09	15.1	30.9	48.8
0.10	0.0833	0.1667	0.417	0.833	1.67	4.17	8.33	16.2	30.1	61.9	97.6
0.20	0.167	0.333	0.833	1.67	3.33	8.33	16.7	32.4	60.2	124	195
0.50	0.417	0.833	2.08	4.17	8.33	20.8	41.7	80.9	150.6	309	488
1.0	0.833	1.67	4.17	8.33	16.7	41.7	83.3	162	301	619	976
2.0	1.67	3.33	8.33	16.7	33.3	83.3	167	324	602	1,238	1,951
5.0	4.17	8.33	20.8	41.7	83.3	208	417	809	1,506	3,094	4,878
10	8.33	16.7	41.7	83.3	167	417	833	1,618	3,012	6,188	9,756
20	16.7	33.3	83.3	167	333	833	1,667	3,236	6,024	12,376	19,512
50	41.7	83.3	208	417	833	2,083	4,167	8,091	15,060	30,941	48,780

Table 5.3 Characteristics of Commercially Available Syringe Injectors

Manufacturer	Model	Number of Speeds	Speed Ratio	Number of Syringes Held	Syringe Sizes Accomodated (mL)	Range of Volume Dispensed (μL/min)
Brinkmann	25 20 060-8	2	2:1	1	0.050 – 0.50	21 – 430
Sage	341 B	11	60:1	1	0.001 – 50	0.01 – 13,000
Sage	351, 352	19	40:1	3	0.0005 – 100	150 – 60,000
Sage	355	Variable	100:1	3	0.0005 – 100	0.0029 – 140,000
Harvard	22	Variable	16,000:1	2	0.010 – 140	0.029 – 55,000

Table 5.4 Syringe Size Characteristics [a]

Manufacturer Materials	Becton Dickinson Plastic	Becton Dickinson Glass	Popper & Sons Glass	Unimetrics Glass	Sherwood Monoject Plastics	Ranfac	Harvard St. steel	Hamilton Glass
Syringe volume (mL)								
0.010				0.460				0.460
0.025				0.729				0.729
0.050				1.031				1.031
0.10				1.460				1.46
0.25			3.45	2.30				2.30
0.50			3.45	3.26				3.26
1.0	4.78	4.64	4.50	4.61	4.65			4.61
2.0		4.64	8.92			9.12		7.28
2.5		8.66						
3.0	8.66		8.99		8.94			
5.0	12.06	11.86	11.70			12.34		10.30
6.0					12.7			
10	14.50	14.34	14.70			14.55		14.57
12					15.9			
20	19.13	19.13	19.58		20.4	19.86	19.13	
25								
30	21.7	22.7	22.7			23.2		23.0
35					23.8			
50	26.7	28.6	29.0			27.6	28.6	32.6
60	26.7				26.6			
100		34.9	35.7				34.9	
140					38.4			

[a] Data taken from Harvard Apparatus Pump 22 Instruction Manual. Publication No. 5381-001, July 29, 1986.

syringes, but once the syringe is empty the operation must be stopped and the syringes refilled if the injection process is to continue. When using syringe pumps, make sure they have a limit or reversing switch, or come equipped with a clutch. If the syringe empties or binds at the extreme end, motor overload and burnout can occur if some method of automatic stoppage is not available.

Motor-driven syringes are extremely versatile and several applications include inhalational toxicity studies,[47,53,62,82,104] gas-stream mixing with nitrogen dioxide,[47] water,[105] cobalt hydrocarbonyl,[106] aromatic materials,[107] ethanol,[56] nickel carbonyl,[50] and sulfuric acid.[54] They have also been used for checking direct-reading air-quality tubes[108] and instruments,[109] producing air pollution standards[37] and gas mixtures for spectral studies,[110] carbon adsorption studies,[59] catalytic feed systems for gas-phase reactions,[111] and for generating test and interfering gases.[58]

These systems have the advantages of little or no waste, such as is seen in the double-dilution techniques. Usually, concentration changes can be made quickly, with the system attaining equilibrium in a few minutes. The concentration range of such systems is large and ranges from the percent to parts per million range for injected liquids to below 1 ppm for gases. The minimum practical input flow is about 0.1 µL/min.[107] The stability and reproducibility at the output concentration should be checked by some alternate technique to ensure best accuracies. Interface liquids, such as water and mercury, should be avoided since they only complicate the reservoir filling operation. Motor-driven systems are fairly expensive and become even more so if they are built from scratch in local shop facilities.[103]

5.2.3 Liquid Pumps

Uninterrupted contaminant injection can be accomplished using piston and peristaltic pumps. The piston pump injects contaminant in short periodic bursts. There is no flow during the filling stroke and hence no liquid is discharged. Peristaltic pumps utilize a carousel of rollers or fingers, which compress the tubing against a raceway. This action traps the fluid and forces it mechanically at a controlled rate through the tubing. Slight variations in pumping rate are observed from surges caused by the periodic depressing and lifting of the rollers. The injection rate depends on several factors, such as pump speed, tubing diameter, number of channels, and tubing material.[73] Several types of peristaltic pumps are summarized in Table 5.5. They can be obtained through most of the major scientific product distributors. All the pumps shown are variable speed, and most use speed-control mechanisms, which compensate for variations in roller resistance. The pumps can be obtained with multiple channels and can dispense at rates from less than a microliter per minute to more than several liters per minute. These pumps can accommodate tubing sizes from 0.8 to 16 mm i.d. Tubing types preferred include Marprene, silicone, Tygon, Viton, Norprene, neoprene, and natural rubber latex. Pisula presents a detailed tubing selection guide.[73] Additional pumps are compared in Table 5.6.

Dynamic Systems for Producing Gas Mixtures

Table 5.5 Characteristics of Commercally Available Peristaltic Pumps[a]

Manufacturer	Model	Speed Ratio	Number of Channels	Tubing Size Accommodated (mm i.d.)	Range of Volume Dispensed (μL/min, mL/min)
Buchler	4262000		4	0.8–4.8	30–21.8
Elmeco	E-50	60:1	1	1–5	700–2.67
Manostat	72500000		20	2.4–4.8	2.5–10.8
Manostat	72300000		1	1.6–7.9	5000–3000
Manostat	72350000		1	1.6–7.9	5000–4000
Manostat	72335000		1	3.2–9.6	3000–5500
Sage	375A		4	0.5–4	20–17
Watson-Marlow	101U/R	>50:1	1	0.5–4.8	0.8–53
Watson-Marlow	202U/AA	100:1	16	0.13–2.8	0.2–10.2
Watson-Marlow	601S/R	10:1	1	6.4–15.9	140–6.2
Watson-Marlow	302F	8:1	1	0.5–8	3–2200
Watson-Marlow	503S	25:1	1	0.5–8	16–1700
Watson-Marlow	501Z/R		1	0.5–8	360–1700

[a] All pumps are variable speed.

5.2.4 Miscellaneous Injection Methods

There are a variety of alternate methods of injecting trace materials into a dynamic system. These include electrolytic, gravity feed, liquid and gas pistons, and pulse diluters. The practicality of some of these systems is suspect. Few of these methods are used today and are presented more for historical purposes.

5.2.4.1 Electrolytic Methods

Both gases and liquids can be introduced into a moving gas stream by using the evolved gas or gases from the electrolytic process as a driving force.[74] Acetone and alcohol, for example, have been introduced into an air stream (100 mL/min) at the rate of 1 mL/min using 89 mA and the electrolytically evolved hydrogen and oxygen from water.[74]

Gases generated electrolytically can also be made to control the rate of advancement of a liquid piston that in turn displaces another gas into the system of interest. One such system is shown in Figure 5.6. Here again, both evolved gases are used to drive a piston of water, oil, or mercury, which in turn delivers helium at a constant rate, depending on the current supplied to a moving stream of nitrogen.

The injected gas reservoir can be refilled by reversing the position of the stopcocks and bleeding the helium in slowly, under a slight positive pressure. The

Table 5.6 Characteristics of Some Low-Flow Pumps

Manufacturer	Model Number	Material Pumped	Range Minimum (µL/min)	Range Maximum (mL/min)	Operating Principle	Number of Speeds
Beckman Instr.	74600	Liquid	40	20	Cam act.	Variable
Buchler Instr.	4267000	Liq, gas	16	47	Peristaltic	Variable
Chromatronix	CMP-2	Liquid	200	100	Dual piston	6
Coleman	102-510	Liq, gas	50	6	Peristaltic	Variable
Exdex	1018	Liquid	200	10	Piston	Variable
Fenet Apparatus	Evenflow	Liq, gas	50	17	Peristaltic	Variable
Harvard Apparatus	22	Liq, gas	0.029	55	Syringe	Variable
Instrum. Special.	380	Liquid	17	53	Piston	18
Manostat Corp.	72300000	Liq, gas	2.5	11	Peristaltic	Variable
Milton Roy Co.	196	Liquid	80	8	Piston	Variable
Phoenix Precis.	4000	Liquid	80	2	Piston	Variable
ProMinent	9121336	Liquid	300	3	Piston	Variable
Quigley-Rochester	59100	Liquid	0.13	0.0833	Recip. syringe	24
Sage Instruments	355	Liq, gas	0.0029	140	Syringe	Variable
Sigmamotor, Inc.	AL-E	Liq, gas	9	1	Peristaltic	Variable

Dynamic Systems for Producing Gas Mixtures

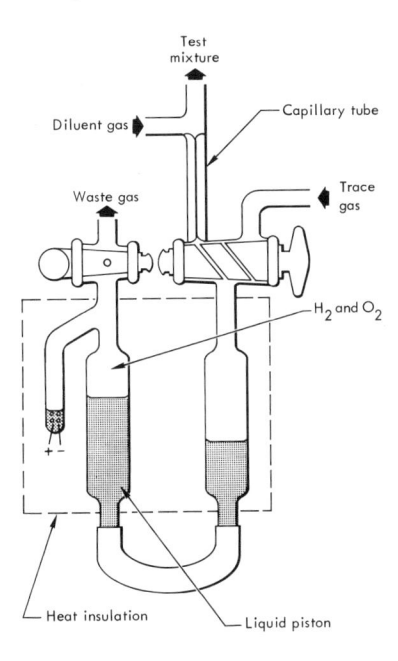

Figure 5.6 Sketch of a method for injecting gas electrolytically. Taken from Hersch et al.[74] with permission.

displacing gas is commonly hydrogen, oxygen, both from aqueous sulfuric acid, or potassium sulfate or potassium hydroxide solutions.[75]

Glass syringes that are actually driven by electrolytically evolved gases are available commercially.[76] They dispense between 0.04 and 4 mL/hr, and will run continuously for as long as 10 days while dispensing 1 mL/day.

5.2.4.2 Gravity-Feed Methods

The pressure created by a column of liquid is sufficient to force it through a restricted opening into the system of interest. Since the flow rate is a function of the height of the liquid, the reservoir level should change by only a small amount during a given time period. Liquid regulation has been achieved using drip regulator clamps[77] or needle valves especially designed for precise flow. Ground-glass stopcocks have been used[78] but are not always successful as flow regulators, since the sealing grease is often flushed away. This, in turn, causes leakage through the stopcock and problems in the system when the grease reappears after the solvents evaporate. This condition can be partially eliminated by using a Teflon stopcock, but precise flow control is still difficult and it should be used only for rough flow adjustments. Direct gravity feed through a capillary has been used,[79,80] but one is limited to just a single flow rate. Micrometer needle valves are generally best, since their flow can be adjusted and reproduced with precision.

Gases can be injected by gravity by using the weight of a ground-glass plunger alone. The plunger's friction with the barrel is minimized by rapid rotation from a tangential blast of air against the vanes[81,82] or a squirrel cage rotor[83] attached directly to the plunger. The plunger then falls under its own mass at a rate that depends on the system pressure, the outlet orifice size, and the nature of the gas.

5.2.4.3 Liquid and Gas Pistons

Various delivery pistons other than glass and Teflon have been used. Yant and Frey, for example, used a dropping water supply to create a steadily increasing pressure head over a piston of mercury.[103] The mercury, in turn, displaces the liquid to be vaporized into the mixing chamber. Gases and liquids have also been delivered by dropping mercury to displace gas through a small capillary tube.

Gas pressure is also used to pneumatically activate a moving liquid piston.[75,84–87] An example is shown in Figure 5.7. The contaminant gas is filled through the stopcock until the liquid piston is nearly to the capillary choke. The stopcock is then reversed, and dilution of the trace material proceeds to the right and empties itself. Other variations on this design have been proposed.[75,85]

5.2.4.4. Pulse Diluters

Continually revolving Teflon stopcocks or the Woesthoff gas-dosing apparatus have been used as the basis of a pulse dilution system.[88] An example is shown in Figure 5.8. The trace gas of interest is bled through a rotating stopcock of precisely known volume, as shown in the left-hand side of Figure 5.8. When the stopcock is turned 90°, as shown in the right-hand side of the figure, the dilution gas purges the stopcock, mixes with more dilution gas, and proceeds to a mixing vessel. The stopcock then returns to the left position to be filled while the dilution gas skirts the stopcock through a bypass and goes to the mixing vessel. Since the gas is added by pulses, rather than smoothly and continuously, special care must be taken to mix the gases and to ensure that the output mixture is as smooth as possible. With this method, pure gases can be diluted to 50 ppm or less if two or more units are used in series.[89,90] According to manufacturers, this technique can be used with such corrosive gases as chlorine, nitrogen dioxide, and sulfur dioxide, since all components in contact with the gas are constructed from Teflon and glass.

Axelrod et al. have used two such systems in series and report dilutions of 10^8 or better and concentrations from 100 ppm to 0.1 ppb at flows of 30 L/min.[91] They have further added mixing flasks after each stage to minimize the pulsing concentration created by the nonuniform introduction of the contaminant material into the diluent gas stream. The major disadvantage is the 3-h time interval required to reach concentration equilibrium.

5.2.5 Injection Ports

Gases are relatively easy to inject into a larger flowing volume of gas, but liquids present somewhat of a problem. They must be injected, evaporated, and mixed smoothly for reliable and invariant gas mixture concentrations. Injectors for either gases or liquids are most often narrow-gauge needles. The smaller the flow rate into the diluent gas, the smaller the needle diameter. This minimizes the equilibrium time. Several injection ports are shown in Figure 5.9. The most common is a glass tee fitted with a rubber serum cap or a septum shown at the top

Dynamic Systems for Producing Gas Mixtures

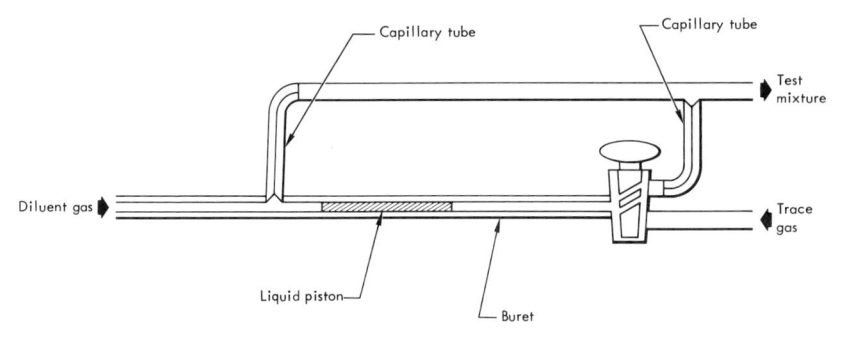

Figure 5.7 Sketch of a method for injecting gas by means of a liquid piston. Taken from Hersch and Whittle[86] with permission.

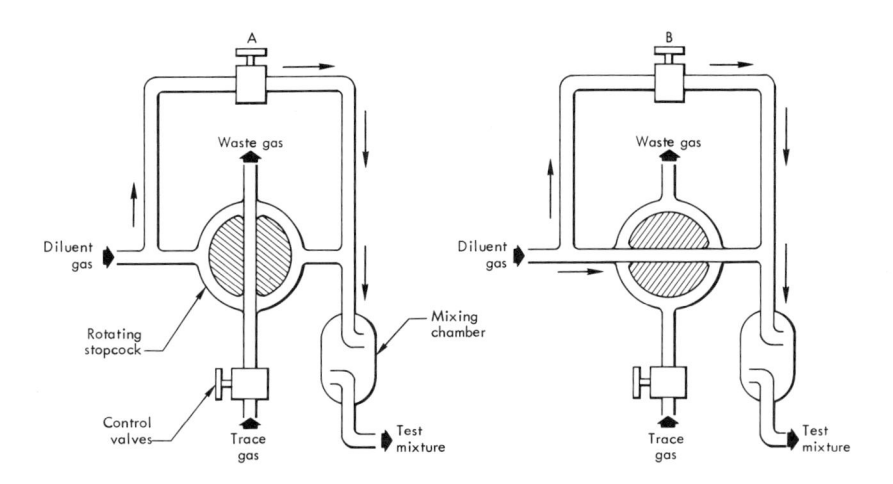

Figure 5.8 Sketch of a pulse diluter or Woesthoff gas dosing apparatus.

of the figure. Here, the gas or liquid of interest can be injected into the air stream without leakage around the needle insertion point. The injection port for liquid, shown in the center of Figure 5.9, is not recommended. The liquid can run down the needle, be trapped, and is slow to evaporate because of the lack of air velocity and turbulence in this region. Some solvents are lost altogether because of leakage and permeability through the septum.

Needles used for injection should be as inert as possible, and both Teflon and 304 stainless steel are commercially available in a wide variety of lengths and diameters. Table 5.7 lists a summary of some typical commercially available needles. The stainless steel or hypodermic needles are quite inexpensive but have the disadvantage of being rigid. Steel needles may be bent if special care is taken not to crimp or close off the needle to normal flow. The Teflon needle, on the other hand, is quite flexible, permitting remote placement of the syringe or reservoir. The most commonly used injection needles are usually 18 to 26 gauge. Platinum needles in selected gauges are also available for corrosive materials. Needles of

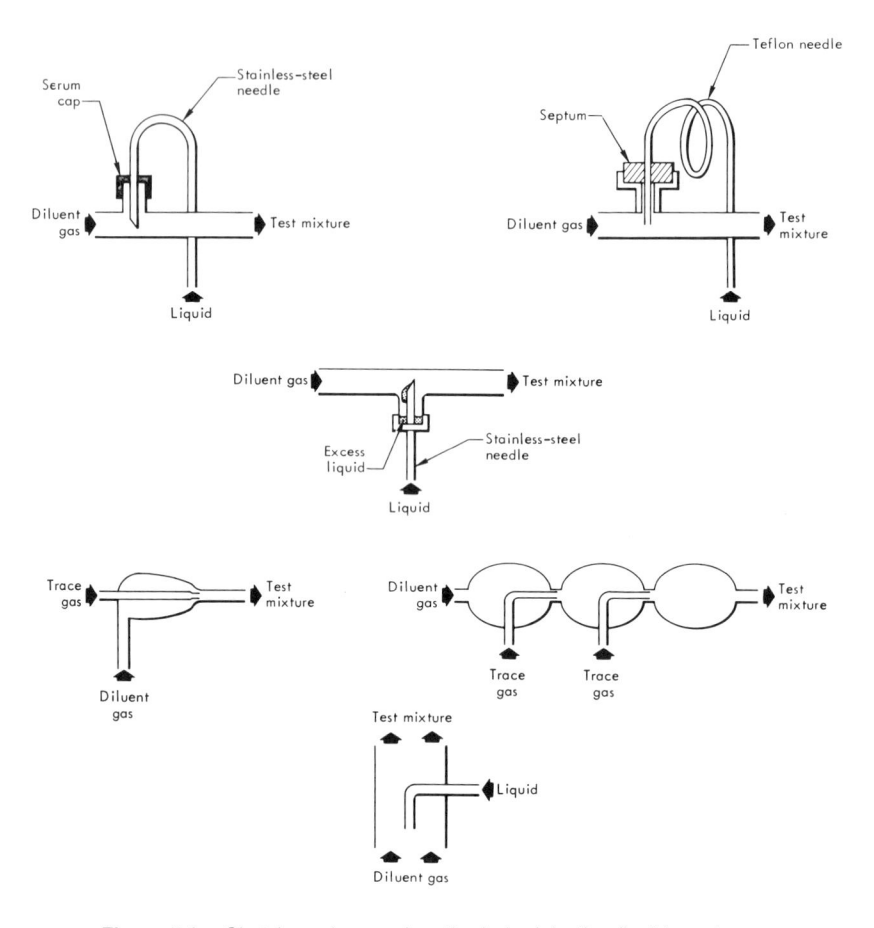

Figure 5.9 Sketches of several methods for injecting liquids and gases.

all types can be connected to support tubing lengths if Luer fittings and adaptors are employed.

5.2.6 Vaporization Techniques

When liquids must be added to the moving gas stream, some means must be provided to evaporate them smoothly so that the controlled addition of vapors will proceed evenly. If the concentration of a more volatile material at 100 ppm or less is desired, the turbulence of the diluent gas is frequently sufficient to evaporate the material satisfactorily. However, as the concentration requirements increase and the solvent volatility decreases, there comes a time when the normal turbulence is not enough and other methods must be employed to aid the vaporization process.

There are three main methods of enhancing the evaporation process: (1) the surface area between the diluent gas and the liquid can be increased, (2) the solvent or surrounding air can be heated, and (3) the amount of turbulence around the incoming solvent can be increased.

Dynamic Systems for Producing Gas Mixtures

Table 5.7 Specifications of Some Commercially Available Needles[a]

Material[b]	Gauge	Outer Diameter Inches	Outer Diameter mm	Inner Diameter Inches	Inner Diameter mm	Needle Volume (μL/inch)	For Use with Syringe Capacities
S, T	33	0.008	0.21	0.004	0.11	0.20	5–100 μL
S, T	32	0.009	0.24	0.004	0.11	0.20	5–100 μL
S, T	31	0.010	0.26	0.005	0.13	0.34	5–100 μL
S, T	30	0.012	0.31	0.006	0.16	0.45	5–100 μL
S, T	28	0.014	0.36	0.007	0.18	0.63	5–100 μL
S, T, P	27	0.016	0.41	0.008	0.21	0.80	
S, T, P	26	0.018	0.46	0.010	0.26	1.25	10 μL–10 mL
S, T	25	0.020	0.52	0.010	0.26	1.25	
S, T	24	0.022	0.57	0.012	0.31	1.80	
S, T	23	0.025	0.64	0.013	0.34	2.17	
S, T, P	22	0.028	0.72	0.016	0.41	3.35	250 μL–10 mL
S, T	21	0.033	0.83	0.020	0.51	5.19	
S, T	20	0.036	0.91	0.024	0.60	6.71	
S, T	19	0.042	1.07	0.027	0.69	9.50	
S, T, P	18	0.050	1.27	0.033	0.84	14.1	
S, T	17	0.058	1.47	0.042	1.07	22.8	
S, T	16	0.065	1.65	0.047	1.19	28.3	
S, T	15	0.072	1.83	0.054	1.37	37.4	
S, T	14	0.083	2.11	0.063	1.60	51.1	
S, T	13	0.095	2.41	0.071	1.80	64.6	
S, T	12	0.109	2.77	0.085	2.16	93.1	
S, T	11	0.120	3.05	0.094	2.39	113.0	
S, T	10	0.134	3.40	0.106	2.69	143.3	

[a] Hamilton Product Catalog, Reno, NV.
[b] S - stainless steel, T - Teflon, P - platinum.

The surface area can be increased in several ways. Plugs of glass wool,[50,74] wicks, filter papers,[75] gauze, and cheese cloth[78,79] have been used. If the vaporization tube is oriented vertically, the liquid will migrate down the gauze and gradually be removed by the diluent gas passing upward. Hence the liquid does not form pools in the bottom of the tube. This has also been accomplished with a column of glass beads.[92] One must be careful when injecting small amounts of liquid, for evaporation sometimes does not proceed as smoothly as one might imagine. When the liquid evaporates, it cools the substrate material, retarding the volatilization of the additive liquid. A pulsing concentration gradient then develops. This can partially be overcome; injecting the liquid into an activated carbon plug[93] or adding a large mixing vessel will dampen the pulsations.

The vaporization of the liquid can be achieved with a wide variety of heaters. Some techniques involve heating the liquid as it travels into the system,[69,93] but the usual procedure is to heat the diluent gas, which in turn vaporizes the solvent. Several of these are shown in Figure 5.10. The diluent gas can be passed through flame-heated coils[94] or tubes heated by combustion furnaces or flames. These represent the brute-force approach and are not particularly energy efficient and quite often lead to contaminant decomposition. Techniques more often employed include wrapping the injection area with heating tape,[59,66,68,78,82] steam,[84] or in-line stainless steel or ceramic heaters.[51,52,95] Figure 5.11 shows data curves for a 200-W heater at various air flow rates. Pressure-sensitive heaters,[96] Electrofilm,[97] and spray-on heating also have applications.

The last alternative is to increase the turbulence at the point of injection. Either the gas turbulence can cause the required vaporization or the liquid can be atomized into an aerosol, which effectively increases the surface area. Figure 5.12 shows a basic type of such a device.[1,56,98,99] The liquid is usually fed into the system through a small stainless-steel needle, around which the diluent gas stream flows at high velocity. The force of the gas striking the emerging liquid, as well as the shape of the nozzle, causes the liquid to disperse into small droplets that are in turn more easily vaporized in the moving gas stream.

There is a commercially available vaporizer from Miller-Nelson Research (Model SV-20) that can vaporize the less volatile materials and generate formaldehyde as well (see Chapter 6).[100] Figure 5.13 shows a schematic of the system. Solvent is injected into the vaporizer using a needle that is directly adjacent to a 100-W heater. The solvent contacts the heater and any excess is held in the retaining cup. The heater controls are adjusted to provide enough heat to just vaporize the material. Excessive heat can, on occasion, however, cause solvent decomposition. The vapor and air mixture is then cooled to room temperature using a fan and cooling coil arrangement.

5.2.7 The Ultimate System

The list of alternatives for assembling such a dynamic systems is indeed large and will, of course, depend on budget and resource considerations, as well as glass blowing and mechanical shop capabilities. There are, however, some viable commercially available items that will make the task of setting up a precision gas and vapor generation system much easier. Figure 5.14 shows a schematic diagram of the ultimate system. Basically it consists of an air and water purification system, an air flow-temperature-humidity controller, a solvent injection system, a trace-gas flow controller, and an analytical system.

Air purification systems have been discussed in Chapter 2. Air entering the system must be free of oil vapor, particulate matter, and any foreign materials that might interfere with the experiments at hand. This is usually done with a series of filters before and after the compressor and at the laboratory source. The air humidity should be at least 20% below the desired level so that the automated controller can function properly. This is normally done using desiccants or a refrigerated air dryer. Purified water at 20 psig is also needed to service the humidification system. Laboratory distilled water or deionized and filtered tap water can be used if the pressure is sufficient.

The purified air and water are then connected to the heart of the system — a Miller-Nelson Research HCS-401 series flow-temperature-humidity controller.[101] Here the air flow, temperature, and relative humidity can be set using direct-reading potentiometers. The actual conditions are read out visually using panel

Dynamic Systems for Producing Gas Mixtures

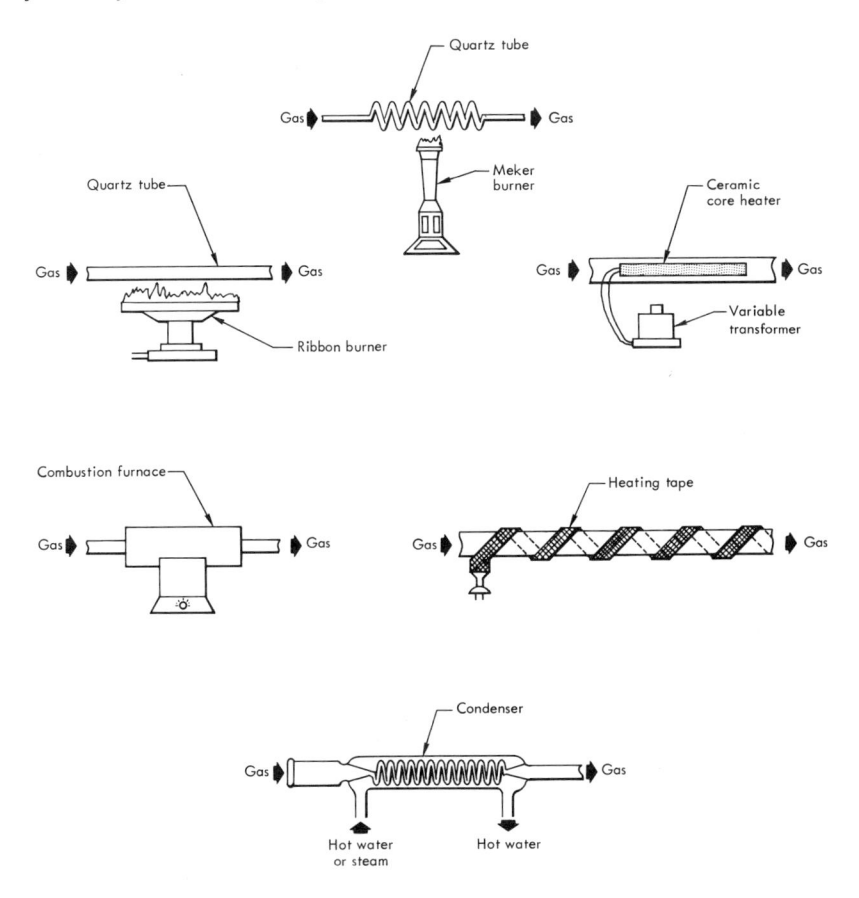

Figure 5.10 Sketches of several methods for heating gases so that they can later vaporize liquids.

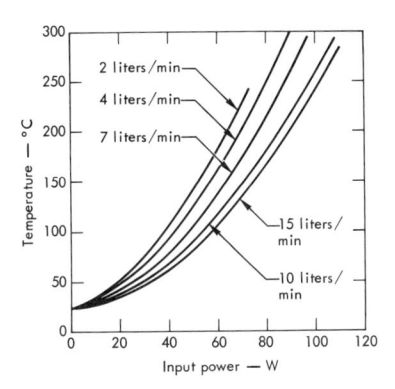

Figure 5.11 Temperature vs power for a 200-W ceramic heater at various air flow rates.

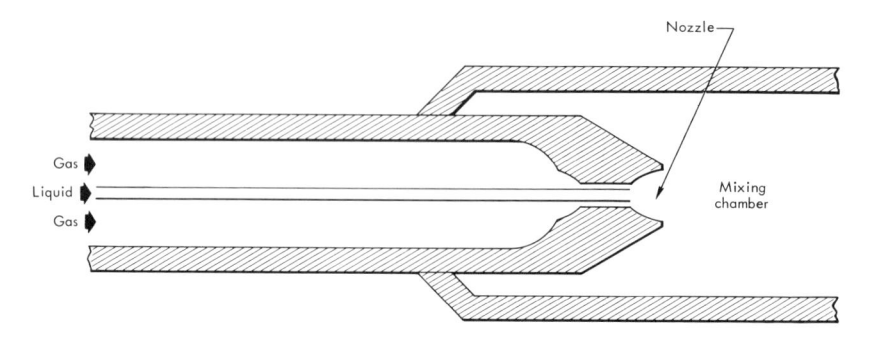

Figure 5.12 Sketch of a typical atomizer for vaporizing liquids.

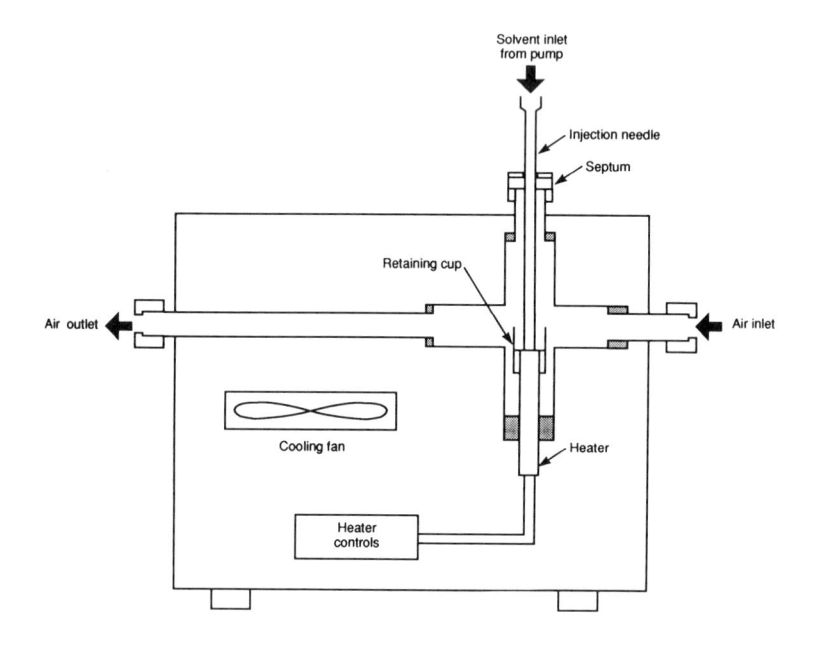

Figure 5.13 Sketch of a commercially available vaporizer. Data taken from Miller-Nelson Research SV-20 brochure with permission.

meters. Control of the temperature and humidity (see Chapter 6) is accomplished by signal comparison and a feedback control system. Air flow control is managed using a mass flow controller, which measures flow at standard conditions (usually 25°C and 1 atm) and automatically compensates for temperature and pressure variations. Available flow ranges are 1–10, 2–20, 5–50, 10–100, 20–200, 25–250, 30–300, 40–400, or 50–500 L/min.

The solvent is injected into the air stream directly and allowed to evaporate naturally or into a Miller-Nelson Research Inc. SV-20 Vaporizer, which was discussed earlier. Here the vapors merge with the preconditioned air stream. The contaminant is supplied for short durations with a syringe pump or with a peristal-

Dynamic Systems for Producing Gas Mixtures

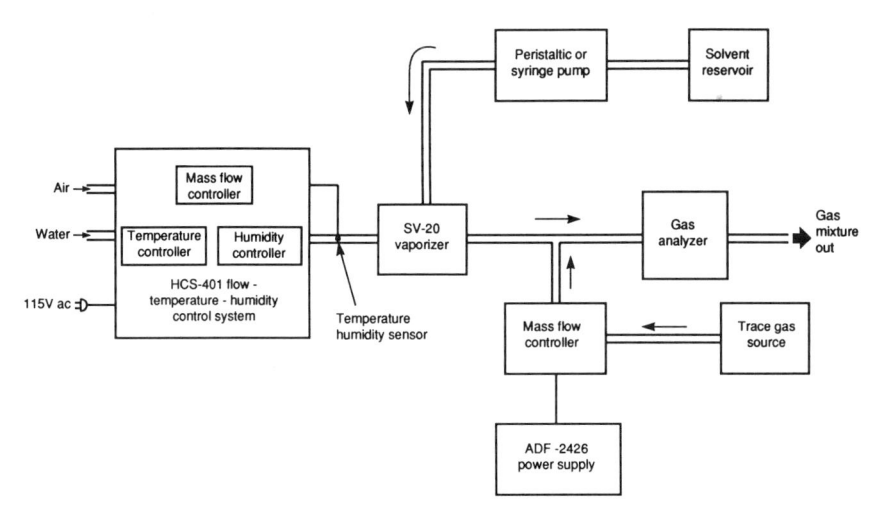

Figure 5.14 Schematic diagram of a precision gas and vapor in air blending system.

tic for longer duration experiments. If low vapor concentrations (less than 10 ppm) are desired, then the diffusion or permeation tubes described later in the chapter may be employed. If a gas contaminant is desired, it is added using a low-range (starting from 1 mL/min) mass flow controller powered by an ADF-2426 power supply.[102]

A final concentration check can be carried out with an appropriate analyzer. The device of choice varies, but a favorite of the author is the MIRAN 1A infrared analyzer with a 20-m gas cell. This analyzer not only measures most vapors nondestructively, but the 5.64-L gas cell provides some concentraton pulse dampening. If irregular concentration-pulses persist, then some sort of mixing vessel might be needed. Other choices of analyzers include flame ionization detectors and various types of single-gas electrochemical detectors.

5.2.8 Calculations

For gases, the concentration produced from a motor-driven syringe, assuming ideality, is

$$C_{ppm} = \frac{10^6 q_G}{q_D} \quad (5.10)$$

where C is the concentration (ppm), q_G is the injection rate (mL/min), and q_D is the dilution gas rate (mL/min).

$$q_G = KR \quad (5.11)$$

where K is the syringe constant (mL/mm) and R is the rate of advance (mm/min). Combining Equations 5.10 and 5.11 yields

$$C_{ppm} = \frac{10^6 \, KR}{q_D} \quad (5.12)$$

The value of K can be derived from

$$K = \frac{V_s}{L} \quad (5.13)$$

where V_s is the volume of the syringe (mL) and L is the length that contains that volume (mm).

Example 5.7. What is the concentration produced when 10 mL/min of methane is injected into a 25-L/min air stream?

Methane flow rate	10 mL/min
Air flow rate	25 L/min
mL/L	1000

From Equation 5.10, the methane concentration is

$$C_{(CH_4)} = \frac{(10)(E+06)}{(25)(1000)} = 400 \text{ ppm}$$

Example 5.8. A syringe contains 20 mL in a 66.5 mm length. Calculate the syringe constant. What advance rate is required to produce a 10 ppm concentration of any gas in an air stream flowing at 30 L/min?

Syringe volume	20 mL
Syringe length	66.5 mm
Desired concentration	10 ppm
Air flow rate	30 L/min
mL/L	1000

From Equation 5.13, the syringe constant is

$$K = \frac{(20)}{(66.5)} = 0.301 \text{ mL/mm}$$

From Equation 5.12, the advance rate is

$$R = \frac{(10)(30)(1000)}{(0.301)(E+06)} = 0.998 \text{ mm/min}$$

For liquids, the concentration produced from a motor-driven syringe or pump is

Dynamic Systems for Producing Gas Mixtures

$$C_{ppm} = \frac{22.4 \times 10^6 \left(\dfrac{T}{273}\right)\left(\dfrac{760}{P}\right) q_L \rho}{q_D M} \quad (5.14)$$

where T is the experimental temperature (°K), P is the experimental pressure (mmHg), q_L is the liquid injection or flow rate (mL/min), r is the density of the liquid (g/mL), q_D is the diluent gas flow rate (L/min), and M is the molecular weight (g/mol). At 25°C and 1 atm, this reduces to

$$C_{ppm} = \frac{24.5 \times 10^6 q_L \rho}{q_D M} \quad (5.15)$$

Combining Equation 5.15 with 5.11, the concentration can be calculated from the injection rate and syringe constant with

$$C_{ppm} = \frac{24.5 \times 10^6 KR\rho}{q_D M} \quad (5.16)$$

Example 5.9. What is the concentration produced at 20°C and 740 mmHg when chlorobenzene (density, 1.10 g/mL; molecular weight, 112.6 g/mol) is continuously fed at the rate of 12 µL/min into an air stream flowing at 15 L/min?

Room temperature	20°C or 293°K
Room pressure	740 mmHg
Chlorobenzene density	1.10 g/mL
Molecular weight	112.6 g/mol
Feed rate	12 µL/min or 0.012 mL/min
Air flow rate	15 L/min

From Equation 5.14, the concentration produced is

$$C = \frac{(22.4)(E+06)\left(\dfrac{293}{298}\right)\left(\dfrac{760}{740}\right)(0.012)(1.10)}{(15)(112.6)} = 177 \text{ ppm}$$

Example 5.10. What would be the required advance rate to produce 1000 ppm carbon tetrachloride in 32 L/min at standard conditions? Use a 20-mL syringe with a syringe constant of 0.301 mL/mm. The carbon tetrachloride density is 1.59 g/mL and the molecular weight is 153.8.

Desired concentration	1000 ppm

Air flow rate	32 L/min
Syringe constant	0.301 mL/mn
Carbon tet. density	1.59 g/mL
Molecular weight	153.8 g/mol

From Equation 5.16, the advance rate would be

$$R = \frac{(1000)(32)(153.8)}{(24.5)(E+06)(0.301)(1.59)} = 0.420 \text{ mm/min}$$

Example 5.11. Calculate the total injection rate to produce a mixture of 1000 ppm each of carbon tetrachloride and carbon disulfide in air flowing at 30 L/min. What would be the weight percent of the solvent mixture to produce these concentrations?

CS_2 concentration	500 ppm
CCl_4 concentration	1500 ppm
Air flow rate	30 L/min
CS_2 density	1.263 g/mL
CCl_4 density	1.594 g/mL
CS_2 molecular weight	76.1 g/mol
CCl_4 molecular weight	153.8 g/mol

From Equation 5.15, compute the injection rates for each liquid.

$$q_{L(CS_2)} = \frac{(500)(30)(76.1)}{(24.5)(E+06)(1.263)} = 0.0369 \text{ mL/min}$$

$$q_{L(CCl_4)} = \frac{(1500)(30)(153.8)}{(24.5)(E+06)(1.594)} = 0.1772 \text{ mL/min}$$

$$q_{L(total)} = 0.0369 + 0.1772 = 0.2141 \text{ mL/min}$$

The weight percent of carbon disulfide would be

$$P_{(CS_2)} = \frac{(0.0369)(1.263)(100)}{(0.0369)(1.263)+(0.1772)(1.594)} = 14.2\%$$

The weight percent of carbon tetrachloride would be

$$P_{(CCl_4)} = \frac{(0.1772)(1.594)(100)}{(0.0369)(1.263)+(0.1772)(1.594)} = 85.8\%$$

Dynamic Systems for Producing Gas Mixtures

Example 5.12. You wish to use a 25-mL Hamilton syringe to inject 50 µL/min into a diluting air stream. What advance rate should you set on the syringe pump? The syringe diameter is 23.0 mm (see Table 5.4).

Syringe diameter	23.0 mm or 2.30 cm
Syringe volume	25 mL
Injection rate	50 µL/min or 0.050 mL/min
mm/cm	10

The syringe length is

$$L = \frac{(25)}{(3.1416)\left(\frac{23.0}{2}\right)^2} = 60.2 \text{ mm}$$

From Equation 5.13, the syringe constant is

$$K = \frac{(25)}{(60.2)} = 0.415 \text{ mL / mm}$$

The required advance rate is

$$R = \frac{(0.050)}{(0.415)} = 0.1203 \text{ mm / min}$$

Table 5.8 illustrates several solvent injection-rate conditions.

5.3 DIFFUSION METHODS

Gases and vapors have the property of diffusing through tubes at a uniform rate if the temperature, concentration gradients, and tube geometry remain unchanged. This phenomenon is a convenient method of producing concentrations of solvent vapors in the low parts per million and parts per billion range.[112-115] This technique has produced mixtures in air of such diverse materials as N,N-dimethylcarbamoyl chloride,[116] benzene,[117] toluene diisocyanate,[118] tetramethyl lead,[119] aniline,[120] and liquified ethylene oxide.[121] Figure 5.15 illustrates a typical diffusion system. A metered gas stream is brought to the desired experimental temperature by passing it through a constant-temperature bath or oven. The preconditioned gas then passes over a diffusion tube and mixes with the vapor of interest at a constant rate. The resultant concentration is controlled by varying the flow rate of either the primary or secondary diluent gas.

Table 5.8 Conditions for Producing Various Concentrations in Dynamic Injection Systems

Contaminant Vapor	Mol. Wt. (g/mol)	Density (g/mL)	Flow Rate (L/min)	Desired Conc. (ppm)	Injection Rate (μL/min)	Syringe Capacity (mL)	Syringe Constant (mL/mm)	Rate of Advance (mm/min)
Acetic acid	60.05	1.049	10	100	2.34	0.50	0.00840	0.278
Acetone	58.08	0.791	10	2000	59.9	10	0.168	0.357
Benzene	78.11	0.879	10	50	1.81	0.25	0.00420	0.432
Butyl acetate	116.16	0.882	10	500	26.9	2.5	0.0420	0.640
Chlorobenzene	112.56	1.106	25	50	5.19	1.0	0.0168	0.309
Chloroform	119.38	1.489	25	100	8.18	1.0	0.0168	0.487
Cyclohexane	84.16	0.779	25	250	27.6	5.0	0.0840	0.328
Decane	142.29	0.730	25	1000	199	10	0.168	1.184
Dibromoethane	187.87	2.055	50	25	4.66	1.0	0.0168	0.278
Diethyl ether	74.12	0.708	50	2000	427	20	0.301	1.420
Hexane	86.18	0.659	50	500	133	10	0.168	0.794
Methanol	32.04	0.792	50	1000	82.6	10	0.168	0.491
Methyl ethyl ketone	72.11	0.805	100	500	183	20	0.301	0.607
Styrene	104.15	0.906	100	25	11.7	1.0	0.0168	0.698
Trichloroethylene	131.39	1.466	100	200	73.2	2.5	0.0420	1.742
Xylene	106.17	0.880	100	100	49.2	2.5	0.0420	1.172

Dynamic Systems for Producing Gas Mixtures

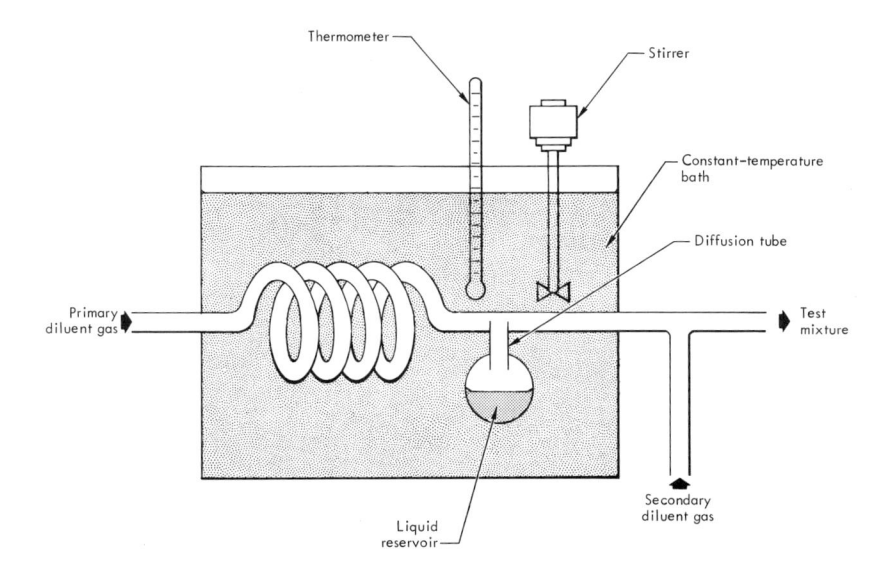

Figure 5.15 Sketch of a dynamic diffusion system.

Diffusion tubes come in a wide variety of configurations. Almost any small vessel that can hold a liquid and allow diffusion through a specified length has been used. The 5-mL capacity volumetric flasks are popular, as well as a number of specially made devices.[122-134] Several of the most common devices are illustrated in Figure 5.16. If the diffusion rates are to be calculated, then both the length and the cross-sectional area must be accurately known. The ratio of area to length should be maintained at less than 0.3 for best results.[135] Tubes should be between 2 and 20 mm in diameter. Diameters smaller than 2 mm make tube filling a problem. Diameters above 20 mm lead to excessive turbulence, which causes additional errors because of the decrease in the effective diffusion path length.[124] For these reasons, types A, B, and C in Figure 5.16 are preferred; the length does not change as the liquid is consumed.[123,136,137] If a large reservoir is available, then the diffusion system can be given ample time to equilibrate, and prolonged runs with the calibrated mixture can be made. Types D and E can be used with success for shorter time periods or else corrected for increasing length.[124,138] If desired, automatic, constant leveling diffusion cells have been designed.[139]

The necessity of proper temperature control cannot be overstated. Generally a change in temperature of 1°C will affect the diffusion rate by 5 to 10%. The temperature should then be controlled to ±0.2°C to maintain an accuracy of ±1%.[5]

If it is desirable to alter the diffusion rate without changing the diluent air flow or temperature, then adjustable diffusion cells can be used. In type F, the effective cross-sectional area is decreased by the partial insertion of a plug affixed to a movable plunger.[137] Type G is controlled by turning the stopcock.[140] However, the most reliable method of controlling the concentration consists of either changing the bath or oven temperature, or subsequently diluting the air downstream from

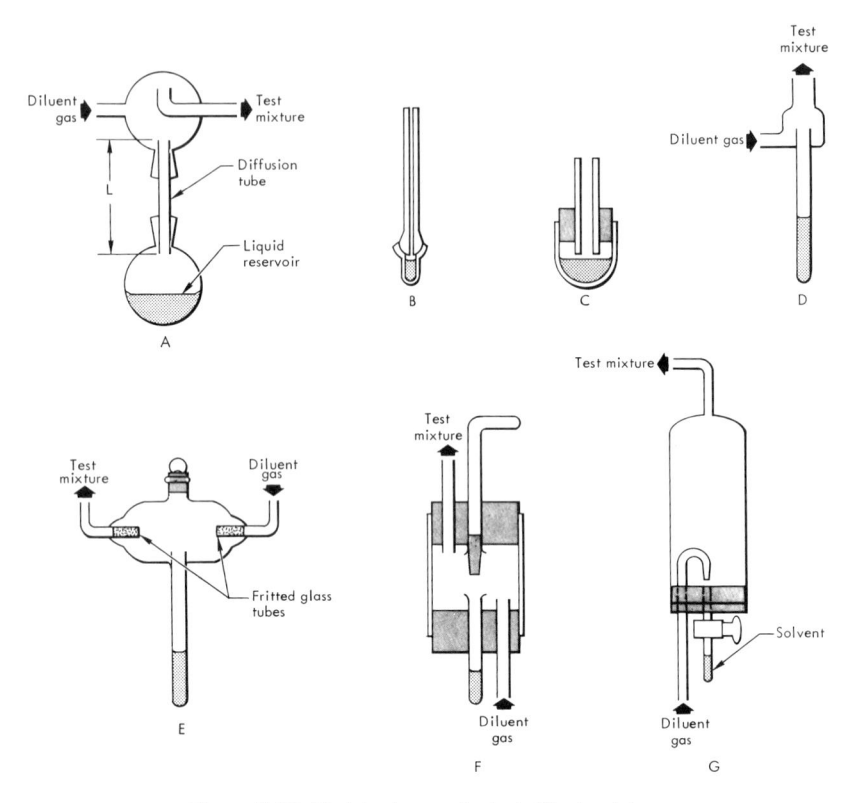

Figure 5.16 Sketch of some typical diffusion tubes.

the diffusion device. Primary gas flow rates through the diffusion chamber usually should not exceed 1 L/min.

Diffusion methods offer one of the least complex methods of generating low parts per million concentration analytical standards. Vapor concentrations from 0.1 to 100 ppm can be routinely achieved[123] for most liquid substances with an appropriately high vapor pressure. This method is not generally useful when concentrations on the order of 1000 ppm or greater are needed above 10 L/min. It should also be mentioned that a single diffusion cell should not contain multicomponent mixtures because of enrichment of the least volatile compound with time. If several components are needed, several diffusion cells should be employed, each with its own pure component.

Calibration of diffusion tubes is usually done gravimetrically.[122,127,128,141-147] The tubes are weighed after a sufficiently long time period of diffusing under the experimental conditions. A total weight loss of 100 mg is desired to minimize weighing errors. Other calibration techniques involve gas chromatograpy,[146,148]

Dynamic Systems for Producing Gas Mixtures

quartz pressure gauges,[149] and the use of a calibrated capillary tube.[124,125,129,150,151] Other techniques will be discussed in the next section on permeation tubes.

The concentration of a dynamic diffusion system can be expressed by

$$C_{ppm} = \frac{10^6 q_d}{Q} \quad (5.17)$$

where C is the concentration (ppm), q_d is the diffusion rate (mL/min), and Q is the dilution flow rate (mL/min).

Example 5.13. What is the resultant concentration of a diffusion system when 10 L/min of air is passed over a tube diffusing 0.20 millimoles/min in a 32°C constant temperature oven? Assume the mixture is cooled back to 25°C after the mixing process.

Diffusion rate	0.20 mmol/min or 0.00020 mol/min
Air flow rate	10 L/min
L/mol	24.5

From Equation 5.17, the concentration is

$$C = \frac{(E+06)(0.00020)(24.5)}{(5.0)} = 490 \text{ ppm}$$

Assuming that the concentration of vapor at the tube exit is maintained at nearly zero by the dilution gas and that the vapor in the tube is saturated,[123] then q_d is usually expressed in mol/sec or g/sec, and can be calculated from

$$q_d = \frac{DMPA \ln\left(\frac{P}{P-p_v}\right)}{LRT} \quad (5.18)$$

where q_d is the diffusion rate (g/sec), D is the diffusion coefficient (cm²/sec), M is the molecular weight of the diffusing vapor (g/mol), P is the pressure — usually atmospheric — in the diffusion cell (mmHg), A is the diffusion tube cross-sectional area (cm²), L is the length of the diffusion tube (cm), R is the molar gas constant (mL-mmHg/mol-°K), T is the absolute temperature (°K), and p_v is the partial pressure of the diffusing vapor (mmHg).

If the volumetric flow rate is desired, then Equation 5.18 reduces to

$$q_d = \frac{DA \ln\left(\frac{P}{(P-p_v)}\right)}{L} \qquad (5.19)$$

where q_d is now expressed in mL/sec.

Example 5.14. What is the diffusion rate of acetic acid at 25°C using a diffusion tube with a 20-mm diameter and an effective length of 5.0 cm? The vapor pressure of acetic acid at 25°C is 15 mmHg. Atmospheric pressure is 730 mmHg and the diffusion coefficient is 0.1235 cm²/sec (see Appendix F). Express the answer in mL/min.

Tube diameter	20.0 mm or 2.00 cm
Tube area	3.14 cm²
Tube length	5.0 cm
Diffusion coefficient	0.124 cm²/sec
Atmospheric pressure	730 mmHg
Partial pressure HAc	15 mmHg
sec/min	60

From Equation 5.19, the diffusion rate is

$$q_d = \left[\frac{(0.1235)(3.14)}{(5.00)}\right] \ln\left[\frac{730}{(730-15)}\right]$$

q_d = 0.00161 mL/sec or 0.0967 mL/min

Example 5.15. Determine the A/L ratio when producing 7.5 ppm of chlorobenzene in 2 L/min of air at 25°C and 1 atm. The vapor pressure and diffusion coefficient of chlorobenzene are 12 mmHg and 0.0747 cm²/sec, respectively. If a maximum diameter of 20 mm is chosen, what tube length will be required?

Concentration	7.5 ppm
Air flow rate	2.0 L/min or 2000 mL/min
Atmospheric pressure	1 atm or 760 mmHg
Chlorobenzene pressure	12 mm
Diffusion coefficient	0.0747 cm²/sec
Tube diameter	20 mm or 2.0 cm
sec/min	60

Dynamic Systems for Producing Gas Mixtures

From Equation 5.17, the required diffusion rate is

$$q_d = \frac{(7.5)(2000)}{(E+06)} = 0.0150 \text{ mL/min}$$

From Equation 5.19, the A/L ratio is

$$\frac{A}{L} = \frac{(0.0150)}{(0.0747)(60)\ln\left[\frac{760}{760-12}\right]} = 0.210 \text{ cm}$$

$$L = \frac{(3.14)\left(\frac{2.0}{2}\right)^2}{(0.210)} = 14.9 \text{ cm}$$

Diffusion coefficients at standard conditions (0°C and 1 atm) can be found in the literature,[152] but more current experimental values at 25°C and 1 atm are shown in Appendix F.[153] Diffusion coefficients at other temperatures and pressures can be calculated from the expression

$$D = D_{298}\left(\frac{T}{298}\right)^n\left(\frac{760}{P}\right) \qquad (5.20)$$

where D is the diffusion coefficient at a new pressure, P, and temperature, T, D_{298} is the diffusion coefficient at 25°C and 760 mmHg (cm²/sec), T is the temperature °K, P is the pressure (mmHg), and n is the number of moles. The value for n, according to kinetic theory, is 1.5, and although experiments indicate that the value is 1.6 to 2.0, [124,154-156] 2.0 is normally used.[157] Also diffusion coefficients vary in the literature and many times cannot be adjusted confidently to the experimental range of interest without introducing further error.

Example 5.16. What is the concentration generated when 0.400 L/min of air is passed over a tube 50 mm long with a 6 mm diameter containing ethyl alcohol? The diffusion coefficient at 25°C is 0.118 cm²/sec, and the experiment is run at 40°C and 740 mmHg. The partial pressure of ethanol is 134 mmHg at 40°C. The air flow is given at 25°C and 740 mmHg.

Air flow rate	0.400 L/min or 400 mL/min
Tube length	50 mm or 5.0 cm
Tube diameter	6 mm or 0.600 cm

Tube area	0.283 cm²
Diff. coeff at 25°C	0.118 cm²/sec
Experimental temperature	40°C or 313°K
Experimental pressure	740 mmHg
EtOH partial pressure	134 mm at 40°C
sec/min	60

From Equation 5.20, the diffusion coefficient at 40°C is

$$D_{313} = (0.118)\left(\frac{313}{298}\right)^2 \left(\frac{760}{740}\right) = 0.134 \text{ cm}^2/\text{sec}$$

From Equation 5.19, the diffusion rate is

$$q_d = \left[\frac{(0.134)(0.283)(60)}{(5.0)}\right] \ln\left[\frac{740}{(740-134)}\right] = 0.0906 \text{ mL/min}$$

This is the flow rate at the elevated temperature, 40°C. The flow rate at the standard temperature, 25°C is

$$q_d = (0.0906)\left(\frac{298}{313}\right) = 0.0863 \text{ mL/min}$$

From Equation 5.17, the final concentration is

$$C = \frac{(E+06)(0.0863)}{400} = 216 \text{ ppm}$$

If diffusion coefficients are not available, they can be calculated from Gilliland's approximation,[158]

$$D = \frac{0.0043\sqrt{T^3}\sqrt{\frac{1}{m_A} + \frac{1}{m_B}}}{\left(v_A^{1/3} + v_B^{1/3}\right)^2 P} \quad (5.21)$$

where P is the pressure (atm), m_A and m_B are the molecular weights of the diluent and diffusing gases (g/mol), and v_A and v_B are the molar volumes at the boiling point (cm³/g-mol). The molar volume usually chosen for air is 29.9 cm³/g-mol.[159] A table of the molar volumes is given in Appendix L.[160]

Dynamic Systems for Producing Gas Mixtures

Example 5.17. Calculate the molecular volume of nitrobenzene using the LeBas approximation shown in Appendix L.

Taking the relevant information from Appendix L:

Number Element	Atomic Volume per Atom (mL/g-mol)	Total Atomic Volume (mL/g-mol)	Atomic Weight (g/mol)	Total Molecular Weight (g/mol)
6 C	14.8	88.8	12.0	72.0
5 H	3.7	18.5	1.0	5.0
1 N	15.6	15.6	14.0	14.0
2 O in union with S,P,N	8.3	16.6	16.0	32.0
1 6 Membered ring	−15.0	−15.0	0.0	0.0
Total		124.5		123.0

The total molecular volume is 124.5 mL/g-mol.

Example 5.18. Calculate the diffusion coefficient of nitrobenzene in air at 25°C. The molecular weight of air is 28.9 g/mol.

Molecular volume NO_2Bz	124.5 mL/g-mol
Molecular volume air	29.9 mL/g-mol
Molecular weight NO_2Bz	123 g/mol
Molecular weight air	28.9 g/mol
Temperature	25°C or 298°K
Pressure	1 atm

From Equation 5.21, the diffusion coefficient is

$$D = \frac{\left[(0.0043)\sqrt{298^3}\right]\left(\frac{A}{B}\right)}{P}$$

$$A = \sqrt{\left(\frac{1}{28.9}\right) + \left(\frac{1}{123}\right)} = 0.207$$

$$B = \left[(29.9)^{1/3} + (124.5)^{1/3}\right]^2 = 65.6$$

$$D = \frac{\left[(0.0043)\sqrt{298^3}\right]\left(\frac{0.207}{65.6}\right)}{1} = 0.0697 \text{ cm}^2/\text{sec}$$

Table 5.9 Comparison of Actual and Calculated Diffusion Coefficients

Compound	Atomic Volume (mL/g-atom)	Diffusion Coefficient at 25°C Experimental	Calculated	Percent Error Calc/Exp
Acetic acid	63.8	0.1235	0.0993	−19.6
Aniline	110.2	0.0735	0.0755	2.7
Benzyl chloride	136.1	0.0713	0.0670	−6.0
Butyl alcohol, n-	103.7	0.0861	0.0797	−7.4
Carbon disulfide	66.0	0.1045	0.0946	−9.5
Carbon tetrachloride	101.2	0.0828	0.0743	−10.3
Chlorotoluene, p-	136.1	0.0621	0.0670	7.9
Diethyl amine	111.9	0.0993	0.0774	−22.1
Dioxane	95.8	0.0922	0.0804	−12.8
Ethyl ether	107.2	0.0918	0.0786	−14.4
Ethylene chlorohydrin	69.7	0.0964	0.0950	−1.5
Hexane, n-	140.6	0.0732	0.0689	−5.9
Mesitylene	162.6	0.0663	0.0625	−5.7
Propyl bromide	97.3	0.0875	0.0770	−12.0
Styrene	133.0	0.0701	0.0690	−1.6
Toluene diisocyanate	199.3	0.0583	0.0554	−5.0
Trichloroethane, 1,1,1-	105.5	0.0794	0.0740	−6.8

Although values of q_d and D can be calculated (Table 5.9), theoretical and experimental diffusion rates can exhibit large variations.[124] The complexity of the diffusion equations is further increased when such terms as the minimum diluent-gas flow rate[137] and steady-state equilibrium time are taken into account.[161] The difference between the actual and theoretical output has been found to be as much as ±10%.[135,137,153] It is strongly suggested that other techniques,[75] such as gas chromatographic, infrared, gravimetric, or volumetric determinations, be used as a primary calibration if there is any doubt about the system. In this manner, a plot of the diffusion rate vs temperature yields a straight line that is easy to use (Figure 5.17, for example) and that allows the concentration to be easily and accurately calculated from Equation 5.17.

5.4 PERMEATION METHODS

One of the best methods for producing low parts per million concentrations of precision gas mixtures involves one of the oldest known phenomena — permeation. Figure 5.18 shows a schematic diagram of a typical system.[162] The heart of the system is the permeation device, usually a tube. Here, the liquid or gas of interest, compressed to the point of liquefaction, is sealed inside of some polymeric container. The chemical then dissolves in and permeates through the walls of the tube at a constant rate for a given temperature and mixes with, and is carried away by, a diluent gas flow.[190]

The gases that work best in this type of system are those that are able to maintain a two-phase equilibrium inside the container walls. Examples include halocarbons,[163,164] 2,4-toluenediisocyanate (TDI),[165] benzene and tetrachloro-

Dynamic Systems for Producing Gas Mixtures 147

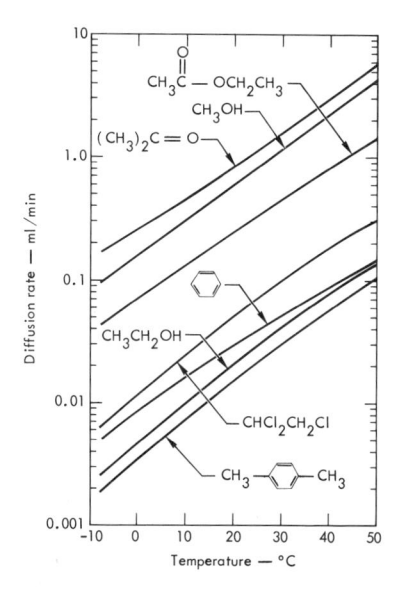

Figure 5.17 Diffusion rate vs temperature for seven organic compounds (tube area, 0.5 cm^2; tube length, 10 cm).

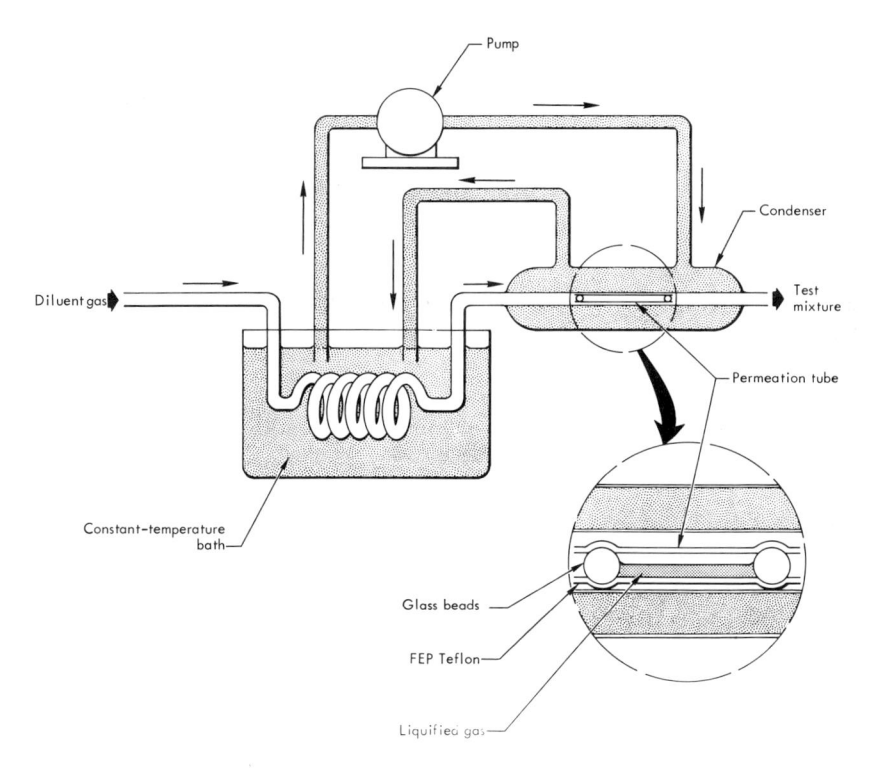

Figure 5.18 Sketch of a method for dynamically producing gas mixtures with a permeation tube.

ethylene,[166] benzene and toluene,[167] organo isocyanates,[168] sulfur dioxide,[169-172] ethylene,[173] vinyl chloride,[174,175] and hydrogen sulfide.[169,176-178] Thus, it is not usual to use materials above their critical temperatures (i.e., acetylene, carbon monoxide, ethane, and methane). There are, however, references in the literature that report the use of one-phase pressurized systems.[179-182] Two-phase solid gas systems have been employed. Formaldehyde has been produced from solid paraformaldehyde and polyoxymethylene.[183-185] Appendix O gives a partial list of permeation materials, typical rates, and some commercial suppliers.

5.4.1 Device Construction

The construction of permeation devices varies but the most common commercially available devices are tubes, microbottles, and wafer devices. These are shown in Figure 5.19. Tubes are made by inserting a glass or stainless steel ball, (or a PTFE or FEP[186] rod usually 1.5 times the internal diameter of the tube), into both ends of the tube. One end of the tube is connected to the liquified gas source. A small pair of plastic pliers pinches the tube at the ball nearest the gas supply and provides a small opening for the gas to enter. The gas supply is turned on and the tube is allowed to fill to approximately 80% capacity. Sometimes immersion in a dry-ice bath is required during this operation. Multiwalled tubes can also be constructed using this technique. Tubes of nesting sizes can be made by stretching the smaller tube and pulling it through the larger tube.[187] Thicker walled tubes can be fabricated using a solid PTFE rod that has been drilled to the desired configuration.[188]

The size of the tube will vary with the concentration sought. The inside diameter of the tube is usually about 1/16 to 1/4 in., and the tube can vary in length from 2 to 30 cm or longer. Tube life will depend on the temperature of use and storage. Sulfur dioxide tubes last about 6 months, while nitrogen dioxide and hydrogen sulfide tubes will last several months if used moderately and stored in a cool place.[189] Microbottles or extended-life devices are made by attaching a permeation tube to an impermeable stainless steel reservoir. These devices will add up to an order of magnitude to the useful life.

Wafer devices were developed to generate gases that have extremely high vapor pressures or low permeation rates. Here, the gas of interest permeates only through a small section of wafer or membrane held in place with a stainless steel cap. These devices have service lifetimes of years. The permeation rate of tubes and wafer devices are shown in Figure 5.20 for butane. These devices are also compared to the emission rate of diffusion tubes discussed in the previous section.

A few of the various polymers used in tube construction are shown in Table 5.10. Although many possibilities exist, fluorinated ethylene propylene resin (FEP Teflon) is usually chosen because of its availability and durability.[190] However, the literature shows that silicon rubber,[191-194] polypropylene,[195] polyester,[195] polyvinyl fluoride,[182,195] polyamide,[182,195] nylon,[196] and polyethylene[183,197,198] have been used.

Other permeation configurations have been proposed, and several are shown in Figure 5.19 and outlined in a summary by Namiesnik.[5] These include Teflon and silicon rubber tubes, tubes immersed in liquid,[191,199,200] large Teflon and PTFE

Dynamic Systems for Producing Gas Mixtures

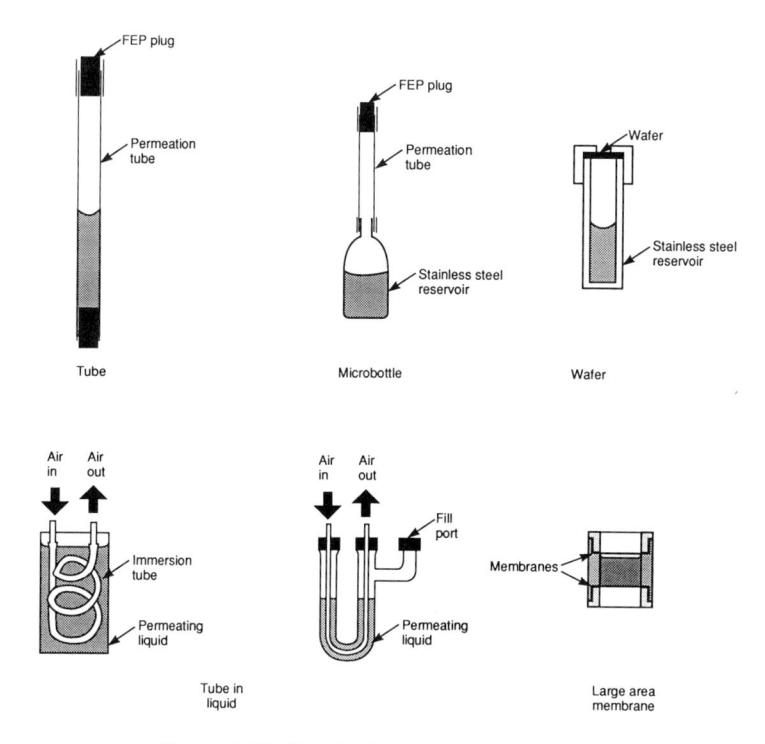

Figure 5.19 Sketch of six permeation devices.

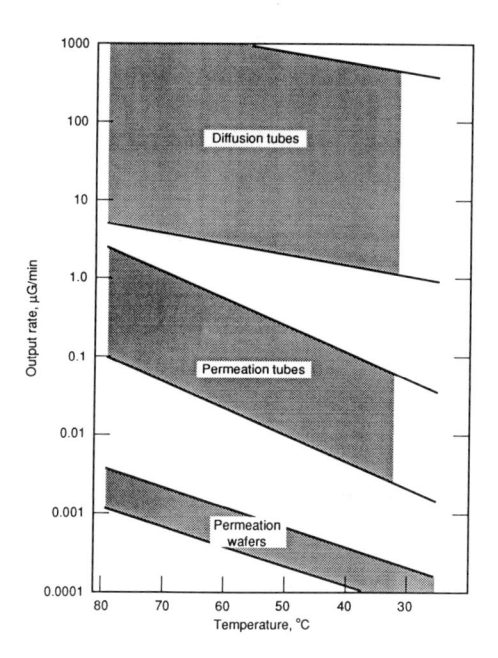

Figure 5.20 Emission rates from diffusion and permeation devices for butane. Output rates extend over seven orders of magnitude. Data taken from Martin et al.[29] with permission.

Table 5.10 Properties of Some Polymers Used in Permeation Devices

Material	Trade Name	Thickness (in.)	References
Fluorinated ethylene propylene	FEP Teflon	0.030	162, 220
Fluorinated ethylene propylene	FEP Teflon	0.012	220
Fluorinated ethylene propylene	FEP Teflon	0.011	217
Polyethylene		0.027–0.526	221
Polyethylene	Alathon	0.025	218
Polyethylene		0.001–0.037	222
Polyvinyl acetate		0.135–1.332	223
Polyvinylidene chloride	Saran wrap	0.025	218
Polyamide	Nylon 6	0.113	218
Polyester	Mylar	0.031	218
Polyethylene terephthalate	Mylar	0.0003–0.006	222
Polythene	Diothene	0.001–0.0635	224
Tetrafluoroethylene	TFE Teflon	0.030	225

membranes,[5] large-area tubes, and membranes for single-phase gas supplies,[203] refillable tubes,[180,182,195,201,202] and thick membranes over glass vials.[173,191,202-205]

5.4.2 Calibration

Calibration of the permeation device can be done several different ways. Gravimetric, volumetric, and manometric techniques have all been employed and are compared in Table 5.11.[206] Before calibration, however, the device is stored at a constant temperature, usually in a nitrogen atmosphere. This prevents reaction with oxygen and water vapor, and allows the device to stabilize structurally.[207] Gravimetric calibration involves weighing the device and storing it in a desiccator at a known temperature and reweighing it after several days. This process is repeated at a new temperature, and plots shown in Figure 5.21 are developed. A semimicrobalance is normally used for best accuracy, since the weight losses are usually on the order of milligrams.[208]

Permeation devices can also be calibrated volumetrically using the microgasometric technique shown in Figure 5.22. A flask housing the permeation tube is attached to a compensated Warburg syringe manometer and brought to temperature equilibrium in a water bath. The meniscus of the manometer fluid (n-nonane when calibrating sulfur dioxide tubes) is brought just to the left of the reference line. As gas permeates through the tube walls and pushes the meniscus to the

Dynamic Systems for Producing Gas Mixtures

Table 5.11 Comparison of Calibration Methods for 2% Precision [a]

Method	Permeation Rate (ng/min)	Minimum Time Hours	Days
Gravimetric — microbalance	2,000	100	4
	200	1,000	40
	20	10,000	400
Volumetric[b]	2,000	2.5	0.1
	200	25	1
	20	250	10
	2	2,500	100
Gravimetric — electrobalance[c]	200	10	0.4
	20	100	4
	2	1,000	40
Partial pressure[a]	200	1	0.04
	20	10	0.4
	2	100	4

[a] Source: Dietz et al.[200]
[b] Source: Saltzman et al.[209]
[c] Source: Purdue and Thompson.[208]

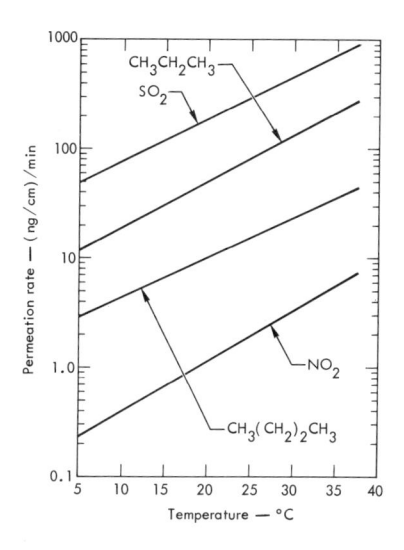

Figure 5.21 Permeation rate vs temperature for four compounds.

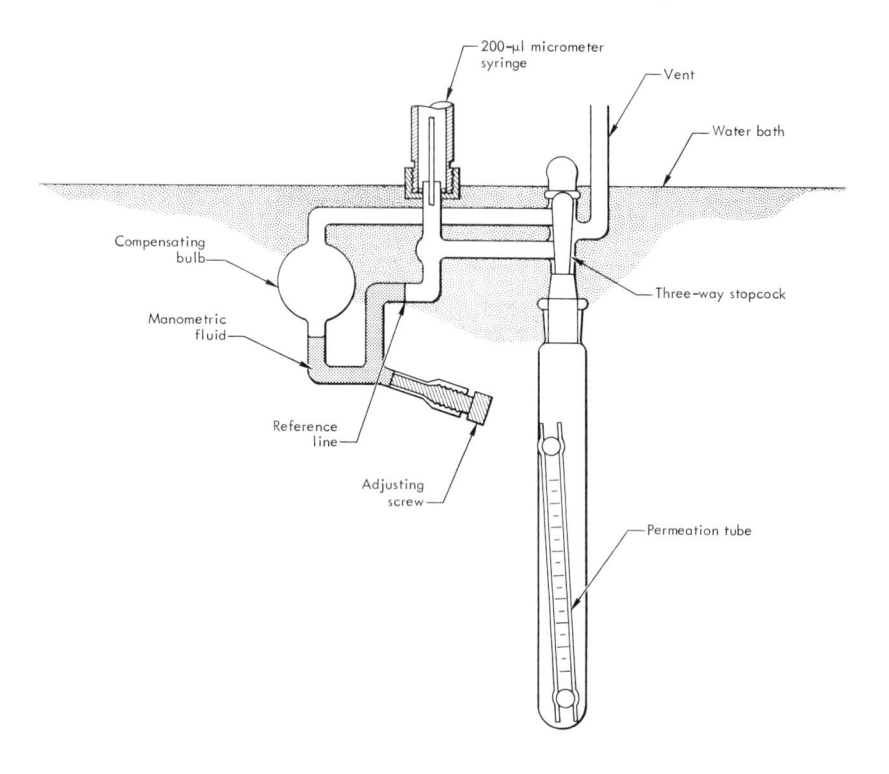

Figure 5.22 Sketch of a microgasometric apparatus for determining permeation rates.

reference line, a timer is started. After 2 hr or less, the micrometer screw is repeatedly adjusted and the volume and timer readings are noted. The permeation rates can then be calculated, given the volume and pressure, by applying a gravimetric temperature correction.[209]

Calibration can also be accomplished using the absolute pressure method of Dietz, Cote, and Smith. This involves monitoring the pressure above the permeation device. This reportedly requires only 1/100 of the time needed to perform a gravimetric calibration (Table 5.11).[206]

5.4.3 Calculations

The permeation of gas through polymeric materials is basically a diffusion process that occurs because of the concentration gradients between the inner and outer surfaces of the membrane wall. The quantity of gas diffusing through the boundaries of a polymer film as expressed by Fick's law is[219]

$$q_d = \frac{DSA(P_1 - P_2)}{L} = \frac{p_G A(P_1 - P_2)}{L} \qquad (5.22)$$

Dynamic Systems for Producing Gas Mixtures

where q_d is the amount diffusing through the polymer per unit area per unit time, D is the diffusion constant, S is the solubility constant of the gas in the polymer, A is the polymer area, $P_{1,2}$ are the pressures on each side of the polymer wall, L is the polymer thickness, and p_G is the gas permeability constant.

Under steady-state conditions, assuming the outside pressure P_2 is negligible and the device is tubular in shape, Equation 5.22 evolves to[188]

$$q_d = \frac{730 \, P_G M P_1}{\log\left(\frac{d_2}{d_1}\right)} \tag{5.23}$$

where q_d is the mass permeation rate (mg/cm of tube length), P_G is the permeation constant (cm²/sec), P is the gas pressure inside the tube (mmHg), d_2 is the outside tube diameter (mm), and d_1 is the inside tube diameter. Equation 5.23 shows that the permeation rate depends on the ratio of the outside to inside diameters and not the absolute diameter. The volume of the enclosed liquified gas (i.e., the lifetime) is proportional to the square of the inside diameter.[188]

The permeation constant is a function of the temperature, the gas characteristics, and the membrane properties, and it is mostly pressure independent.[210,211] The permeation constant is given by

$$P_G = P_G^\circ e^{-E_P/RT} \tag{5.24}$$

where P_G° is a constant, E_P is the permeation activation energy, R is the molar gas constant, and T is the absolute temperature. In any permeation tube system with a fixed geometry, all the variables should remain essentially constant, except for the temperature. The dependence of permeation rate is derived from Equation 5.24 and is given as

$$\log\left(\frac{q_{d_2}}{q_{d_1}}\right) = \left(\frac{E_P}{2.303R}\right)\left(\frac{1}{T_1} - \frac{1}{T_2}\right) \tag{5.25}$$

where $q_{d_{1,2}}$ are the gravimetric permeation rates at the two temperatures (mg/cm), $T_{1,2}$ are the temperatures (°K), E_P is the activation energy, and R is the molar gas constant (1.99 cal/g-mol-°K). If the experimental permeation rate is measured at one temperature, the rate at a new temperature can be calculated from the empirical equation derived from Equation 5.25, which states

$$\log q_{d_1} = \log q_{d_2} - 2950\left(\frac{1}{T_1} - \frac{1}{T_2}\right) \tag{5.26}$$

Experience has shown this equation to be accurate to ±5% for a 10°C temperature change and should be used as an approximation only. Equation 5.26, however, shows the important dependence of the permeation rate on temperature. For every 1°C change in temperature, the permeation rate will change approximately 10%.[212] Therefore, if 1% accuracy is required, the temperature must be controlled to 0.1°C. If a temperature control of ±0.02°C is maintained, average deviation of ±0.5 % from 1.00 ppm of sulfur dioxide has been reported.[213] The temperature stability, however, can be improved through the use of a two-membrane system. Permeation rate changes of 3%/°C have been reported.[214] To preserve the thermal stability, the diluent gas flow should be kept at 1 L/min or less.

Example 5.19. The permeation rate of an ammonia tube is 200 ng/min at 25°C. What is the new rate if the temperature is raised to 40°C?

Permeation rate   200 ng/min at 25°C or 298°K
New temperature   40°C or 313°K

From Equation 5.26, the permeation rate at 40°C is

$$\text{Log } q_{d_1} = \log 200 - 2950\left[\left(\frac{1}{313}\right) - \left(\frac{1}{298}\right)\right]$$

$$\text{Log } q_{d_1} = 2.7754$$

$$q_{d_1} = 596 \text{ ng / min}$$

Often it is desirable to calculate the size of a permeation device required to yield a known concentration. A K factor has been developed for this purpose and is listed for each material in Appendix O. The permeation rate needed to generate a known concentration level in a moving gas stream is

$$q_d = \frac{CQ}{K} \tag{5.27}$$

where q_d is the permeation rate (ng/min), C is the concentration (ppm), Q is the flow rate (mL/min), and K is the gas constant (L/g). The constant K at 25°C is calculated from

Dynamic Systems for Producing Gas Mixtures

$$K = \left(\frac{22.4}{M}\right)\left(\frac{T}{273}\right)\left(\frac{760}{P}\right) \qquad (5.28)$$

At 25°C and 760 mmHg, this reduces to

$$K = \frac{24.5}{M} \qquad (5.29)$$

where M is the molecular weight (g/mol).

Example 5.20. What is the active length of a tube required to produce 1 ppm benzene in an air stream flowing at 0.6 L/min at 25°C? The permeation rate of the tube is 250 ng/min/cm.

Concentration	1 ppm
Flow rate	0.6 L/min or 600 mL/min
Molecular weight	78.1 g/mol
Permeation rate	250 ng/min/cm

From Equation 5.29, the K at 25°C is

$$K = \frac{(24.5)}{(78.1)} = 0.314 \, L/g$$

From Equation 5.27, the total permeation rate is

$$q_d = \frac{(1)(600)}{(0.314)} = 1913 \, ng/min$$

$$\text{Tube length} = \frac{(1913)}{(250)} = 7.65 \, cm$$

The concentration yielded by a permeation device can be calculated from

$$C = \frac{22.4 \times 10^6 \left(\frac{T}{273}\right)\left(\frac{760}{P}\right) q_d}{q_D M} \qquad (5.30)$$

where T is the temperature (°K), P is the pressure (mmHg), q_d is the permeation rate (g/min), q_D is the flow rate of diluent gas (L/min), and M is the molecular weight (g/mol).

Example 5.21. What permeation rate is required to produce a concentration of 0.5 ppm sulfur dioxide in 1 L/min air at 25°C and 1 atm? Express the answer in ng/min.

Concentration	0.5 ppm
Molecular weight	64.1 g/mol
Flow rate	1 L/min
Temperature	25°C or 298°K
Pressure	1 atm or 760 mmHg
ng/g	1.0E + 09

From Equation 5.30, the permeation rate is

$$q_d = \frac{(0.5)(1)(64.1)}{(22.4)(E+06)\left(\frac{298}{273}\right)\left(\frac{760}{760}\right)}$$

q_d = 1.31E – 06 g/min or 1311 ng/min

Since most permeation devices have a finite lifetime and are not refillable, it is sometimes useful to estimate the tube service life. Under normal conditions, a permeation device experiences three stages: saturation, steady state, and depletion.[215] The steady-state lifetime depends simply on the volume of the tube, the weight of the material inside the tube, and the permeation rate. For standard tubes with wall thicknesses of 0.062 and 0.125 in., the lifetime can be calculated from[216]

$$L_s = \frac{1465\rho}{q_d} \quad (5.31)$$

where L_s is the service lifetime (months), ρ is the density (g/mL), and q_d is the permeation rate (ng/min/cm). For thin-wall tubes with a 0.030-in. wall thickness, the service lifetime in months is

$$L_s = \frac{3386\rho}{q_d} \quad (5.32)$$

For wafer devices, the service lifetime is

Dynamic Systems for Producing Gas Mixtures

$$L_s = \frac{11{,}000\rho}{q_d} \qquad (5.33)$$

and for extended-life microbottles, the service lifetime is

$$L_s = \frac{23{,}000\rho}{q_d L} \qquad (5.34)$$

where L is the active length (cm). All these service lifetime formulas assume that the device will lose 75% of total tube volume.

Example 5.22. Estimate the useful life of a standard acetaldehyde tube that has a wall thickness of 0.125 and has a permeation rate of 45 ng/min/cm.

Permeation rate   45 ng/min/cm
Density   0.861 g/mL

From Equation 5.31, the service life is

$$L_s = \frac{(1465)(0.861)}{45} = 28 \text{ months}$$

5.4.4 Discussion

Permeation devices offer several unusual advantages over their dynamic counterparts. The low degree of complexity and low range (0.2 to 200 ppm)[189] make them ideal for field instrument and laboratory standards. They are especially useful as a primary standard for calibration of air pollution analyzers, as well as flame photometric detectors. High precision and an accuracy of 1% can be obtained if a temperature control of 0.1°C is maintained.[190] They come in a wide range of materials, and small, portable generators are available.

Some undesirable characteristics do exist, however. There are relatively few materials that permeate fast enough to be useful, even at low flow rates. Liquids tend to diffuse so slowly (10^{-5} to 10^{-7} mg/min/cm) that they are sometimes below the useful range. In addition, only liquifiable or solid materials can be used with permeation devices unless gas sources are equipped with precise pressure controls. Also, some gases, such as nitrogen dioxide, have been known to weaken and rupture the walls of tubes.[91] Usually, tubes can be stored in the refrigerator when not in use to prolong the useful life. However, it has sometimes been observed that cooling the tube in low temperatures and then reheating it to the experimental conditions produces some sort of thermal hysteresis and invalidates the expected permeation rate.[217]

5.5 EVAPORATION METHODS

Evaporation techniques comprise one of the most generally useful dynamic methods. The diluent gas is passed in close proximity to the liquid to be vaporized by either dispersing it through or passing it over the liquid of interest. The gas stream can be saturated and further diluted, but partial saturation can be used and indeed is usually most desirable. The diluent gas can also be passed through a volatile solid as well as a liquid. Gas streams totally saturated with vapor from low-boiling materials are not normally used because of the explosive hazard and because the high concentrations require extensive dilution of the mixture if the part per million range is desired. However, some saturated systems such as water vapor[251] and mercury (see Chapter 6) in air, have been successful.

The classical types of evaporation units are shown in Figure 5.23. Usually a two-piece vessel with a capacity of 100 mL to 2 L connected by a ground-glass joint is chosen. The gas normally enters through a centrally located gas dispersion tube and is dispersed into minute bubbles, the size of which depend on the tube porosity. Dispersion tubes, shown in Figure 5.24, are available in the arbitrary porosity designations of extra fine, fine, medium coarse, and extra course. The finer the porosity, the greater the pressure required to force the gas through the tube, and the better the efficiency of evaporation. Small-volume bubblers with replaceable dispersion tubes (lower right in Figure 5.23) have also been specially designed.[226] The amount of liquid evaporated will depend on the bubble size, the rate of ascent, the temperature of the liquid, and the height of the liquid column, as well as its boiling point, vapor pressure, and viscosity. It should also be noted that the carrier or diluent gas should be purified and free from dust, oil mists, water, and other organic materials. Accumulations of such materials will alter the evaporation characteristics of vapor generation systems.

The rate of evaporation will slowly decrease with time as the liquid is carried away and the column of liquid above the dispersion tube slowly recedes. This can be minimized by attaching a large auxiliary tank or eliminated altogether by using a reservoir of the type shown in Figure 5.25. Hence, if the gas pressure in the dispersion bottle remains constant, which it should at a given flow, liquid from the reservoir will flow to the bubbler as it is needed[227] until the pressure difference between the bubbler and the reservoir just equals the pressure exerted by the difference in the liquid heights.

If saturation is not required, the bubbler should be enclosed in a constant-temperature bath, since the vapor pressure of the solvent is logarithmically dependent on the temperature. As the solvent evaporates, the liquid cools, causing a rapid decrease in the vapor output. A circulating temperature bath will help ensure more uniform evaporation and reduce the time required to reach temperature equilibrium. The vaporizer output can also be monitored with an appropriate

Dynamic Systems for Producing Gas Mixtures

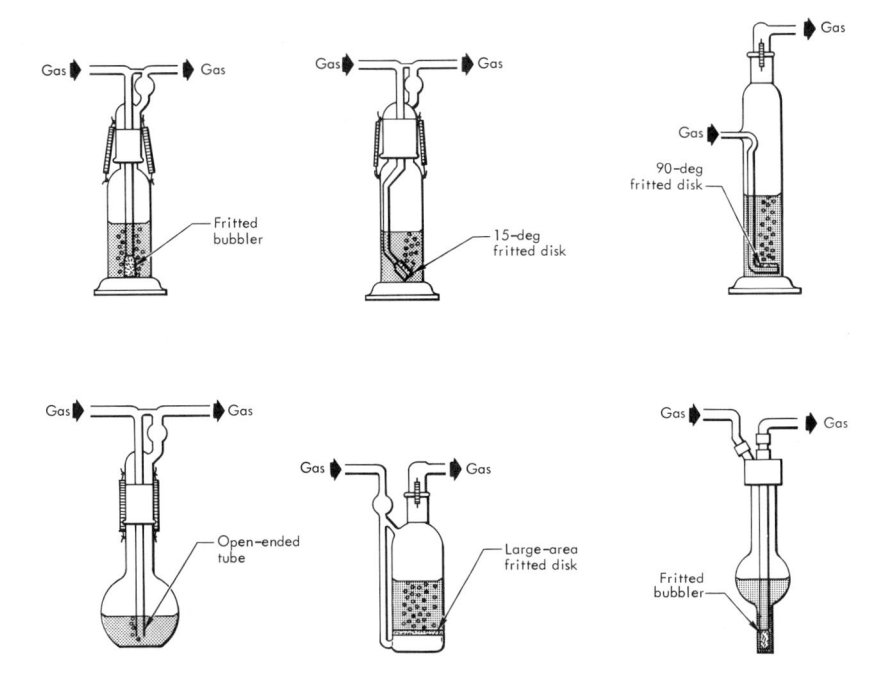

Figure 5.23 Sketches of six typical gas dispersion bottles.

analyzer. If the desired concentration begins to decrease, small incremental bath-temperature changes can return the concentration to the desired level.

The dispersed gas is seldom saturated with vapor, even at reduced flow rates and with an extra-fine gas dispersion apparatus. Saturation will only occur if condensation like that shown in Figure 5.26 takes place. Here, the carrier gas is swept over hot, refluxing vapors up the vertical condenser and is further diluted with diluent gas. The total system flow is now the sum of the contributions from the carrier, dilution, and solvent vapor flow rates. The vapor concentration can be approximated by knowing the temperature at the point where the solvent vapors meet the diluent gas.

Production of contaminants in air streams is often initiated by bubbling nitrogen through the material. This prevents further oxidation and chemical reactions caused by prolonged contact with oxygen, water vapor, or carbon dioxide. Figure 5.27 gives an example of such a system. Here, hydrazine is evaporated and further diluted with additional nitrogen. The primary concentration is measured with an infrared analyzer. Concentration adjustments can be made by varying the nitrogen flow ratios. The final dilution is made with a Miller-Nelson Research Model 401 flow-temperature-humidity control system.[101] Variations of this technique have been used to produce concentrations of 1,1-dimethyl hydrazine,[228,229] trichloroethylene,[230,231] acetic acid,[232] nitro olefins,[228,233] perchloroethylene,[234] tributyl phosphate,[235] carbon tetrachloride,[236] carbon disulfide,[237] phosphorous halides,[238] tributyl phosphene,[239] cyanates,[240,241] and other miscellaneous materials.[242-245]

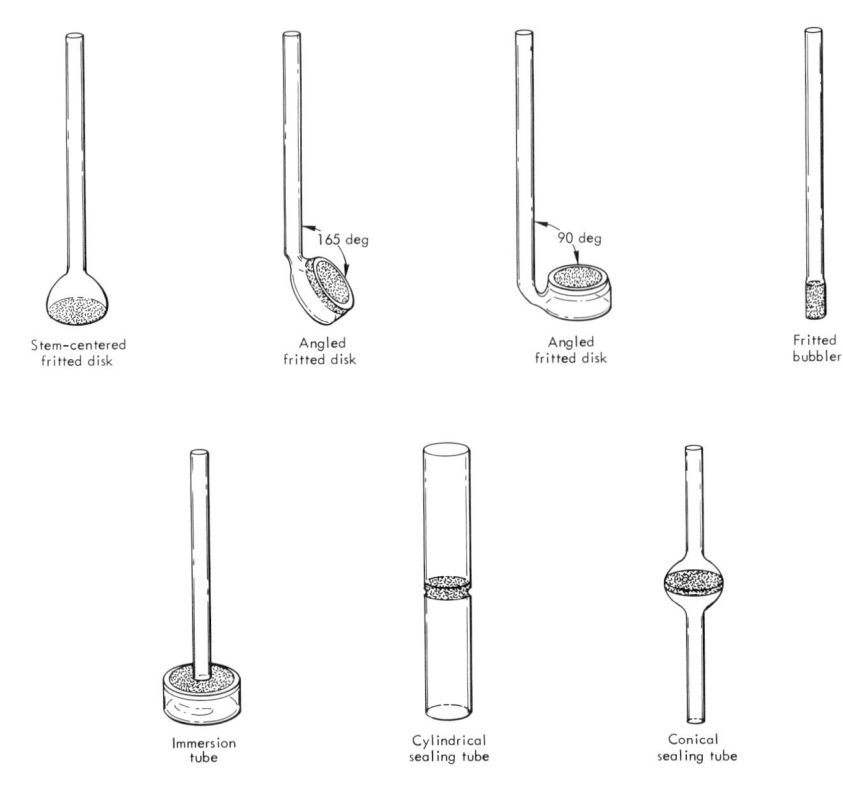

Figure 5.24 Sketches of seven kinds of gas dispersion tubes.

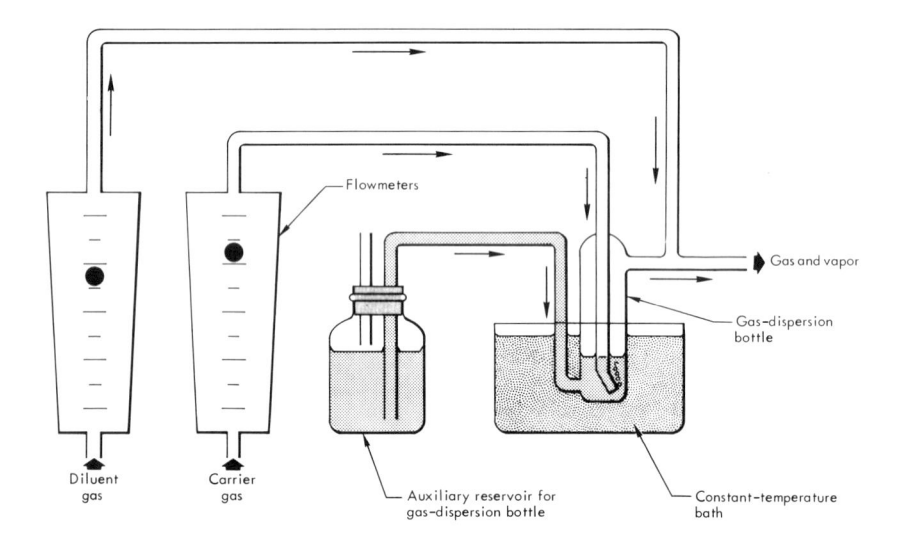

Figure 5.25 Sketch of a constant-level evaporation unit with an auxiliary reservoir.

Dynamic Systems for Producing Gas Mixtures

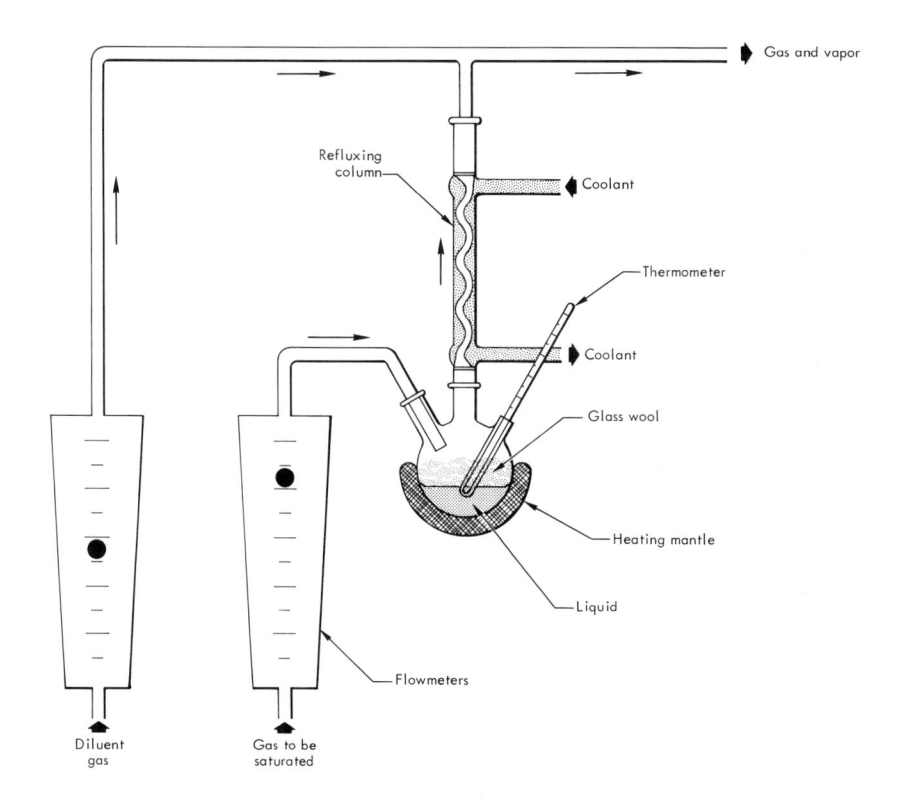

Figure 5.26 Sketch of the apparatus used to produce saturated vapor mixtures via condensation.

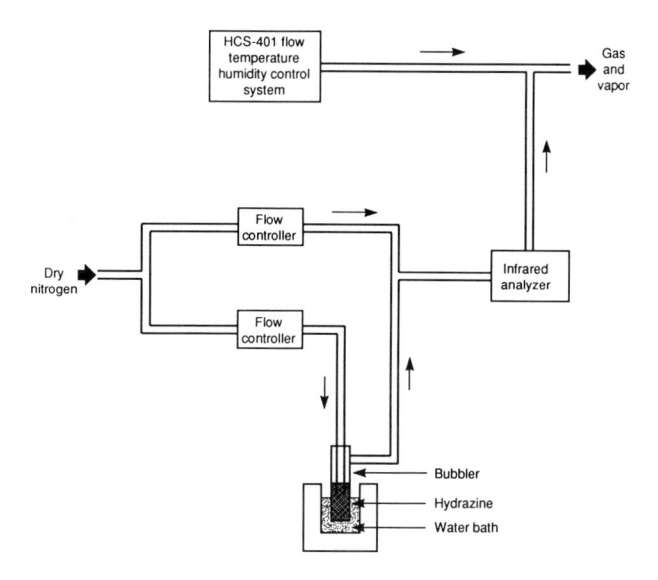

Figure 5.27 Sketch of a hydrazine generation system using nitrogen as the primary diluent gas.

Usually, liquid mixtures cannot be evaporated simultaneously because the mole fraction of the vapor phase is quite different from the mole fraction of the liquid phase, unless a minimum boiling azeotropic mixture is present. Such azeotropes are useful for handling reaction systems of hydrofluoric acid[246-248] and formaldehyde.[246] Nevertheless, gases dissolved in very heavy liquids can be used as the volatile liquid, with only the gas being selectively removed from solution. Trimethyl borane[249] and phosgene[250] dissolved in oil are prime examples.

Evaporation methods have been used for checking analytical methods,[231,239,252254] inhalation studies,[228,230,237,238,249,255-257] carbon evaluations,[236] and vegetation experiments,[246] but they have found only limited use in producing standards for gas or vapor analytical instrumentation. The method is excellent for adding a single volatile liquid to a relatively high-volume (10 to 100 L/min) gas stream, but it has the major disadvantage that the effluent gas mixture must be independently analyzed by some alternate analytical method. The usual methods that have been employed include gravimetric, polarographic,[258] flame ionizing,[230] conductometric,[232] and spectrophotometric[259] techniques.

The concentration of vapor in an air stream is calculated from

$$C_{ppm} = \frac{22.4 \times 10^6 \left(\frac{T}{273}\right)\left(\frac{760}{P}\right) W_t}{M(q_D + q_v)} \quad (5.35)$$

where C is the average concentration (ppm), T is the experimental temperature °K, P is the experimental pressure (mmHg), W_t is the weight lost per unit time from the evaporation vessel (g/min), M is the molecular weight (g/mol), q_D is flow rate of the diluent gas (L/min), and q_v is the flow of the contaminant vapor (L/min). At 25°C and 1 atm, Equation 5.35 reduces to

$$C = \frac{24.5 \times 10^6 W_t}{M(q_d + q_v)} \quad (5.36)$$

The flow of contaminant vapor, q_v, can be calculated from

$$q_v = \frac{24.5 W_t}{M} \quad (5.37)$$

Example 5.23. What is the concentration generated when 0.2 g/min of benzene

Dynamic Systems for Producing Gas Mixtures

is evaporated into a stream of air flowing at 1 ft³/min at 25°C and 1 atm?

Molecular weight	78.0 g/mol
Evaporation rate	0.2 g/min
Flow rate	1.00 ft³/min or 28.3 L/min
L/ft³	28.3

From Equation 5.37, the flow rate of benzene vapor is

$$q_v = \frac{(24.5)(0.2)}{78.0} = 0.06 \text{ L/min}$$

From Equation 5.36, the resultant concentration is

$$C = \frac{(24.5)(E+06)(0.2)}{(78.0)(28.3+0.06)} = 2215 \text{ ppm}$$

Concentrations may also be calculated directly from vapor pressure data if it is certain that the diluent gas stream has reached theoretical saturation at a known temperature. If the total pressure, P, is expressed as the sum of the partial vapor pressures at a specified temperature, then

$$P = p_a + p_b + \cdots + p_n \tag{5.38}$$

The concentration in percent by volume is then

$$C_\% = \frac{10^2 p_n}{P} \tag{5.39}$$

Expressed in parts per million, Equation 5.39 becomes

$$C_{ppm} = \frac{10^6 p_n}{P} \tag{5.40}$$

Example 5.24. The vapor pressure of hexane is 150 mmHg at 25°C. What is the concentration in percent at saturation at 1 atm?

Vapor pressure	150 mmHg
Atmospheric pressure	1 atm or 760 mmHg
mmHg/atm	760

From Equation 5.39, the concentration is

$$C_\% = \frac{(100)(150)}{760} = 19.7\%$$

The vapor pressure exerted by a liquid at a given temperature is obtained from the Antoine relationship:

$$\log p = A - \frac{B}{(C+t)} \tag{5.41}$$

where p is the vapor pressure (mmHg), t is the temperature (°C), and A, B, and C are constants characteristics of each liquid. These constants are experimentally determined and are tabulated in Appendix G for a few of the more common liquids.

Example 5.25. Calculate the vapor pressure of bromocyclohexane at 25°C.

From Appendix G, the Antoine constants are

A	7.3414
B	1778.8
C	235
Temperature	25°C

From Equation 5.41, the vapor pressure is

$$\log p = 7.3414 - \left[\frac{1778.8}{(235+25)}\right] = 0.4999$$

$$p = 3.16 \text{ mmHg}$$

Example 5.26. What is the concentration of dimethyl formamide if it saturates a stream of nitrogen gas at 40°C and 1 atm? Express the concentration in percent.

From Appendix G, the Antoine constants are

A	7.3438
B	1624.7
C	216.2
Temperature	40°C
Atm. pressure	1 atm or 760 mmHg
mmHg/atm	760

Dynamic Systems for Producing Gas Mixtures

From Equation 5.41, the vapor pressure is

$$\log p = 7.3438 - \left[\frac{1624.7}{(216.2+40)}\right] = 1.0023$$

$$p = 10.1 \text{ mmHg}$$

From Equation 5.39, the concentration is

$$C_\% = \frac{(100)(10.1)}{760} = 1.32\%$$

Example 5.27. An air flow is saturated with acetone and is diluted 100-fold by further downstream dilution. At what temperature must the saturation unit be held to maintain the effluent concentration at 3000 ppm? Atmospheric pressure is 740 mmHg.

From Appendix G the Antoine constants are

A	7.0245
B	1161
C	224
Final concentration	3000 ppm or 0.30%
Atm. pressure	740 mmHg
Dilution	100

The concentration exiting the saturation unit is

$$C = (100)(0.30) = 30.0\%$$

From Equation 5.39, the required vapor pressure is

$$p = \frac{(30.0)(740)}{100} = 222 \text{ mmHg}$$

From Equation 5.41, the required temperature is

$$t = \left[\frac{-1161}{(\log 222 - 7.0245)}\right] - 224 = 24.2°C$$

These calculations provide only a rough estimate of concentrations as they really exist in an experimental test apparatus, and errors of ±15% are not uncommon, especially when working with the more volatile materials in saturated gas streams. Better success is noted when working with liquids whose partial pressures are less than 1 mmHg. Here, the amount of evaporated material is so small that temperature equilibrium and large-scale dilution are much less of a problem.

5.6 ELECTROLYTIC METHODS

Many gases can be synthesized on a laboratory scale using electrolytic or coulometric methods. This method has long been used in analytical chemistry to electrodeposit metal ions from solution. The gases that are normally evolved in side reactions are of special interest in producing several types of dynamic systems.

The electrolysis of any solution takes place when enough potential is applied to two electrodes mutually arranged in a solution so that the current can flow. As the voltage across the solution is raised from zero, a significant current increase is noticed at a certain point (the decomposition potential), and the amount of gas evolved from that point on is theoretically a linear function of the applied potential. The electrodes, normally constructed from an inert metal such as platinum, are either deposited with a metal ion from solution or evolve certain gaseous species depending on the electrolyte employed. The gas generated can subsequently be mixed with some diluting carrier gas, and a mixture of known concentration is thus produced.

Selection of an appropriate electrolyte involves the consideration of four requirements outlined by Hersch.[74] First, no side reactions must occur at the electrode of interest. This can be avoided by assuring that a completely pure and plentiful supply of electrolyte is maintained. There must be more electrolyte at the electrode that can be potentially used or serious deviation from expected gas evolution rates will be observed.

Second, gases at the other electrode, if they will interfere, must be diverted from the carrier gas stream. This can usually be done by shaping the reaction vesssel in the form of a U and having the unwanted materials sent to the atmosphere, as shown in Figure 5.28. Another alternate method of electrode separation is the use of a stopcock sealed with a highly conductive electrolyte, such as sulfuric acid, as shown in Figure 5.29. A 45-V source will drive several milliamperes through such a stopcock. If both gases produced from the electrodes do not interfere with one another, then an electrolysis cell of the type shown in Figure 5.30 can be employed. If the arrangement shown in Figure 5.31 is used, the unwanted gas can be suppressed entirely. For example, oxygen will not evolve at the anode of a platinum-screen electrode buried in a bed of activated carbon. Other examples are cited by Hersch[74,260] and Harmon.[261]

Third, the time to attain equilibrium concentration should be minimized. This can be accomplished within seconds if the electrode is either in the shape of a thin vertical platinum wire emerging from an inverted glass "icicle", as shown in Figure 5.29, or hanging down with a glass bead fused to the tip, as shown in Figure 5.31. The solution level should only contact about 1 mm of exposed electrode wire. This configuration favors the almost immediate saturation of the surrounding solution so that the diluent gas thus receives all the evolved gas.

Fourth, the concentration produced should be as smooth as possible. If the electrode is small and if at least a 45-V source is used, small, barely visible bubbles should be continuously generated. The smallest current provided should not be lower than 10 µA, since smooth gas evolution then becomes difficult.[74]

Dynamic Systems for Producing Gas Mixtures

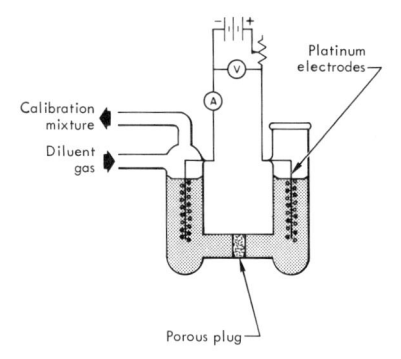

Figure 5.28 Sketch of a U-shaped electrolytic reaction vessel in which the unwanted gas generated by the right hand electrode is vented to the atmosphere.

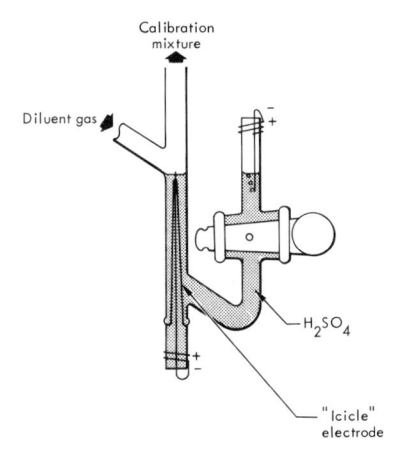

Figure 5.29 Sketch of an electrolytic reaction vessel with an "icicle" electrode.

Excess heating caused by high currents (above 0.5 A) must be avoided as well. Examples of various types of electrolytic systems are shown in Table 5.12. The arsine and stibine methods discussed by Saltzman are particularly useful.[23]

The concentration of gas mixtures produced electrolytically are based on calculations using a variation of Faraday's Law,[17] which states that

$$I = \frac{znF}{\tau} = zFq_E \qquad (5.42)$$

where I is the current (A), z is the number of electrons required to liberate 1 mol of gas, n is the number of moles liberated, F is Faraday's constant (96,489 Coul/mol), τ is the time (sec), and q_E is the rate of gas production (mol/sec).

The concentration, C, produced in parts per million when diluted with a carrier

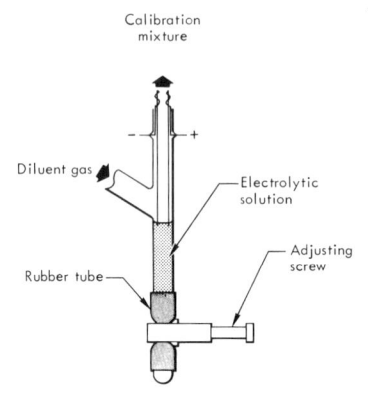

Figure 5.30 Sketch of an electrolytic reaction vessel with unseparated electrodes. Both generated gases are combined in the diluent gas. The adjusting screw is used to adjust the level of the electrolyte.

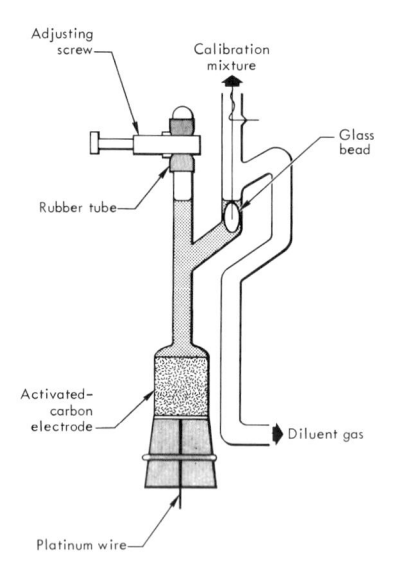

Figure 5.31 Sketch of an electrolytic reaction vessel with an activated carbon electrode. The glass bead provides fast equilibrium times.

gas is

$$C = \frac{10^6 \, q_E}{q_D} = \frac{10^6 \, I}{zFq_D} \tag{5.43}$$

where q_D is the carrier-gas flow rate (mol/sec).

Dynamic Systems for Producing Gas Mixtures

Table 5.12 Some Common Electrolytic Systems

Gas Produced	Elecrolyte and Electrodes	Gas Produced From Alternate Electrode	References
AsH_3	Pt, Na_3AsO_3 + H_2SO_4	O_2	1, 23
SbH_3	Pt, $KSbC_4H_4O_7$, Pt	O_2	1, 23
$O_3 + O_2$	Pt, H_2SO_4	H_2	74,260,262–265
$C_2H_6 + CO_2$	Pt, CH_3CONa, Pt	H_2	266
Cl_2	Pt, NaCl, Pt	H_2	74, 260
NO	Pt, $NOHSO_4$ + H_2SO_4	O_2	75, 267
NO_2	Pt, $NOHSO_4$ + H_2SO_4	(a)	75, 267, 268
$O_2(H_2)$	Pt, H_2SO_4, Pt Pt, KOH, Pt	$H_2(O_2)$	74, 260
$O_2(H_2)$	Pt, H_2SO_4, Pt Pt, H_2O, C Pt, KOH + HgO + C, Hg Pt, KOH + CdO, Cd	None	74, 260
N_2	Pt, N_2H_4 + HCl, Pt Pt, N_2H_4 + H_2SO_4, Pt	H_2	74, 75, 260 269
Air	Pt, N_2H_4 + HCl, Pt[a] Pt, KOH, Pt[b]	H_2	75
D_2	Pt, K_2SO_4 + D_2O, Pt	O_2	74, 260
CO_2	Pt, $H_2C_2O_4$, Pt Pt, CH_3CO_2Na + HCO_3^-, Pt	H_2 H_2	74, 75, 260 270
T_2	Pt, Ni, KOH + TOH, Ni	O_2	75

[a] Both gases evolved react to form NO_2.
[b] Two separate electrolytic cells. The anodic gases N_2 and O_2 are generated and mixed in a 4:1 ratio.

If 1 mol of gas contains 24,060 mL at 20°C and 1 atm, and if the current, I, is expressed in milliamps, then Equation 5.43 becomes

$$C = \frac{14.96\, I}{zq_D} \tag{5.44}$$

Example 5.28. What is the concentration in ppm of carbon dioxide produced when 200 μA are supplied to a solution of oxalic acid at room temperature and pressure? The flow rate of the diluent gas is 0.3 L/min, and the anode reaction is $C_2O_4^= = 2CO_2 + 2e^-$.

Supply current	200 µA or 0.200 mA
Air flow rate	0.3 L/min
Number of electrons	2
µA/mA	1000

From Equation 5.44, the concentration is

$$C = \frac{(14.96)(0.200)}{(2)(0.3)} = 4.99 \text{ ppm}$$

Example 5.29. What current must be supplied to a solution of sodium chloride to produce a concentration of 1 ppm chlorine in nitrogen flowing at 0.5 ft³/min? The anode reaction is $2Cl^- = Cl_2 + 2e^-$.

Concentration	1 ppm
Flow rate	0.5 ft³/min or 14.2 L/min
Number of electrons	2
L/ft³	28.3

From Equation 5.44, the current is

$$I = \frac{(1)(2)(14.2)}{14.96} = 1.89 \text{ mA}$$

where C is the concentration (ppm), I is the supply current (mA), and q_D is the diluent gas flow rate (L/min).

The electrolytic process does not always proceed as theoretically planned. Usually, a correction or yield factor, N, must be inserted. Equation 5.44 then becomes

$$C = \frac{14.96 \text{ IN}}{zq_D} \tag{5.45}$$

If conditions other than 20°C and 1 atm are maintained for both the diluent and electrolytically produced gases, no temperature or pressure correction is needed. Thus the concentration is a ratio of the two corrected gas flow rates. However, if conditions change in the electrolytic cell, but a corresponding change is not felt by the diluent gas, Equation 5.45 becomes

$$C = \frac{14.96 \text{ IN}}{zq_D}\left[1 + \frac{(T-293)}{293} + \frac{(760-P)}{760}\right] \tag{5.46}$$

Example 5.30. What current is required to produce a concentration of 10 ppm of arsine in 1 L/min of nitrogen? The yield factor is 45% and the cell

Dynamic Systems for Producing Gas Mixtures

temperature is 35°C at 1 atm. The cell reaction is $AsO^{3-} + 9H^+ + 6e^- = 3H_2O + AsH_3$.

Concentration	10 ppm
Flow rate	1 L/min
Number of electrons	6
Yield factor	0.45
Temperature	35°C or 308°K
Pressure	1 atm or 760 mmHg
mmHg/atm	760

From Equation 5.46, the current is

$$I = \frac{(10)(6)(1)}{(0.45)(14.96)\left[1 + \frac{(308-293)}{293} + \frac{(760-760)}{760}\right]} = 8.48 \text{ mA}$$

The electrolytic method of adding trace contaminants has the advantage of producing very low concentrations with the flick of a switch. It requires no moving parts, but the construction of some systems requires the services of a first-rate glass blower. Small electrolytic units are not generally available commercially and are somewhat complicated. The small volumes generated (a few microliters per minute) make this an ideal method of checking instruments where continuous low part per million concentrations are needed. However, many trace contaminants cannot be produced by electrolysis or in sufficient volumes to be useful in the percent range of concentration. If a particular reaction of interest is desired but has not been well investigated, the yield factor must be checked with some alternate method to ensure reliability. This factor is commonly affected by electrode discoloration, electrolytic impurities, and insufficient electrolyte.

5.7 CHEMICAL METHODS

Controlled addition of a contaminant gas to a moving gas stream can be accomplished with chemical reaction techniques. If the material of interest is unstable, extraordinarily reactive, commercially unavailable, or prohibitively expensive, it may be desirable to continually produce certain types of gases on a small scale in the laboratory.

The apparatus involved is almost as diverse as the number of possible reactions themselves. Most systems involve a gas-liquid or gas-solid reaction system. An example of a gas liquid involves bubbling nitric oxide through a solution of permanganate to produce nitrogen dioxide. An example of a gas-solid system uses dilute chlorine in the presence of sodium chlorite to produce chlorine dioxide. Both of these reactions are discussed in more detail in Chapter 6.

A number of possibilities for using reaction systems are shown in Table 5.13, but only a few have been practiced.[75] As noted, the additive may be oxidative or reductive or a catalytic gas, liquid, or vapor. Tables 5.14 and 5.15 list some additional reactions that have been tried. Ozone, as produced from the irradiation

Table 5.13 Conversion Reactions Used to Produce Contaminant Gases[a]

Reactants	Products
C_2H_5OH	$(Al_2O_3) = C_2H_4 + H_2O$ (in N_2)
C_6H_5OH	$= CO + 5C + 3H_2$ (in N_2)
$Fe(CO)_5$	$= 5CO + Fe$ (in N_2)
$2H_2 + O_2$	$(Pt) = 2H_2O$
$C_{10}H_8 + 12O_2$	$= 10CO_2 + H_2O$
$Cl_2 + H_2$	$(UV) = 2HCl$
$R-NH_2 + H_2$	$(Ni) = RH + NH_3$ (in N_2)
$O_2 + 2C$	$= 2CO$ (in N_2)
$H_2O + C$	$= CO + H_2$ (in N_2)
$H_2 + Ag_2S$	$(500°C) = H_2S + 2Ag$ (in N_2)
$H_2 + NiCl_2$	$(600°C) = 2HCl + Ni$ (in N_2)
$H_2 + 2CoF_3$	$(300°C) = 2HF = 2CoF_2$ (in N_2)
$6H_2O + Mg_3N_2$	$= 2NH_3 + 3Mg(OH)_2$
$6H_2O + Al_2S_3$	$= 2Al(OH)_3$
$3H_2O + Al(OR)_3$	$= 3ROH + Al(OH)_3$
$HCl + Na_2S_2O_5$	$= SO_2 + NaHSO_3 + NaCl$
$2CF_3COOH + CaC_2$	$= C_2H_2 + (CF_3COO)_2Ca$
$2Cl_2 + N_2H_4 \cdot H_2SO_4$	$= 4HCl + H_2SO_4 + N_2$
$Cl_2 + C_6H_4(OH)_2 Kr^{85}$	$= Kr^{85} + C_6H_4O_2 + 2HCl$
$Cl_2 + 2NaClO_2$	$= 2ClO_2 + 2NaCl$
$SO_2 + 3NaClO_2$	$= 2ClO_2 + NaCl + Na_2SO_4$
$3NO + 2CrO_3$	$(200°C) = 3NO_2 + Cr_2O_3$
$3NO + 2HMnO_4(aq)$	$= 3NO_2 + 2MnO_2 + H_2O$

[a] Data taken from Hersch.[75]

Dynamic Systems for Producing Gas Mixtures

Table 5.14 Additional Reactions Used to Produce Gas and Vapor Mixtures

Reactants	Products	References
$4H^+ + 2NO_2^- + 2I^-$	$= I_2 + 2NO + H_2O$	271
$NO + O_3$	$= NO_2 + O_2$	272
$NaF + H_2O$	$= HF + NaOH$	246, 273
$H_2SO_4 + 2NaF$	$= 2HF + Na_2SO_4$	274
$3O_2 + (UV)$	$= 2O_3$	Chapter 6
$H_2C_2O_4 + (UV)$	$= CO + CO_2 + H_2O$	75
$HCOOH + (heat, H_2SO_4)$	$= CO + H_2O$	2, 274, 275
$2NO + O_2$	$= 2NO_2$	Chapter 6
$2Pb(NO_3)_2 + (heat)$	$= 4NO_2 + 2PbO + O_2$	277, 278
$C_5H_5NHCo(CO)_4 + H^+$	$= HCo(CO)_4 + C_5H_5NH^+$	106, 279
$KCN + H_2O$	$= HCN + KOH$	23
$2KCN + H_2SO_4$	$= 2HCN + K_2SO_4$	280, 281
$(CH_3)_2CO + (heat)$	$= CH_2=C=O + CH_4$	282
$-CH=CH(CH_2)_4- + (heat)$	$= CH_2=CHCH=CH_2 + CH_2=CH_2$	282
$H_2O + CaC_2$	$= C_2H_2 + CaO$	5
$H_2O + CH_3MgI$	$= CH_4 + MgOHI$	5
$H_2O + PCl_5$	$= 2HCl$	5
$HCl + NaHCO_3$	$= CO_2 + NaCl + H_2O$	5
$H_2O + 2RCOCl$	$= 2HCl + (RCO)_2O_3$	5
$5H_2 + Ag_2SO_4 + heat$	$= H_2S + 4H_2O + Ag$	5
$CH_3CCl_3 + heat$	$= CH_2=CCl_2 + HCl$	283
$2AlP + 3H_2O$	$= 2PH_3 + Al_2O_3$	276
$NaHSO_3 + H^+$	$= SO_2 + H_2O + Na^+$	280
$NaNO_2 + H^+$	$= NO_2 + NO + H_2O + Na^+$	280

Gas Mixtures: Preparation and Control

Table 5.15 Generaton of Gases by Thermal Decomposition[a]

Reactants	Temp (°C)	Products
7KMnO$_4$	250	= K$_3$MnO$_4$ + 2 K$_2$MnO$_4$ + 4MnO$_2$ + 3O$_2$
NH$_4$NO$_2$	70	= N$_2$ + 2H$_2$O
Ba(N$_3$)$_2$	150	= 3N$_2$ + Ba
NH$_4$NO$_3$	200	= N$_2$O + 2H$_2$O
4NH$_4$NO$_3$	200	= 2NO + 3N$_2$ + 8H$_2$O
2Pb(NO$_3$)$_2$	(O$_2$)	= 2PbO + 4NO$_2$ + O$_2$
CaC$_2$O$_4$	440	= CaCO$_3$ + CO
Ni(CO)$_4$	200	= Ni + 4CO
2NaHCO$_3$	200	= Na$_2$CO$_3$ + H$_2$O + CO$_2$
MgCO$_3$	540	= MgO + CO$_2$
2AgCN	400	= 2Ag + (CN)$_2$
2AuCl	Heat	= 2Au + Cl$_2$
CH$_3$COONa	400	= CH$_4$ + Na$_2$CO$_3$

Source: Data taken from Namiesnik.[5]

of oxygen by ultraviolet light, is particularly useful and is also discussed in Chapter 6.

Most of the calculations covered previously in this chapter can be applied to systems involving chemical reactions. Examples 5.31 and 5.32 are typical.

Example 5.31. At what rate must hydrogen be passed over heated nickelous chloride to produce a concentration of 5.0% hydrogen chloride when diluted with air flowing at 10 L/min? Assume that the gases are mixed at room temperature and that the conversion reaction of hydrogen chloride proceeds stoichiometrically to completion according to the equation in Table 5.13.

  Concentration   5% or 0.05
  Flow rate   10 L/min
  HCl/H$_2$ 2

From Equation 5.1, the flow of hydrogen chloride is

Dynamic Systems for Producing Gas Mixtures

$$q_{HCl} = \frac{(0.05)10}{(1-0.05)} = 0.526 \text{ L/min}$$

$$q_{H_2} = \frac{0.526}{2} = 0.263 \text{ L/min}$$

Example 5.32. Cyclohexene is injected and evaporated in a 1 ft³/min stream of nitrogen at the rate of 1.2 g/hr. It is subsequently converted to 1,3-butadiene according to the equation in Table 5.14. What concentration is generated at 21°C and 740 mmHg?

Injection rate	1.2 g/hr or 0.020 g/min
Flow rate	1 ft³/min or 28.3 L/min
MW of cyclohexene	82.1 g/mol
MW of 1,3-butadiene	54.1 g/mol
Temperature	21°C or 294°K
Pressure	740 mmHg
L/ft³	28.3
min/hr	60

The rate of 1,3-butadiene produced is

$$W = \frac{(0.020)(54.1)}{82.1} = 0.0132 \text{ g/min}$$

From Equation 5.35, the concentration of 1,3-butadiene produced is

$$C = \frac{(22.4)(E+06)(0.0132)\left(\frac{294}{273}\right)\left(\frac{760}{740}\right)}{(28.3)(54.1)} = 213 \text{ ppm}$$

Although the chemical reaction technique seems straightforward enough, especially in the classical sense, a number of difficulties arise. The reactant feed mechanisms required are normally more complex than the usual dynamic system, since two or more reactants are often required in the gas phase or some carefully regulated reaction vessel must be provided when liquids are used. Characteristically, the rate of reaction is a logarithmic function of temperature, and the usual rule of doubling the reaction rate with every 10°C rise in temperature applies.

Although the number of reactions available is practically limitless, two factors must be considered before selecting an appropriate reaction: first, are the products formed within a reasonable length of time after mixing? Second, does the reaction proceed stoichiometrically to completion? The rate of most reactions can be found in the literature from experimentally determined reaction rate constants, and it is expressed in moles per unit time. The degree of quantitative completion (i.e., the ratio of products to reactants) is also an experimental quantity given by the equilibrium constant. If the reaction system under study does not proceed to completion, or if doubt exists as to the character of the effluent gases and vapors, then further painstaking and time-comsuming analysis is required to assess the true nature of the gas mixtures.

As expected, many reactions that produce a selected contaminant also produce additional products or generate unwanted side reactions. These by-products, along with any unreacted starting material, must be selectively removed by filtration, absorption, or adsorption, which further complicate the system.

It is desirable, except in a few selected instances, to purchase the gas of interest, either in its pure form or diluted by an inert gas. Then the gas can be analyzed and further diluted as necessary via the conventional, commercially available flow-metering devices.

REFERENCES

1. Kusnetz, H. L., B. E. Saltzman, and M. E. Lanier. *Am. Ind. Hyg. Assoc. J.* 21:361 (1960).
2. Caplan, P. E. "Calibration of Air Sampling Instruments," in *Air Sampling Instruments for Evaluation of Atmospheric Contaminants* (Cincinnati, OH: American Conference of Governmental Industrial Hygienists, 1966).
3. Smith, A. F. "Standard Atmospheres," in *Detection and Measurement of Hazardous Gases,* C. F. Cullis and J. G Firth, Eds. (Exeter, England: Heineman Educational Books, 1981), chap. 7.
4. Axelrod, H. D. and J. P. Lodge. "Sampling and Calibration of Gaseous Pollutants," in *Air Pollution,* Vol. III, A. C. Stern, Ed. (San Francisco: Academic Press, 1976).
5. Namiesnik, J. *J. Chromatog.* 300:79 (1984).
6. Woodfin, W. *J. Am. Ind. Hyg. Assoc. J.* 45:138 (1984).
7. Dixon, S. W., J. F. Vasta, L. T. Freeland, D. J. Calvo, and R. E. Hemingway. *Am. Ind. Hyg. Assoc. J.* 45:99 (1984).
8. Bownik, M., J. Namiesnik, and L. Torres. *Chromatographia* 17:503 (1983).
9. Berardinelli, S. P., E. S. Moyer, and R. C. Hall. *Am. Ind. Hyg. Assoc. J.* 51:595 (1990).
10. Irish, D. D. and E. M. Adams. *Ind. Med. Surg.* 1:1 (1940).
11. Box, W. D. *Am. Ind. Hyg. Assoc. J.* 24:618 (1963).
12. Jellinek, H. H. G. and F. J. Kryman. *Environ. Sci. Technol.* 1:658 (1967).
13. Birks, N. and T. Flatley. *J. Sci. Instr.* 2:463 (1969).
14. Bate, G. C., A. D'Aoust, and D. T. Canvin. *Plant Physiol.* 44:1122 (1969).
15. Adams, D. F. *J. Air Pollut. Control Assoc.* 13:88 (1963).
16. Mastromatteo, E., A. M. Fisher, H. Christie, and H. Danzinger. *Am. Ind. Hyg. Assoc. J.* 21:394 (1960).
17. Angely, L., E. Levart, G. Guiochon, and G. Peslerbe. *Anal. Chem.* 41:1456 (1969).
18. Haun, C. C., E. H. Vernot, D. L. Geiger, and J. M. McMerney. *Am. Ind. Hyg. Assoc. J.* 30:551 (1969).

19. Pecora, L. *J. Am. Ind. Hyg. Assoc. J.* 20:235 (1959).
20. Lester, D., L. A. Greenberg, and W. R. Adams. *Am. Ind. Hyg. Assoc. J.* 24:265 (1963).
21. Zocchi, F. *J. Gas Chromatog.* 6:100 (1968).
22. Levinskas, G. J., M. R. Paslian, and W. R. Bleckman. *Am. Ind. Hyg. Assoc. J.* 19:46 (1958).
23. Saltzman, B. E. *Anal. Chem.* 33:1100 (1961).
24. Murphy, S. D., C. E. Ulrich, S. H. Frankowtz, and C. Xintaras. *Am. Ind. Hyg. Assoc. J.* 25:246 (1964).
25. Duckworth, S., D. Levaggi, and J. Lim. *J. Air Pollut. Control Assoc.* 13:429 (1963).
26. Dalhamn, T. and L. Strandberg. *Int. J. Air Water Pollut.* 4:154 (1961).
27. Malanchuk, M. *Am. Ind. Hyg. Assoc. J.* 28:76 (1967).
28. Barr, G. *J. Sci. Instr.* 11:321 (1934).
29. Wilby, F. V. *J. Air Pollut. Control Assoc.* 19:96 (1969).
30. Zolty, S. and M. J. Prager. *J. Gas Chromatog.* 5:533 (1967).
31. Matheson Gas Products Catalog No. 88, East Rutherford, NJ, 1987.
32. Brochure, Calibrated Instruments, Inc., New York, NY.
33. McQuaker, N. R., H. Haboosheh, and W. Best. *Am. Lab.* 13:105 (1981).
34. Axelrod, H. D., J. B. Pate, W. R. Barchet, and J. P. Lodge. *Atmos. Environ.* 4:209 (1970).
35. Sordelli, D. *Rev. Combust.* 13:122 (1959).
36. Schnelle, P. D. *Instr. Soc. Am. J.* 4:128 (1957).
37. Troy, D. *J. Anal. Chem.* 27:1217 (1955).
38. Cummings, W. G. and M. W. Redfearn. *Chem. Ind. (London)* 1950:809 (1957).
39. Namiesnik, J. *J. Chromatog.* 300:79 (1984).
40. McArthur, B. R. *Am. Ind. Hyg. Assoc. J.* 41:151 (1980).
41. Goldan, P. D., W. C. Kuster, and D. L. Albritton. *Atmos. Environ.* 20:1203 (1986).
42. Dixon, S. W., J. F. Vasta, L. T. Freeland, D. J. Calvo, and R. E. Hemingway "A Multiconcentration Controlled Test Atmosphere System for Efficient Passive Dosimeter/Charcoal Tube Studies," E. I. du Pont Company (1985).
43. Kapila, S., R. K. Malhotra, and C. R. Vogt. *Am. Chem. Soc.* 149:533 (1981).
44. Butterfield, K. K. Private communication, University of California, Berkeley, CA.
45. Lodge, J. P. "Production of Controlled Test Atmospheres," in *Air Pollution*, A. C. Stern, Ed. (New York: Academic Press, 1962).
46. Rogoff, J. M. *J. Lab. Clin. Med.* 25:853 (1939).
47. Saltzman, B. E. *Am. Ind. Hyg. Assoc. J.* 16:121 (1955).
48. Leichnitz, K. *Pure & Appl. Chem. (London)* 55:1239 (1983).
49. Bulletin No. 5075m, Davis Instruments, North Charlottesville, VA.
50. Sunderman, F. W., J. F. Kincaid, W. Kooch, and E. A Bermelin. *Am. J. Pathol.* 26:1211 (1956).
51. Nelson, G. O. and K. S. Griggs. "A Precision Dynamic Method for Producing Known Concentrations of Gas and Solvent Vapor in Air," Report No. UCRL-70394, (Livermore, CA: Lawrence Livermore National Laboratory, 1967).
52. Nelson, G. O. and K. S. Griggs. *Rev. Sci. Instr.* 39:927 (1968).
53. Lester, D. and W. R. Adams. *Am. Ind. Hyg. Assoc. J.* 26:562 (1965).
54. Amdur, M. O., R. Z. Schulz, and P. Drinker. *Arch Ind. Hyg. Occup. Med.* 5:318 (1952).
55. Jones, B. W., S. A. Jones, and M. B. Newworth. *Ind. Eng. Chem.* 44:2233 (1952).
56. Hill, D. W. and H. A. Newell. *J. Sci Instr.* 42:783 (1965).
57. Torkelson, T. R., F. Oyen, and V. K. Rowe. *Am. Ind. Hyg. Assoc. J.* 22:354 (1961).

58. Saltzman, B. E. *Anal. Chem.* 26:1949 (1954).
59. Fraust, C. L. and E. R. Hermann.. *Am. Ind. Hyg. Assoc. J.* 27:68 (1966).
60. Kuczynski, E. R. *Environ. Sci. Technol.* 1:68 (1967).
61. Rowe, V. K., T. Wujkowski, M. A. Wolf, S. E. Sadek, and R. D. Stewart. *Am. Ind. Hyg. Assoc. J.* 24:541 (1963).
62. Fultyn, R. V. *Am. Ind. Hyg. Assoc. J.* 22:49 (1961).
63. Bulletin No. 900, Harvard Apparatus Company, Inc., Millis, MA.
64. Sherberger, R. F., G. P. Happ, F. A. Miller, and D. W. Fassett. *Am. Ind. Hyg. Assoc. J.* 19:494 (1958).
65. Henry, N. W. *Am. Ind. Hyg. Assoc. J.* 40:1017 (1979).
66. Brooks, B. I. *Analyst* 106:403 (1981).
67. Kristiansen, U., L. F. Hanson, and G. D. Nielsen. *App. Ind. Hyg.* 4:171 (1989).
68. Nishimoto, K. K., R. L. Schumacher, L. Monteith, and P. Breysse. *Am. Ind. Hyg. Assoc. J.* 41:223 (1980).
69. Decker, J. R., O. R. Moss, and B. L. Kay. *Am. Ind. Hyg. Assoc. J.* 43:400 (1982).
70. Moyer, E. S. and S. P. Berardenelli. *Am. Ind. Hyg. Assoc. J.* 48:315 (1987).
71. Dharmarajan, V., R. D. Ling, and D. R. Hackathorn. *Am. Ind. Hyg. Assoc. J.* 47:393 (1986).
72. Swearengen, P. M. and S. C. Weaver. *Am. Ind. Hyg. Assoc. J.* 49:70 (1988).
73. Pisula, J. D. *Res. Develop* 31:78 (1989).
74. Hersch, P. A., C. J. Sambucetti, and R. Deuringer. "Electrolytic Calibration of Gas Monitors," paper presented at the 9th National Instrument Symposium, Houston, TX, (1963).
75. Hersch, P. A. *J. Air Pollut. Control Assoc.* 19:164 (1969).
76. Brochure, Sage Instruments, Inc., White Plains, NY.
77. Bulletin No. 540, Harvard Apparatus Company, Inc., Millis, MA.
78. Weeks, M. H., T. O. Downing, N. P. Musselman, T. R. Carson, and W. A. Groff. *Am. Ind. Hyg. Assoc. J.* 21:374 (1960).
79. Jacobson, K. H., W. E. Rinehart, H. J. Wheelwright, M. A. Ross, J. L. Papin, R. C. Daly, E. A. Greene, and W. A. Groff. *Am. Ind. Hyg. Assoc. J.* 19:91 (1958).
80. Sayers, R. R., W. P. Yant, C. P. Waite, and F. A. Patty. Public Health Report (U. S.) 45:225 (1930).
81. Brubach, H. F. *Rev. Sci. Instr.* 18:363 (1947).
82. Goetz, A. and T. Kallai. *Am Ind Hyg Assoc. J.* 24:453 (1963).
83. Rossano, A. T. and H. B. H. Cooper. *J. Air Pollut. Control Assoc.* 13:518 (1963).
84. Charnley, A., G. L. Isles, and J. S. Rowlinson. *J. Sci. Instr.* 31:145 (1954).
85. Calcote, H. F. *Anal. Chem.* 22:1058 (1950).
86. Hersch, P. and J. E. Whittle. *J. Sci. Instr.* 35:32 (1958).
87. Lundsted, L. G., A. B. Ash, and N. L. Koslin. *Anal. Chem.* 22:626 (1950).
88. Brochure, Calibrated Instruments, Inc., New York, NY.
89. Tokiwa, Y. and E. R. DeVera. "The Woesthoff Gas Dosing Apparatus," Paper presented at the 8th Conference on Methods in Air Pollution and Industrial Hygiene Studies, Oakland, CA (1967).
90. Axelrod, H. D., J. H. Carey, J. E. Nonelli, and J. P. Lodge. *Anal. Chem.* 41:1856 (1969).
91. Axelrod, H. D., J. B. Pate, W. R. Barchet, and J. P. Lodge. *Atmos. Environ.* 4:209 (1970).
92. Miller, R. R., R. L. Letts, W. J. Potts, and M. J. McKenna. *Am. Ind. Hyg. Assoc. J.* 41:844 (1980).
93. Nelson, G. O. and D. J. Hodgkins. *Am. Ind. Hyg. Assoc. J.* 33:100 (1972).

94. Nelson, G. O., W. Van Sandt, and P. E. Barry. *Am. Ind. Hyg. Assoc. J.* 26:388 (1965).
95. Nelson, G. O. "A Simplified Method for Producing Known Concentrations of Mercury Vapor in Air," Report No. UCRL-71481 (Livermore, CA: Lawrence Livermore National Laboratory, 1969).
96. Brochure, Clayborn Labs Inc., Santa Ana, CA.
97. Brochure, Electrofilm, Inc., North Hollywood, CA.
98. Diggle, W. M. and J. C. Gage. *Analyst* 78:473 (1953).
99. Gage, J. C. *J. Sci. Instr.* 30:25 (1953).
100. Brochure, Miller-Nelson Research, SV-20 Solvent Vaporizer, Monterey, CA.
101. Brochure, Miller-Nelson Research, HCS-401 Series Flow-Temperature-Humidity Control Systems, Monterey, CA.
102. Brochure, Miller-Nelson Research, ADF-2426 Power Supply, Monterey, CA.
103. Yant, W. P. and F. E. Frey. *Ind. Eng. Chem.* 17:692 (1925).
104. Stewart, R. D. *Am. Ind. Hyg. Assoc. J.* 25:560 (1964).
105. Hill, D. W. and H. A. Newell. *Nature* 205:593 (1965).
106. Fanney, J. H. *Am. Ind. Hyg. Assoc. J.* 24:245 (1963).
107. Dambrauskas, T. and W. A. Cook. *Am. Ind. Hyg. Assoc. J.* 24:568 (1963).
108. Silverman, L. and G. R. Gardner. *Am. Ind. Hyg. Assoc. J.* 26:97 (1965).
109. Nelson, G. O. *Am. Ind. Hyg. Assoc. J.* 29:586 (1968).
110. Nelson, G. O. *Appl. Spectr.* 23:133 (1969).
111. Priante, S. J. Miller-Nelson Research Inc. private communication.
112. Smith, S. B. and R. J. Grant. "A Non-Selective Collector for Sampling Gaseous Air Pollutants," Report No. NP-8-193 (Cincinnati, OH: U. S. Public Health Service, 1960).
113. Nader, J. S. *Anal. Chim. Acta* 15:521 (1956).
114. Devaux, P. and G. Guiochon, *Bull. Soc. Chim. France* 1404 (1966).
115. Dravnieks, A. and B. K. Krotoszynski. *Gas Chromatog.* 6:144 (1968).
116. Matienzo, L. J. and C. J. Hensler. *Am. Ind. Hyg. Assoc. J.* 43:838 (1982).
117. Levine, S. P., J. A. Gonzalez, and E. V. King. *Am. Ind. Hyg. Assoc. J.* 47:347 (1986).
118. Neilson, A. and K. S. Booth. *Am. Ind. Hyg. Assoc. J.* 35:169 (1975).
119. Fielden, P. R. and G. M. Greenway. *Anal. Chem.* 61:1993 (1989).
120. Campbell, J. E. and R. B. Konzen. *Am. Ind. Hyg. Assoc. J.* 41:180 (1980).
121. Kring, E. V., P. D. McGibney, and G. D. Thornley. *Am. Ind. Hyg. Assoc. J.* 46:620 (1985).
122. Debbrecht, F. J., D. T. Daugherty, and T. M. Neel. Nat. Bur. Stand. Spec. Publ. No. 519, 761 (1979).
123. McKelvey, J. M. and H. E. Hoelscher. *Anal. Chem.* 29:123 (1957).
124. Altshuller, A. P. and J. R. Cohen. *Anal. Chem.* 32:802 (1960).
125. Stiepanienko, V. E. and S. I. Kriczmar. *Zh. Anal. Khim.* 26:147 (1971).
126. Sentik, A. *Prace Centr. Inst. Ochr. Pr.* 17:97 (1967).
127. Kozlowski, E. and J. Namiesnik, *Mikrochim. Acta* 11:435 (1978).
128. Miquel, A. H. and D. F. S. Natusch. *Anal. Chem.* 47:1705 (1975).
129. Goldup, A. and M. T. Westaway. *Anal. Chem.* 38:1657 (1966).
130. Arito, H. and R. Soda. *Ind. Health* 6:120 (1968).
131. Lipera, J. Lawrence Livermore National Laboratory, Livermore, CA. private communication.
132. Johnson, J. S. Lawrence Livermore National Laboratory, Livermore, CA. private communication.
133. Savitsky, A. C. and S. Siggia. *Anal. Chem.* 44:1712 (1972).

134. Freeland, L. T. *Am. Ind. Hyg. Assoc. J.* 38:712 (1977).
135. Fortuin, J. M. H. *Anal. Chim. Acta* 15:521 (1956).
136. Mgalobleschvili, K. D. *Geofiz. Apparat.* 35, 164 (1968).
137. Avera, C. B. U. S. Public Health Service, Cincinnati, OH. private communication.
138. Lovelock, J. E. *Anal. Chem.* 33:162 (1961).
139. Lugg, G. A. *Anal. Chem.* 41:1911 (1969).
140. Gisclard, J. B. *Ind. Eng. Chem.* 15:582 (1943).
141. Namiesnik, J., L. Torres, E. Kozlowski, and J. Mathieu. *J. Chromatogr.* 208:239 (1981).
142. Devaux, P. and G. Guiochon. *Bull. Soc. Chim. Fr.* 29:123 (1957).
143. Torres, L., J. Mathieu, M. Frikha, and J. Namiesnik. *Chromatographia,* 14:712 (1981).
144. Torres, L., M. Frikha, J. Mathieu, M. L. Riba, and J. Namiesnik. *Int. J. Environ. Anal. Chim.* 13:155 (1983).
145. Namiesnik, J. and E. Kozlowski. *Chem. Anal. (Warsaw)* 25:999 (1981).
146. Garcia Sanz, M. R. and M. M. Perez Garcia. *Junta Energ. Nucl. (Rep.), J.E.N.* 445 (1979).
147. Blacker, J. H. and R. S. Brief. *Am. Ind. Hyg. Assoc. J.* 32:668 (1971).
148. Kowalski, W. J., L. Ogierman, and J. Rzepa. *Ochr. Powietrza* 9:111 (1975).
149. Dietz, R. N., E. A. Cote, and J. D. Smith. *Anal. Chem.* 46:315 (1974).
150. Altshuller, A. P. and C. A. Clemons. *Anal. Chem.* 34:466 (1962).
151. Sawitsky, A. C. and S. Sigga. *Anal. Chem.* 44:1712 (1972).
152. *International Critical Tables* (New York: McGraw-Hill Book Company, Inc., 1929).
153. Lugg, G. A. *Anal. Chem.* 40:1072 (1968).
154. Chen, N. H. and D. F. Othmer. *J. Chem. Eng. Data* 7:37 (1962).
155. Nafikov, E. M. and A. G. Usmanov. *Inzh. Fiz. Zh. Akad. Nauk Belorussk. SSR* 17:530 (1969).
156. Fuller, E. N., P. D. Schettler, and J. C. Giddings. *Ind. Eng. Chem.* 58:19 (1966).
157. Roberts, R. C. *American Institute of Physics Handbook* (New York: McGraw-Hill Book Company, Inc., 1957).
158. Gilliland, E. R. *Ind. Eng. Chem.* 26:681 (1934).
159. Treybal, R. E. *Mass Transfer Operations* (New York: McGraw-Hill Book Company, Inc., 1955).
160. LeBas, F. S. *The Molecular Volumes of Liquid Chemical Compounds,* (London: Longmans, 1915).
161. Lee, C. Y. and C. R. Wilke. *Ind. Eng. Chem.* 46:238 (1954).
162. Brochure, Polyscience Corporation, Evanston, IL.
163. Singh, H. B. and L. Salas. *Environ. Sci. & Tech.* 11:511 (1977).
164. Crescentini, G., F. Mangani, A. R. Mastrogiacomo, and F. Bruner. *J. Chromatogr.* 204:445 (1981).
165. Burg, W. R. and C. Shau-Nong. *Am. Ind. Hyg. Assoc. J.* 42:426 (1981).
166. Schmidt, W. P. and H. L. Rook. *Anal. Chem.* 55:290 (1983).
167. Nano, G., A. Borroni, and B. Massa. *Am. Ind. Hyg. Assoc. J.* 48:814 (1987).
168. Dharmarajan, V. and R. J. Rando. *Am. Ind. Hyg. Assoc. J.* 40:870 (1979).
169. Goldan, P. D., W. C. Kuster, and D. L. Albritton. *Atmos. Environ.* 20:1203 (1986).
170. Bruner, F., C. Canulli, and M. Possanzini. *Anal. Chem.* 45:1790 (1973).
171. Ash, R. M. and J. R. Lynch. *Am. Ind. Hyg. Assoc. J.* 32:490 (1971).
172. DeSouza, T. L. C. and S. P. Bhatia. *Anal. Chem.* 48:2234 (1976).
173. Crosset, R. N. and D. J. Campbell. *Lab. Practice* 43:369 (1973).

174. Bankovich, P. W. and R. W. Modrell. *Am. Ind. Hyg. Assoc. J.* 37:640 (1976).
175. Burg, W. R., S. R. Birch, J. E. Cuddeback, and B. E. Saltzman. *Environ. Sci. & Tech.* 13:1233 (1976).
176. Stevens, R. K., A. E. O'Keeffe, and G. C. Ortman. *Environ. Sci. & Technol.* 3:652 (1969).
177. McKee, E. S. and P. W. McConnaughey. *Am. Ind. Hyg. Assoc. J.* 47:475 (1986).
178. Blacker, J. H. and R. S. Brief. *Am. Ind. Hyg. Assoc. J.* 32:668 (1971).
179. Ibusuki, T., F. Toyokawa, and K. Imagami. *Bull. Chem. Soc. Jpn..* 52:2105 (1979).
180. Namiesnik, J., M. Bownik, and E. Kozlowski. *Pomiary Autom. Kontrola,* 28:67 (1982).
181. Ibusuki, T., M. Sakuma, and K. Imagami. *Nippon Kagaku Kaishi* 6:882 (1978).
182. Ibusuki, T. and K. Imagami. *Jpn. Kogai,* 14:79 (1979).
183. Godin, J., G. Bouley, and C. Boudene. *Anal. Lett.* A11:319 (1978).
184. Beasley, R. K., C. E. Hoffman, M. L. Rueppel, and J. W. Worley. *Anal. Chem.* 52:1110, (1980).
185. Geisling, K. L., R. R. Miksch, and S. M. Rappaport. *Anal Chem.* 54:140 (1982).
186. O'Keeffe, A. E. *Anal. Chem.* 49:1278 (1977).
187. Scaringelli, F. P., A. E. O'Keeffe, E. Rosenberg, and J. P. Bell. *Anal. Chem.* 42:871 (1970).
188. Stellmack, M. L. and K. W. Street. *Am. Lab.* 14:25 (1982).
189. Bulletin No. 30-76, Metronics Associates, Inc., Palo Alto, CA.
190. O'Keeffe, A. E. and G. C. Ortman. *Anal. Chem.* 38:760 (1966).
191. Godin, J. and C. Boudene. *Anal. Chim. Acta* 96:221 (1978).
192. Dharmarajan, V. and R. J. Rando. *Am. Ind. Hyg. Assoc. J.* 41:437 (1980).
193. Shou, Y. H. and P. O. Schlect. *Am. Ind. Hyg. Assoc. J.* 42:70 (1981).
194. Brocco, D. and M. Possanzini, *Anal. Lett.* 7:153 (1974).
195. Ibusuki, T., M. Sakuma, T. Hirasawa, and K. Imagami. *Jpn. Kogai* 13:101 (1978).
196. Calibration Standard Notebook, Analytical Instrument Development Inc., Avondale, PA (1977).
197. Andrew, P. and R. Wood, *Chem. Ind. (London)* 1836 (1968).
198. Andrew, P., A. F. Smith, and R. Wood, *Analyst (London)* 96:528 (1971).
199. Namiesnik, J. *Chromatographia* 17:47 (1983).
200. Langmaier, J. and F. Opekar. *Anal. Chim. Acta* 166:305 (1984).
201. Lindqvist, F. and R. W. Lanting. *Atmospheric Environ.* 6:943 (1972).
202. Teckentrup, A. and D. Klockow. *Anal. Chem.* 50:1728 (1978).
203. Stellmack, M. L. and K. W. Street. *Analyt. Lett.* 16:77 (1983).
204. Raschdorf, F. *Chimia* 32:478 (1978).
205. Hartkamp, H. and A. Ionescu. *Staub. Reinhalt. Luft* 40:151 (1980).
206. Dietz, R. N., E. A. Cote, and J. D. Smith. *Anal. Chem.* 46:315 (1974).
207. Quinn, F. A., D. E. Roberts, and R. N. Work. *J. Appl. Phys.* 22:1085 (1951).
208. Purdue, L. J. and R. J. Thompson. *Anal. Chem.* 44:1034 (1972).
209. Saltzman, B. E., C. R. Feldman, and A. E. O'Keeffe. *Environ. Sci. & Tech.* 3:1275 (1969).
210. Stern, S. A., J. T. Mullhaupt, and P. J. Gareis. *Am. Inst. Chem. Eng. J.* 15:64 (1969).
211. Li, N. N. and R. B. Long. *Am. Inst. Chem. Eng. J.* 15:73 (1969).
212. Saltzman, B. E. *J. Air Pollut. Control Assoc.* 18:326 (1968).
213. Rodes, C. E., H. F. Palmer, L A. Elfers, and C. H. Norris. *J. Air Polut. Control Assoc.* 19:575 (1969).
214. Brochure, 11-87, G. C. Industries, Chatsworth, CA.

215. Lucero, D. P. *Anal. Chem.* 43:1744 (1971).
216. Calibration Standards Seminar, Analytical Instrument Development, Inc., Avondale, PA.
217. Stern, S. A., T. F. Sinclair, P. J. Gareis, N. P. Vahldieck, and P. H. Mohr. *Ind. Eng. Chem.* 57:49 (1965).
218. Waack, R., N. H. Alex, H. L. Frisch, V. Annett, and M. Szwarc. *Ind. Eng. Chem.* 47:2524 (1955).
219. Martin, A. J., F. J. Debbrecht, and G. R. Umbreit, "Devices for Preparing Low-Level Gas Mixtures," Analytical Instrument Development, Inc., West Chester, PA.
220. Scaringelli, F. P., S. A. Frey, and B. E. Saltzman. *Am. Ind. Hyg. Assoc. J.* 28:260 (1967).
221. Meyer, J. A., C. Rogers, V. Stannett, and M. Szware. *Tappi* 40:142 (1957).
222. Burbaker, D. W. and K. Kammermeyer. *Ind. Eng. Chem.* 45:1148 (1953).
223. Meares, P. *J. Am. Chem. Soc.* 76:3415 (1954).
224. Andrew, P. and R. Wood. *Chem. Ind. (London)* 1936 (1968).
225. DeMaio, L. *Instrument. Technol.* 19:37 (1972).
226. Wartburg, A. F., J. B. Pate, and J. P. Lodge. *Environ. Sci. & Tech.* 3:767 (1969).
227. Heppel, L. A., P. A. Neal, T. L. Perrin, M. L. Orr, and V. T. Porterfield. *J. Ind. Hyg. Tox.* 26:8 (1944).
228. Weeks, M. H., G. C. Maxey, M. E. Sicks, and E. A. Greene. *Am. Ind. Hyg. Assoc. J.* 24:137 (1963).
229. Rinehart, W. E., R. C. Garbers, E. A. Greene, and R. M. Stoufer. *Am. Ind. Hyg. Assoc. J.* 19:80 (1958).
230. Stopps, G. J. and M. McLaughlin. *Am. Ind. Hyg. Assoc. J.* 28:43 (1967).
231. Campbell, E. E., M. F. Milligan, and H. M. Miller. *Am. Ind. Hyg. Assoc. J.* 20:138 (1959).
232. Amdur, M. O. *Am. Ind. Hyg. Assoc. J.* 22:1 (1961).
233. MacDonald, W. E., W. B. Deichmann, and E. Bernal. *Am. Ind. Hyg. Assoc. J.* 24:539 (1963).
234. Roper, C. P. *Am. Ind. Hyg. Assoc. J.* 32:847 (1971).
235. Parker, G. B. *Am. Ind. Hyg. Assoc. J.* 41:220 (1980).
236. Cohen, H. J. and R. P. Garrison. *Am. Ind. Hyg. Assoc. J.* 50:486 (1989).
237. Cohen, A. E. *Am. Ind. Hyg. Assoc. J.* 20:303 (1959).
238. Weeks, M. H., N. P. Musselman, P. P. Yevich, K. H. Jacobson, and F. W. Oberst. *Am. Ind. Hyg. Assoc. J.* 25:470 (1964).
239. Bolton, N. E., J. B. Johnson, W. H. McDermott, and V. T. Stack, *Am. Ind. Hyg. Assoc. J.* 20, 32 (1959).
240. Brochhagen, F. K. and H. P. Schal. *Am. Ind. Hyg. Assoc. J.* 47:225 (1986).
241. Dharmarajan, V. and R. J. Rando. *Am. Ind. Hyg. Assoc. J.* 41:437 (1980).
242. Urano, K., S. Omori, and E. Yamamoto. *Environ. Sci. Technol.* 16:10 (1982).
243. Potts, W. J. and E. C. Steiner. *Am. Ind. Hyg. Assoc. J.* 41:141 (1980).
244. Weber, H., H. Stenner, and A. Kettrup. *Fresenius Z. Anal. Chem.* 325:64 (1986).
245. Voelte, D. R. and F. W. Weir. *Am. Ind. Hyg. Assoc. J.* 42:845 (1981).
246. Hill, A. C., L. G. Transtrum, M. R. Pack, and A. Holloman. *J. Air Pollut. Control Assoc.* 9:22 (1959).
247. Hill, A. C., L. G. Transtrum, M. R. Pack, and W. S. Winters. *Am. Soc. Agron. J.* 50:562 (1958).
248. Wilson, W. L., M. W. Campbell, L. D. Eddy, and W. H. Poppe. *Am. Ind. Hyg. Assoc. J.* 28:254 (1967).

249. Rinehart, W. E. *Am. Ind. Hyg. Assoc. J.* 21:389 (1960).
250. Rinehart, W. E. and T. Hatch. *Am. Ind. Hyg. Assoc. J.* 25:545 (1964).
251. Gelizunas, V. L. *Anal. Chem.* 41:1400 (1969).
252. Jahn, R. E. *Analyst* 80:700 (1955).
253. Kacy, H. W. and R. W. Cope. *Am. Ind. Hyg. Assoc. J.* 16:55 (1955).
254. Prager, M. J. *Am. Ind. Hyg. Assoc. J.* 27:272 (1966).
255. Marshal, E. K. and A. C. Kolls. *J. Pharmacol. Exp. Therap.* 12:385 (1918).
256. Sandage, C. "Tolerance Criteria for Continuous Inhalation Exposure to Toxic Material," Report No. ASD-61-519(I), U. S. Air Force, Wright-Patterson Air Force Base, Ohio (1961).
257. Rinehart, W. E. *Am. Ind. Hyg. Assoc. J.* 28:561 (1967).
258. Bryan, F. A. and V. Silis. *Am. Ind. Hyg. Assoc. J.* 21:423 (1960).
259. Priante, S. J. Lawrence Livermore National Laboratory, Livermore, CA. private communication.
260. Hersch, P., C. J. Sambucetti, and R. Deuringer. *Chim. Anal. (Paris)* 46:31 (1964).
261. Harmon, J. N. "Electrochemical Generation of Pollutant Standards," ASTM Special Technical Publication 598, p. 282, Philadelphia, PA (1975).
262. Pring, J. N. and G. M. Westrip. *Nature* 170:530 (1952).
263. Boer, H. and E. C. Kooyman. *Anal. Chim. Acta* 5:550 (1951).
264. Boer, H. *Rec. Trav. Chim.* 70:1020 (1951).
265. Boer, H. *Rev. Trav. Chim.* 217 (1948).
266. Noller, C. R. *Chemistry of Organic Compounds* (Philadelphia: W. B. Saunders Company, 1951).
267. Hersch, P. and R. Deuringer. *J. Air Pollution Control Assoc.* 13:538 (1963).
268. Shaw, J. T. *Atmos. Environ.* 1:81 (1967).
269. Page, J. A. and J. J. Lingane. *Anal. Chim. Acta* 16:175 (1957).
270. Waclawik, J. *Chim. Anal. (Paris)* 40:247 (1958).
271. Ludwick, J. D. *Anal. Chem.* 41:1907 (1969).
272. Singh, T., R. F. Sawyer, E. S. Starkman, and L. S. Caretto. *J. Air Pollut. Control Assoc.* 18:102 (1968).
273. Thompson, C. R. and J. O. Ivie. *Int. J. Air Water Pollut.* 9:799 (1965).
274. Silverman, L., "Experimental Test Methods," in *Air Pollution Handbook*, P. L. Magill, F. R. Holden, and C. Ackley, Eds. (New York: McGraw-Hill Book Company, Inc., 1956).
275. Powell, C. H. and A. D. Hosey, Eds. "The Industrial Environment — Its Evaluation and Control" (Washington, D. C.: U.S. Government Printing Office, 1965).
276. Hughes, J. G. and A. T. Jones. *Am. Ind. Hyg. Assoc. J.* 24:164 (1963).
277. Gill, W. E. *Am. Ind. Hyg. Assoc. J.* 21:87 (1960).
278. Stratman, H. and M. Buck. *Int. J. Air Water Pollut.* 10:313 (1966).
279. Palmes, E. D., N. Nelson, S. Laskin, and M. Kuschner. *Am. Ind. Hyg. Assoc. J.* 20:453 (1959).
280. Feldner, A. C., G. G. Oberfell, M. C. Teague, and J. N. Lawrence. *Ind. Eng. Chem.* 11:519 (1919).
281. Hashimoto, Y. and S. Tanaka. *Environ. Science & Technol.* 14:413 (1980).
282. Williams, J. W. and C. D. Hurd. *J. Org. Chem.* 5, 122 (1940).
283. B. T. Glisson, B. F. Craft, J. H. Nelson, and H. L. C. Meuzelaar. *Am. Ind. Hyg. Assoc. J.* 47:427 (1986).

CHAPTER 6

SPECIALIZED SYSTEMS

There are several materials that must be considered separately when formulating procedures for both static and dynamic methods. Because of their unusual reactivities, toxicities, concentrations, vapor pressures, or calculation considerations, they must be treated as special cases. Such systems include ozone, nitrogen dioxide, mercury, hydrogen cyanide, hydrogen fluoride, formaldehyde, chlorine dioxide, miscellaneous halogens, and water vapor.

6.1 OZONE

The production of ozone must be done dynamically if continuous, constant, and reproducible mixtures with a diluent gas are to be achieved. This is due primarily to ozone's extraordinary reactivity and decomposition to oxygen. If static systems are attempted, one quickly discovers that the concentration can decay from $1/3$[1] to $1/100$ of the initial concentration, depending on the circumstances.[2]

6.1.1 Methods of Generation

Ozone is produced by a variety of methods. The most common include ultraviolet illumination and electric discharge. Electrolytic techniques have also been tried.[3] Ozone may also be obtained commercially, but only in a semipurified state.

Probably the most preferred method of laboratory ozone production is the use of ultraviolet irradiation shown in Figure 6.1.[4-16] Here, air or oxygen is admitted to the chamber at a known rate, where the oxygen is partially converted to ozone with an ultraviolet lamp. The mixture is then further diluted and the test mixture is ready for use. The concentration of ozone can be varied by adjusting the air flow or by altering the input voltage to the ultraviolet lamp with a variable transformer. The ultraviolet lamp should be run for several hours to allow for stabilization. In

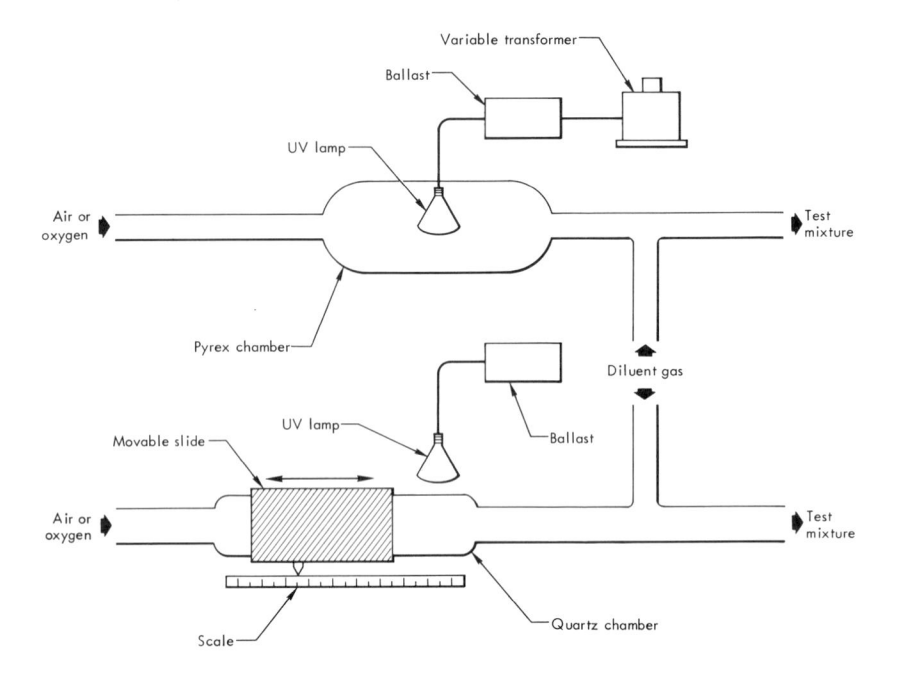

Figure 6.1 Sketch of two systems for continuously generating ozone with an internal (top) and external (bottom) lamp.

some cases, it is desirable to mount the lamp outside the chamber. If this is done, a quartz-walled chamber must be used to permit entry of the radiation, and concentration changes can be made with a movable slide.[11,17]

All downstream apparatus should be made of glass or Teflon, since these materials have the least effect on ozone.[6] If polyethylene or polyvinyl tubing must be used, then connections should be as short as possible. Rubber tubing should be avoided.

The other major method of ozone production is the electrical discharge.[18-21] The kinetics and mechanism of this method have been discussed extensively.[22] Usually, a silent discharge is preferred,[23] because actual spark discharges produce excessive quantities of nitrogen oxides, which constitute a primary interference in many ozone analyses.

Figure 6.2 shows a typical system.[24] Air or oxygen is passed between two concentrically oriented electrodes separated by Teflon spacers. The electrodes are activated with a high-voltage transformer (a 10 to 15 kV neon sign transformer is usually sufficient). A rheostat alters the ozone concentration by appropriate adjustments. An electrostatic precipitator functions on the same principle, except that the high-voltage output is usually rectified. This type of precipitator can also be used as an ozone source.

Ozone can be purchased commercially in liter-sized stainless steel cylinders. It is obtained as 8 mol% in the liquid phase with the balance being an inert gas,

Specialized Systems

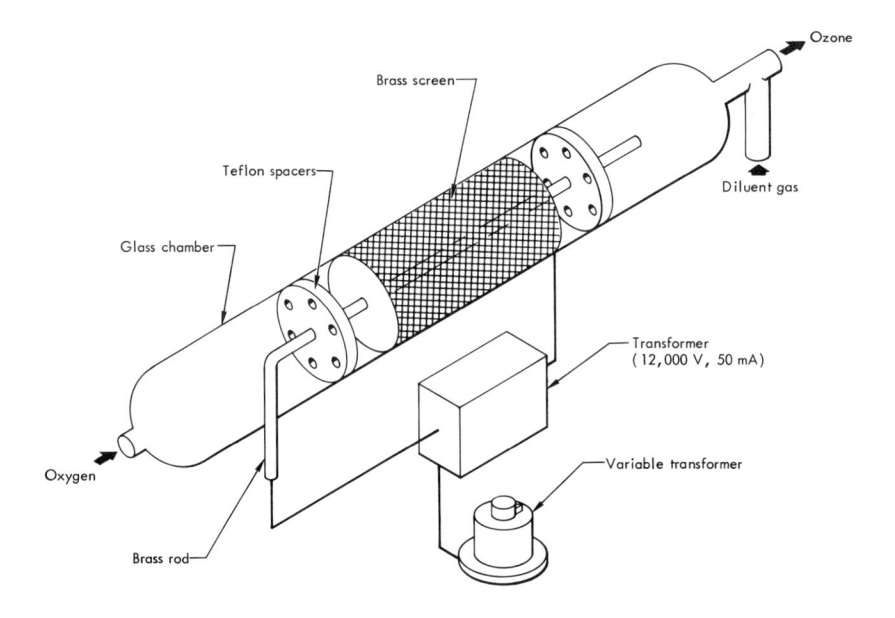

Figure 6.2 Sketch of an apparatus for generating ozone using a silent discharge.

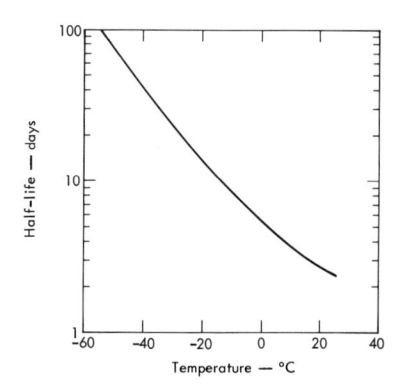

Figure 6.3 Half-life of ozone vs temperature. Ozone decomposes to liquid oxygen.

usually Freon 13 (chlorotrifluoromethane). This method, however, has three serious drawbacks. First, the decomposition of ozone to oxygen, as shown in Figure 6.3, is extremely rapid above −20°C; hence the shelf life is limited. Second, if the ozone is withdrawn as a vapor, the concentration gradually decreases due to the presence of its oxygen decomposition product. Thus, the ozone must be expelled from inverted cylinders as a liquid. This necessitates the use of large dilutions to achieve the concentrations normally needed for experimental work, especially in animal toxicology, plant, and air pollution studies. Third, ozone, especially in a highly concentrated form, is an extremely powerful oxidant and can detonate with a variety of fuels.

6.1.2 Methods of Analysis

Although ozone mixtures can be generated in the parts per million range, there is no absolute method of predicting with accuracy the concentration of ozone produced. System geometry, gas flow rates, and the intensity of the arc discharge or ultraviolet light all influence the final concentration. It is therefore necessary to analyze the gas mixture at the various experimental conditions to ascertain the exact ozone concentration.

There are numerous wet chemical and instrumental methods available. Most of these are reliable when dealing exclusively with ozone, but many methods deviate sharply from the expected values in the presence of interfering gases. Nitrogen dioxide, hydrogen sulfide, and sulfur dioxide are a few examples.

The methods commonly in use include potassium iodide (both wet chemical and coulometric),[5,8,25] phenolphthalein, diacetyldihydrolutedine,[26] polymer chain scission (rubber decomposition),[27] calibrated detector tubes,[28] and a host of others.

The most widely used and generally accepted wet chemical method is the potassium iodide method. Ozone in the air, when bubbled through neutral iodide solution, reacts as follows:[29]

$$O_3 + H_2O + 2I^- = I_2 + 2OH^- + O_2 \tag{6.1}$$

In an excess of potassium iodide,

$$I_2 + I^- = I_3^- \tag{6.2}$$

The concentration can then be determined spectrophotometrically by the intensity of the triiodide ion at 352 Nm. In an alkaline solution, a somewhat different reaction takes place.[29]

$$3O_3 + I^- = 3O_2 + IO_3^- \tag{6.3}$$

When acidified,

$$IO_3^- + 6H^+ + 8I^- = 3I_3^- + 3H_2O \tag{6.4}$$

In reality, the interaction of ozone with potassium iodide is more complex than the equations indicate and the stoichiometry is often under discussion.[30-32]

The phenolphthalein method is extremely sensitive, but there is also some question as to the reaction stoichiometry at low concentrations.[24] Other wet chemical techniques include sodium diphenylamine sulfonate,[33] 4,4'-dimethoxystilbene,[34] boric acid, buffered potassium iodide,[35] peroxy isocyanate,[36] and the nitrogen dioxide equivalent method.[37]

Specialized Systems

Several instrumental methods are available for continuously monitoring ozone.[38] One method involves applying a small potential across the electrodes. This potential is, in turn, influenced by the reaction of ozone drawn through a potassium iodide solution, as shown in Equation 6.1. Here, hydrogen, formed at the cathode, combines with the iodine produced by ozone oxidation of iodide. As ozone enters the system, a current is produced that is directly proportional to the ozone concentration. Such instruments have a full-scale deflection of 1 ppm, with reported sensitivities of ±0.01 ppm.[39] Additional performance criteria are given by Potter and Duckworth.[40]

Other measurement techniques include photometric, chemiluminescent, clathrate, and galvanic-type analyzers. Figure 6.4 illustrates the photometric method. Here, the ozone can either be scrubbed of ozone using a catalytic converter or measured directly using 254-nm ultraviolet light source. The difference in measurement is displayed as parts per million ozone. The detection limit is approximately 0.02 ppm at a full scale of 10 ppm ozone.[39] In the chemiluminescent method, the gas sample is drawn over a material containing Rhodamine B, which, in turn, is viewed with a sensitive photocell.[41-49] The ozone causes the material to emit a quantity of light proportional to its concentration. This technique enjoys a fast response time and low sensitivity. However, the light output-concentration relationship must be determined chemically.

The clathrate method, on the other hand, can be used to measure extremely low concentrations directly. Ozone, passing over Kr^{85} (containing quinol clathrate), releases radioactivity[50] that can be measured. This is done when the ozone oxidizes the quinol to quinoline, which in turn liberates the krypton as follows:

$$(C_6H_4(OH)_2)_3 \cdot Kr^{85} + O_3 = 3C_6H_4O_2 + 3H_2O + Kr^{85} \quad (6.5)$$

This technique can be used to measure concentrations on the order of one part per 10 billion or less. However, the method is limited by its relatively long response time.

Galvanic monitoring of ozone can be achieved by using bromide as the principal electrolytic constituent.[13] The ozone and bromide interact in the following manner:

$$O_3 + 2Br^- = O_2 + O^= + Br_2 \quad (6.6)$$

$$2e^- + Br_2 = 2Br^- \quad (6.7)$$

$$C + O^= = CO + 2e^- \quad (6.8)$$

The anode reaction represents the oxidation of the relatively large charcoal electrode. This method has the advantage of using no applied potential and of continually reusing and regenerating the electrolyte.

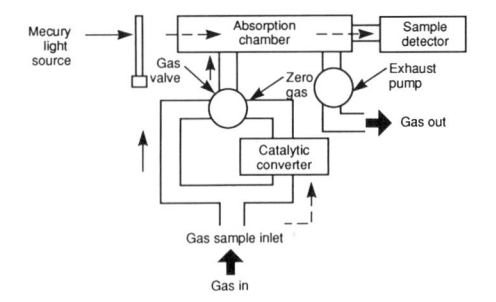

Figure 6.4 Sketch of a photometric ozone analyzer.

6.2 NITROGEN DIOXIDE

Nitrogen dioxide and its mixtures with other gases, usually air, are commonly required in low concentrations, especially in air pollution studies. There are, however, several problems that make accurate and precise production of such a test atmosphere difficult to control and maintain.

One of the primary difficulties when handling nitrogen dioxide is its propensity for dimerizing by the reversible reaction

$$2NO_2 = N_2O_4 \tag{6.9}$$

As expected, both temperature and pressure affect this equilibrium. Some typical results showing the contribution of each variable over the usual range of interest are shown in Figure 6.5. This graph was constructed from the equilibrium constants given by Verhoek and Daniels.[53] For example, a sample of pure nitrogen dioxide taken at 40°C and 1 atm would contain 58% nitrogen tetroxide and 42% nitrogen dioxide. There are numerous other values given for the dissociation constants and heats of vaporization that are involved in these calculations, which would cause the N_2O_4/NO_2 ratio to be shifted slightly.[54-56]

A further difficulty with nitrogen dioxide is its boiling point of 21.3°C. Many times it is desirable to conduct experiments at room temperature. Often one will observe the nitrogen dioxide alternating between the gas and liquid phase if the temperature is in the neighborhood of 20°C. If the temperature falls below 15°C, air will actually enter the cylinder of nitrogen dioxide when the valve is opened, rather than the nitrogen dioxide flowing to the atmosphere.

6.2.1 Static Systems

Known concentrations of nitrogen dioxide can be prepared in closed vessels, either by direct addition or by a chemical reaction within the vessel. If direct addition is used, then Figure 6.5 must be used to estimate and compensate for the partial dimerization.

Specialized Systems

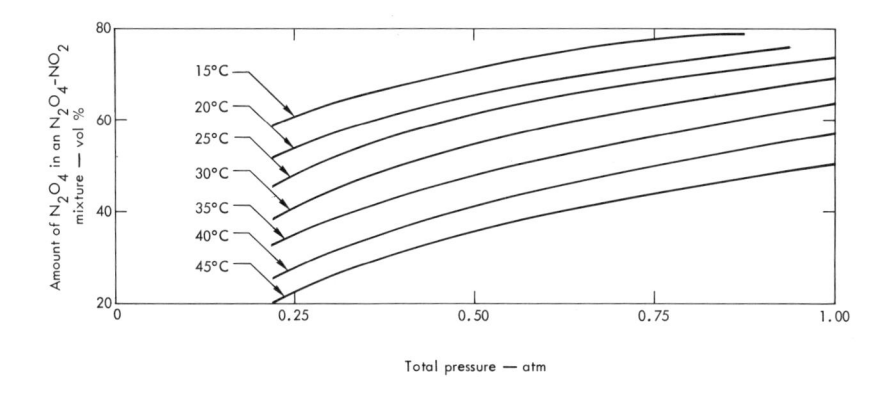

Figure 6.5 Dimerization of nitrogen dioxide vs pressure at seven isotherms.

Example 6.1. Estimate the volume of nitrogen dioxide/nitrogen tetroxide required to produce 5 ppm nitrogen dioxide in a 40-L vessel at 25°C and 1 atm. Assume the volume contains 26% nitrogen dioxide.

Concentration	5 ppm
Vessel size	40 L or 40,000 mL
NO_2/N_2O_4 concentration	26% or 0.26 decimal percent
mL/L	1,000

From Equation 4.2, the total volume of nitrogen dioxide required is

$$V_{(NO_2)} = \frac{(5)(40,000)}{(E+06)} = 0.200 \text{ mL}$$

If the nitrogen tetroxide fully dissociates into nitrogen dioxide, each milliliter contains the following amount of nitrogen dioxide

$$V_{\left(\frac{NO_2}{mL}\right)} = 0.26 + (2)(1.00 - 0.26) = 1.74 \text{ mL}$$

The total volume of the mixture required is then

$$V = \frac{0.200}{1.74} = 0.115 \text{ mL}$$

Concentrations of nitrogen dioxide can also be conveniently prepared by chemical reaction. Oxidation of nitric oxide by air can be used if high concentrations are desired or if the nitric oxide is first reacted with oxygen[57,58] and then injected into the system. Low parts per million concentrations are not usually created by direct injection of pure nitric oxide because of the relatively slow rate of reaction.[59-61] This reaction rate is normally considered third order.[62] If the initial oxygen concentration is sufficiently high and does not change appreciably as the reaction nears completion, it becomes second order.[63] Here

$$\frac{d(NO_2)}{dt} = -K(NO)^2 \qquad (6.10)$$

where K, the rate constant, is 3×10^{-4} ppm/min.[64] Figure 6.6 illustrates the reaction process with time. Note that even after 1 hr only 15% of an original concentration of 10 ppm of nitric oxide reacted to form nitrogen dioxide.

The thermal decomposition of lead nitrate has also been used successfully to produce nitrogen dioxide.[57,60] The reaction is

$$2Pb(NO_3)_2 = 2PbO + 4NO_2 + O_2 \qquad (6.11)$$

Figure 6.7 shows the apparatus required to operate such a system. Known weights of lead nitrate crystals are placed between glass-wool plugs. The chamber is partially evacuated and the lead nitrate is heated with a small burner. When the brown nitrogen dioxide becomes evident, a small amount of dilution gas is allowed to enter and sweep the evolved gases into the chamber. This process is repeated several times until no more nitrogen dioxide is evident, even after extensive heating. The test chamber is then brought back to atmospheric pressure. If desired, a combustion tube furnace may be substituted as a heat source, with a maximum temperature of about 450°C.[65]

Example 6.2. What weight of lead nitrate is needed to produce 200 ppm nitrogen dioxide in a 100-L chamber at 25°C and 1 atm?

Concentration	200 ppm
Chamber volume	100 L
MW of NO_2	46.1 g/mol
MW of $Pb(NO_3)_2$	331 g/mol
$Pb(NO_3)_2/NO_2$	0.5

From Equation 4.6, the weight of nitrogen dioxide required is

$$W_{(NO_2)} = \frac{(200)(100)(46.1)}{(24.5E+06)} = 0.0376 \text{ g}$$

If Equation 6.11 proceeds to completion, then

Specialized Systems 193

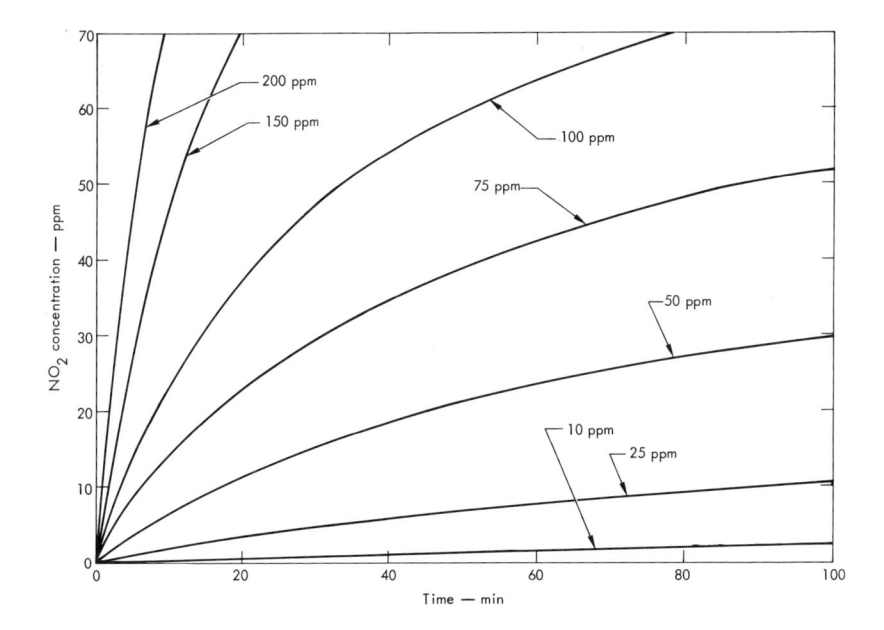

Figure 6.6 Concentration of nitrogen dioxide vs time at seven initial nitric oxide concentrations.

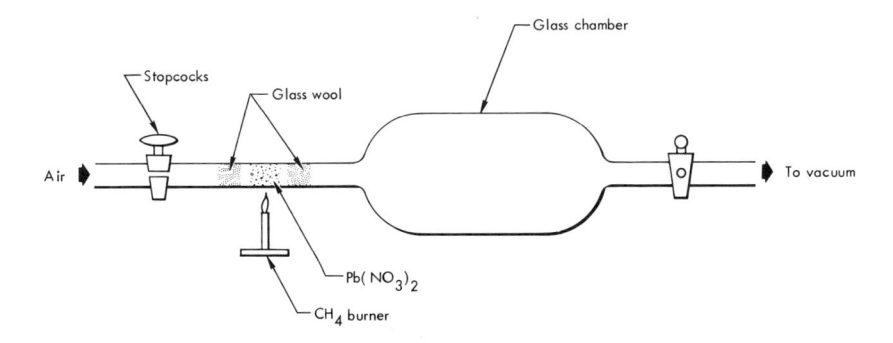

Figure 6.7 Sketch of an apparatus for producing static nitrogen dioxide mixtures in air by decomposing lead nitrate.

$$W_{\left[Pb(NO_3)_2\right]} = \frac{(0.0376)(0.5)(331)}{46.1} = 0.135 \text{ g}$$

6.2.2 Dynamic Systems

Low parts per million concentrations of nitrogen dioxide can be made in the liter per minute range using the permeation tubes described in Chapter 5.[66-72] There has, however, been a problem with errors due to the presence of water vapor.

Nitrogen dioxide concentrations in the parts per million range can be generated by most of the methods outlined in Chapter 5 only if a preliminary dilution is made to obtain a usable stock mixture.[73-75] Nitrogen dioxide in air, helium, and nitrogen is available commercially in concentrations as high as 1%. Concentrations greater than 0.65% will vary proportionally with pressure because of the condensation problems explained in Chapter 4.[76] Any attempt to meter the pure gas at normal cylinder pressure and room temperature or to produce mixtures by single or double dilution will meet with little or no success. Asbestos-plug flow meters can be successfully used,[77,78] but the pure substance must again be diluted to prevent fouling of the plug.[79] Motor-driven syringes have also been used to dispense a previously diluted amount of nitrogen dioxide.[80]

An interesting method for producing mixtures electrolytically has been designed by Shaw.[81] There seems to be no suitable reaction to evolve nitrogen dioxide directly. However, nitric oxide is easily generated quantitatively by the electrolysis of nitrosyl hydrogen sulfate ($NOHSO_4$) dissolved in concentrated sulfuric acid. The reaction using a platinum electrode under an atmosphere of nitrogen is

$$NO^+ + e^- = NO \tag{6.12}$$

The anode reaction produces oxygen from the sulfate by the reaction

$$4SO_4^= = 2S_2O_7^= + O_2 + 4e^- \tag{6.13}$$

If the oxygen and nitric oxide are allowed to mix before dilution for a sufficient time interval, the yield is almost quantitative.

Another, lesser known method of generation involves the reaction of nitric oxide with excess ozone by the following reaction:[82]

$$NO + O_3 = NO_2 + O_2 \tag{6.14}$$

Additional byproducts, such as nitrogen pentoxide, are formed, but they, along with excess ozone, can be thermally decomposed by passing the gas mixture through a chamber filled with glass beads and heated to 150°C. Here the concentrations can be varied from 100 to 5000 ppm, with a reproducibility of 10% at the low concentrations and 3% at the high concentrations.

Specialized Systems

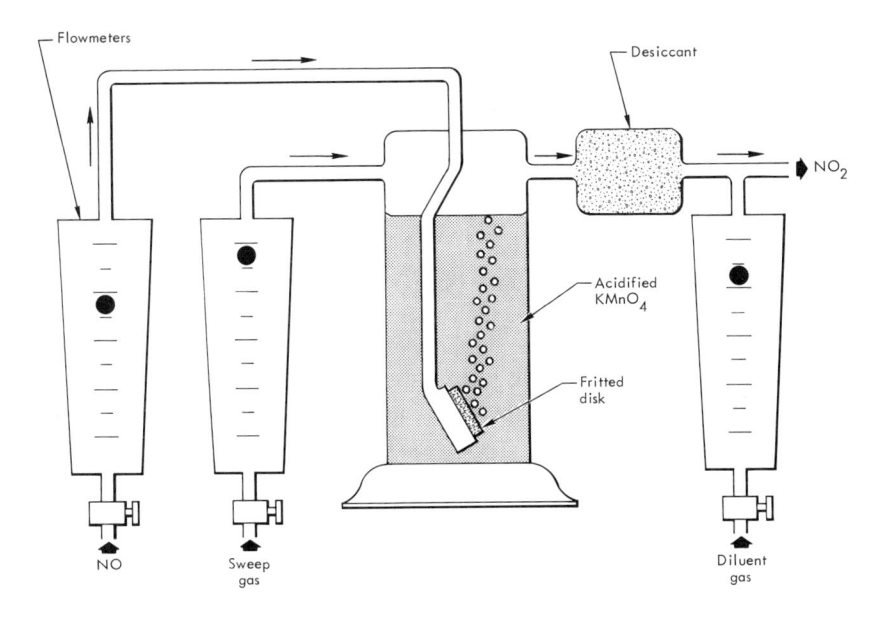

Figure 6.8 Sketch of an apparatus for continuously producing nitrogen dioxide from nitric oxide and acidified potassium permanganate.

Figure 6.8 illustrates another method of nitrogen dioxide production. A metered stream of nitric oxide is bubbled into a solution of acidified permanganate solution, and nitrogen dioxide is generated by the reaction[83]

$$3NO + 2HMnO_4 = 3NO_2 + 2MnO_2 + H_2O \qquad (6.15)$$

As the nitrogen dioxide bubbles to the surface in the reaction flask, it is swept away by an appropriate diluent gas, normally nitrogen. A desiccant may be provided to remove unwanted water vapor from the mixture. If desired, further dilution can be accomplished downstream. The system must equilibrate for several hours before it can be used because of the solubility of the resultant nitrogen dioxide in the permanganate solution and because of the adsorption of some of the gas in the desiccant.

Nitrogen dioxide can also be produced by reacting stoichiometric quantities of nitric oxide and oxygen in the presence of a platinum catalyst. This can be mixed with humidified air and passed through chromium trioxide impregnated carbon to completely convert all the nitric oxide to nitrogen dioxide.[84]

Oxides of nitrogen can be measured by a variety of techniques. Wet chemical techniques predominated before 1970.[78,80,85-90] Current instrumental methods involve chemiluminescent,[91] ion selective electrode,[92] infrared spectrometric,[93] and electrochemical techniques.

6.3 MERCURY

Mercury is a most unusual metal because of its relatively high (for a metal) vapor pressure (0.0012 mmHg at 20°C). It therefore requires special techniques to produce known concentrations. Injecting the pure liquid into an air stream is not practical because of the extremely low flow rates and the volatilization problems involved. However, it does lend itself readily to the evaporation technique discussed in Chapter 5.

Figure 6.9 shows a typical dynamic system for mercury vapor production.[94-98] Basically, it entails diluting a saturated air stream to obtain the desired concentration. Here, filtered compressed air is split into two streams. One line goes through a flow meter (about 40 L/min) and is used as the dilution air. The other line flows through a lower range flow device (about 1 L/min) into the saturation unit. Valves or flow controllers govern the flow rate in each flow meter.

In the saturation unit, a ceramic core heater raises the air temperature to about 60°C. The air then impinges upon a pool of warm mercury. The mercury-laden air then travels to two equilibrium vessels, which also contain pools of mercury. Here, the air cools rapidly, and any excess mercury condenses. The air becomes saturated during this cooling phase. After the air passes from the equilibrium bottles, an accurate temperature reading (to the nearest 0.1°C) is taken just before the saturated air stream meets the dilution air.

A stopcock placed after the thermometer provides an easy method of discontinuing the flow of mercury vapor and also facilitates checking for leaks in the saturation unit. No flow should be observed in the low-flow flow meter when its valve is open and the stopcock is closed. Various concentrations are made by adjusting the air and mercury vapor in air dilution ratios. The measured temperature should be near ambient temperature if the air flows through the saturation unit at less than 1 L/min. If the temperature begins to rise rapidly, or if a greater flow rate through the saturator is required, the equilibrium bottles should be enlarged and maintained in a constant temperature bath for the best results.

An accurate measurement of the temperature is the most crucial part of the calibration operation. Since the vapor pressure of mercury is so dependent on the temperature (Table 6.1), an error of 1°C causes a deviation from the expected concentration of 10%. For this reason, the thermometer should not be handled and should be placed as far as possible from the heating equipment.

Mercury permeation tubes have been used for concentrations in the $\mu g/m^3$ range.[99] These tubes must first be calibrated using a photometric technique since the gravimetric weight lost is so low (1 mg every 37 days). The photometer is calibrated by filling a syringe with saturated mercury vapor and injecting it at a known rate into an air stream.

Several instruments have been used to detect mercury vapor. The photometric type is similar to the ozone analyzer shown in Figure 6.9. Another analyzer involves measuring the resistance change of a gold film in the presence of mercury vapor.[100]

Concentrations in parts per million by volume of the saturated air stream can be calculated from

Specialized Systems

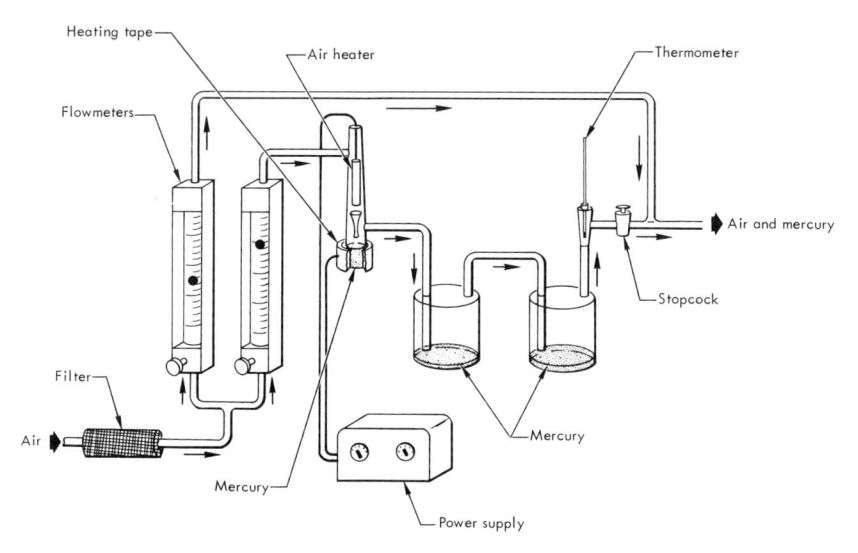

Figure 6.9 Sketch of an apparatus for dynamically generating mercury vapor.

Table 6.1 Calculated Concentrations of Air Saturated with Mercury[a]

Temp (°C)	Vapor Pressure (mmHg)	Concentration of Saturated Air (mg/m³)	Antoine Coefficients
15	0.000767	8.58	A = 8.274427
16	0.000840	9.36	B = 3280.205
17	0.000919	10.2	C = 273
18	0.00101	11.1	
19	0.00110	12.1	
20	0.00120	13.2	
21	0.00131	14.3	
22	0.00143	15.6	
23	0.00156	17.0	
24	0.00170	18.4	
25	0.00185	20.0	
26	0.00201	21.7	
27	0.00219	23.5	
28	0.00238	25.5	
29	0.00259	27.6	
30	0.00281	29.9	

[a] Calculated from Equation 6.17.

$$C_{ppm} = \frac{p_{(Hg)} \times 10^6}{P} = \frac{W}{V} \left[\frac{22.4 \times 10^6 \left(\frac{T_m}{273}\right)\left(\frac{760}{P}\right)}{M_{(Hg)}} \right] \quad (6.16)$$

where C is the concentration (ppm), p_{Hg} is the vapor pressure of mercury (mmHg), T_m is the measured temperature of air saturated with mercury (°K), P is the atmospheric pressure (mmHg), and M_{Hg} is the molecular weight of mercury (200.6 g/mol).

Assuming atmospheric pressure to be 760 mmHg and substituting in the molecular weight of mercury, the concentration in mg/m³ becomes

$$C = \frac{3.22 \times 10^6 \, p_{(Hg)}}{T_m} \tag{6.17}$$

The values of the p_{Hg} are well documented and some are shown in Table 6.1.

When air dilutes the saturated mercury vapor, Equation 6.17 becomes

$$C = \left[\frac{q_{(Hg)}}{q_{(Hg)} + q_{air}} \right] \left[\frac{3.22 \times 10^6 \, p_{(Hg)}}{T_m} \right] \tag{6.18}$$

where q_{Hg} is the flow rate of the air saturated with mercury (L/min) and q_{air} is the flow rate of the dilution air (L/min). Table 6.2 shows some sample calculations using Equation 6.18.

6.4 HYDROGEN CYANIDE

Because of its extreme toxicity and room-temperature boiling point (25.7°C, see Table 6.3), hydrogen cyanide is one of the most difficult of all materials to handle in the purified state. It is also dangerous to store for extended time periods, for it sometimes undergoes rapid polymerization, causing sudden, violent explosions. Thus the shelf life of hydrogen cyanide cylinders is limited. The generation of the gas in the laboratory as it is needed is generally more acceptable than storing it in cylinders, most of which, in all probability, will go unused.

6.4.1 Static Systems

The classical method of small-scale laboratory production is illustrated in Figure 6.10. Two reagent reservoirs are made in an ordinary glass test tube by heating the tube to red heat and creating a slight positive pressure within the tube. Excess sulfuric acid is added to one reservoir, and a small predetermined volume of sodium cyanide is added to the other, as shown. The end of the tube is then sealed with a torch and inverted, as shown in the right-hand side of Figure 6.10. As the solutions mix, a stoichiometric quantity of hydrogen cyanide gas is liberated into the air space above the mixed solution. Only a small percentage (usually less than 5%) remains dissolved. The glass ampoule can then be trans-

Specialized Systems

Table 6.2 Calculated Concentrations of Mercury Vapor in Air[a]

Flow Rate of Dilution Air (L/min)	Flow Rate of Hg Sat'd Air (L/min)	\multicolumn{5}{c}{Concentration of Hg Vapor (mg/m^3) at the Specified Temperature (°C)}
		20	22	24	26	28
0	All	13.2	15.6	18.4	21.7	25.5
18.0	0.0180	0.0132	0.0156	0.0184	0.0217	0.0254
18.0	0.0360	0.0263	0.0311	0.0367	0.0433	0.0508
14.0	0.0500	0.0469	0.0555	0.0655	0.0771	0.0906
10.0	0.0500	0.0656	0.0776	0.0916	0.108	0.127
10.0	0.100	0.131	0.154	0.182	0.215	0.252
10.0	0.200	0.259	0.306	0.361	0.425	0.499
10.0	0.300	0.384	0.454	0.536	0.631	0.742
10.0	0.400	0.507	0.600	0.708	0.834	0.980
10.0	0.500	0.628	0.743	0.877	1.03	1.21
10.0	0.900	1.09	1.29	1.52	1.79	2.10
5.00	1.00	2.20	2.60	3.07	3.61	4.24

[a] Calculated from Equation 6.18.

Table 6.3 Hydrogen Cyanide Vapor Pressure

Temp (°C)	Pressure (mmHg)	Gauge Pressure (psi)
10	406	−6.8
15	499	−5.0
20	609	−2.9
25	737	−0.4
25.8	760	−0.0
30	888	2.5
35	1062	5.8
40	1264	9.7
45	1496	14.2
50	1761	19.4
55	2062	25.2
60	2404	31.8

Antoine vapor pressure constants: $A = 7.752$, $B = 1456$, $C = 273.1$, $\log p = A - [B/(C + t)]$.

ferred to a larger test chamber, broken, and the gas dispersed throughout a larger volume of diluent gas.[101]

Under static conditions, the concentration of hydrogen cyanide can be determined from a calculation based on Equation 4.5. If the gas is generated from the reaction of a standard cyanide solution with excess acid, the concentration of hydrogen cyanide is given by

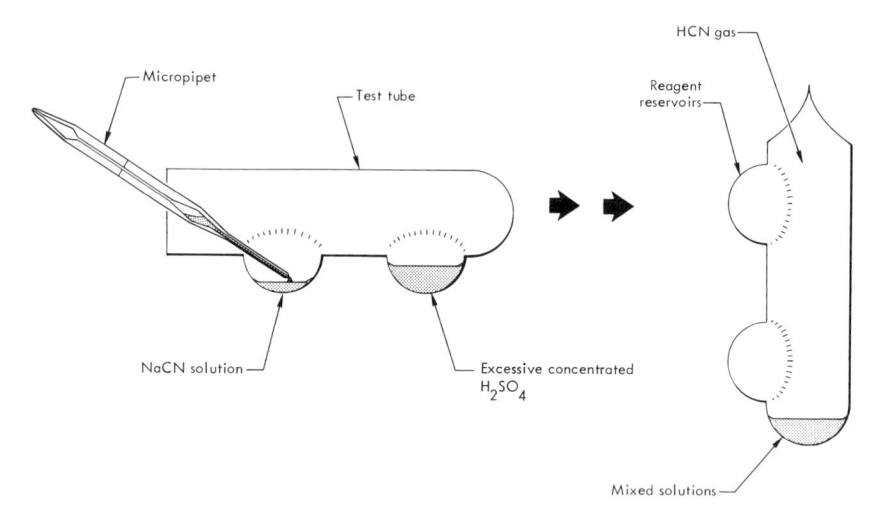

Figure 6.10 Sketch of an apparatus for producing hydrogen cyanide.

$$C_{ppm} = \frac{10^6 \, v_{(HCN)}}{v_d} = \frac{22.4 \times 10^6 \left(\dfrac{T}{273}\right)\left(\dfrac{760}{P}\right) v_s c_s}{v_d} \qquad (6.19)$$

where C is the concentration (ppm), v_{HCN} is the volume of hydrogen cyanide produced from the acid reaction (L), v_d is the volume of diluent gas (L), v_s is the volume of a standard cyanide solution (L), c_s is the concentration of a standard cyanide solution (mol/L), T is the temperature (°K), and P is the pressure (mmHg).

Example 6.3. How many microliters of a 1 M solution of sodium cyanide are required to create a concentration of 100 ppm in a 2 ft³ chamber? The final temperature and pressure are 25°C and 745 mmHg, respectively.

Concentration, air	100 ppm
Concentration, liquid	1 mol/L
Vessel volume	2 ft³
Temperature	25°C or 298°K
Pressure	745 mmHg
μL/L	E + 06
L/ft³	28.3

From Equation 6.19, the required liquid volume is

$$C = \frac{(100)(28.3)(2)(E+06)}{(22.4E+06)\left(\dfrac{760}{745}\right)\left(\dfrac{298}{273}\right)} = 227 \; \mu L$$

6.4.2 Dynamic Systems

Low concentrations in the parts per million range of hydrogen cyanide can be generated using the permeation tubes discussed in Chapter 5. The flow rates, however, will be in the liter per minute range. If higher flows or concentrations are required, standard gas mixtures are available commercially in helium and nitrogen that are as high as 1%. However, concentrations greater than 0.65 % will vary proportionally with pressure because of the condensation problems explained in Chapter 4.[76]

Saltzman has suggested a technique shown in Figure 6.11 for producing 100 ppm at flows of 0.25 to 0.5 L/min.[59] Here, the carbon dioxide in air is removed with Ascarite. The cleansed air then flows through a condenser and a midget impinger containing a 30% solution of potassium cyanide. As the air bubbles through, it picks up the hydrogen cyanide from the hydrolyzed cyanide ion and is further diluted with more air if necessary. The water bath and gas supply are maintained at about 29 to 30°C to stabilize the hydrogen cyanide production. Using this technique, 100 ppm can be maintained for up to 10 hr. An elaboration of this method is also discussed by Hashimoto.[102]

High concentrations at the 1000 ppm level can be produced by feeding a cyanide salt into excess acid over a chamber filled with glass beads. The dilution air also passes over the beads and carries away the hydrogen cyanide as it is formed. The resultant liquid is periodically drained away.[103]

Another technique for producing long-term, high concentrations is shown in Figure 6.12. Here, a small cylinder (100 mL capacity), complete with regulator, is placed in an oven at 35°C (do not heat above this point!). This will vaporize the hydrogen cyanide liquid and create a gauge pressure of about 6 psig. The flow rate of gas can then be controlled by the mass flow controller, which is also inside the oven. The dilution air supply also enters and exits the oven. Once the hydrogen cyanide gas merges with the dilution air stream, condensation is no longer a problem and the resultant gas mixture is ready for use.

There are several techniques used to measure hydrogen cyanide. Wet chemical methods include the colorimetric method by Aldridge[104] and ion-specific electrodes. Instrumental methods include infrared, electrochemical, and flame ionization techniques.

6.5 HYDROGEN FLUORIDE

Hydrogen fluoride is one of the most reactive compounds to handle in the gas phase. Its room-temperature boiling point of 19.5°C (see Table 6.4) and chemical reactivity make this material one of the most difficult to work with in the laboratory. It can, however, be dispersed from aqueous solutions with somewhat less difficulty.

Static systems have been used but must be fashioned from inert polymeric materials, such as polyethylene, polypropylene, or Teflon. Hydrogen fluoride reacts with glass and most metallic materials. Hydrofluoric acid is commercially available in solutions of 48 to 51% by weight. It can be added to vessels in either

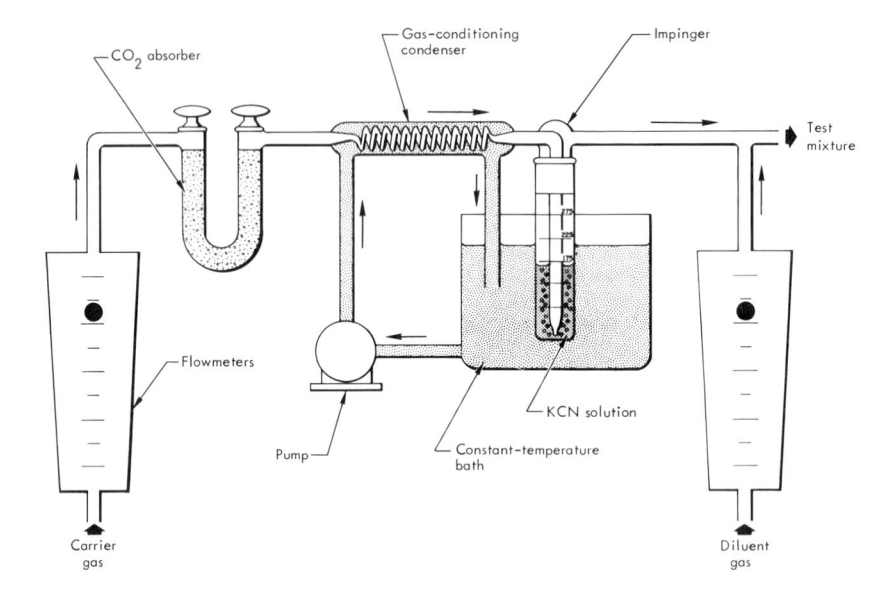

Figure 6.11 Sketch of an apparatus for dynamically producing hydrogen cyanide by hydrolyzing cyanide ions.

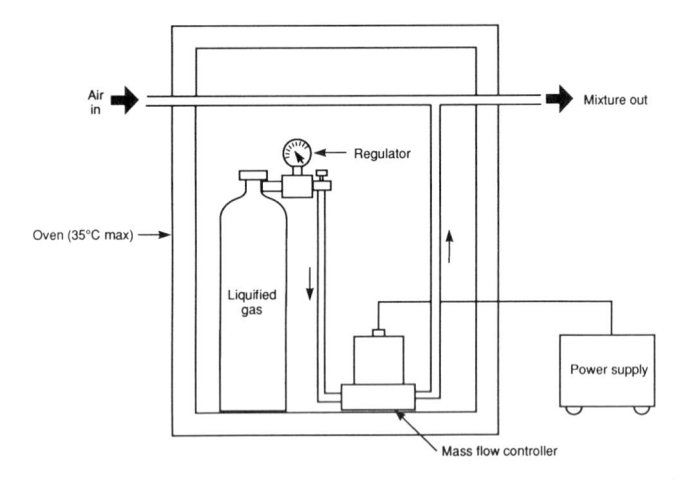

Figure 6.12 Sketch of an apparatus for generating hydrogen cyanide from the liquified gas.

its pure or diluted form, using plastic syringes with Teflon-tipped plungers. The reaction of sodium fluoride with excess sulfuric acid and its subsequent evaporation has also been used.[105] In addition, the partial pressure technique has been employed.[106] The desired gas mixture should equilibrate, be withdrawn, and again be added to season the vessel walls and to help increase the accuracy. This seasoning process is further described in Chapter 4.

Specialized Systems

Table 6.4 Hydrogen Fluoride Vapor Pressure

Temp (°C)	Pressure (mmHg)	Gauge Pressure (psi)
10	534	−4.4
15	644	−2.2
19.5	760	0.0
20	773	0.3
25	922	3.1
30	1094	6.5
35	1291	10.3
40	1515	14.6
45	1770	19.5
50	2059	25.1
55	2384	31.4
60	2749	38.5

Antoine vapor pressure constants: $A = 7.685$, $B = 1478.55$, $C = 288.22$, $\log p = A - [B/(C + t)]$.

Dynamic methods are generally employed for hydrogen fluoride mixtures. Concentrations at the parts per million level at the liter per minute flow range can be accomplished using the permeation tubes discussed in Chapter 5.[107-109] Hersch uses the continuous conversion of hydrogen passing over cobalt trifluoride at 300°C to produce hydrogen fluoride.[110] Nebulizers containing dilute solutions have also been used.[111] Higher concentrations and flows can be obtained from the system shown in Figure 6.13.[112,113] Here, air or the gas of interest is metered through two vessels surrounded by a constant-temperature bath. The first vessel contains water only, while the second contains the selected concentration of aqueous hydrofluoric acid. Preconditioning water vessels helps to stabilize the concentration produced over prolonged time intervals. A Miller-Nelson Research Model HCS-401 flow-temperature-humidity control system can also be used for this purpose.[114] The temperature of the water bath plays an important role in the concentrations generated. Hill has found that an increase in the bath temperature from 27 to 48°C triples the output at the same air flow.[112]

Another technique that is useful for producing concentrations in the 50 to 100 ppm range at 50 L/min is injecting the aqueous hydrofluoric acid directly into the gas stream. Figure 6.14 shows a special injection port used for this purpose. Here, the aqueous hydrofluoric acid is injected into a moving air stream, through a septum, using a polyethylene syringe with a Teflon tip. As the liquid exits the needle, droplets form at the tip or are collected on the platinum coil, which centers the Teflon needle in the air stream. A glass tube viewing port can be used if it is protected from the hydrofluoric acid using a specially machined Teflon sleeve. Heating the air slightly may be required to help stabilize the concentration output. A 5-gal plastic vessel may be required for mixing and dampening any concentration irregularities.

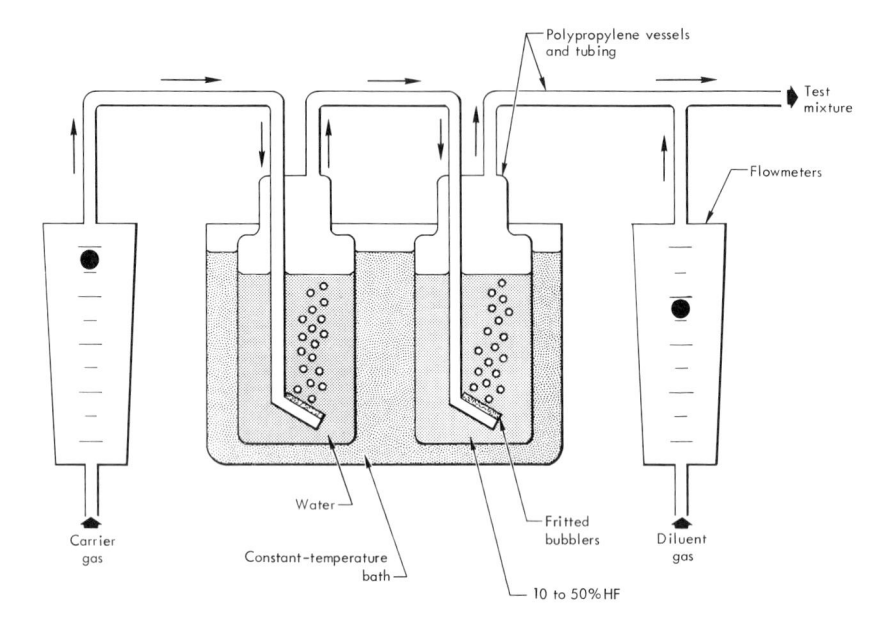

Figure 6.13 Sketch of an apparatus for dynamically producing hydrogen fluoride.

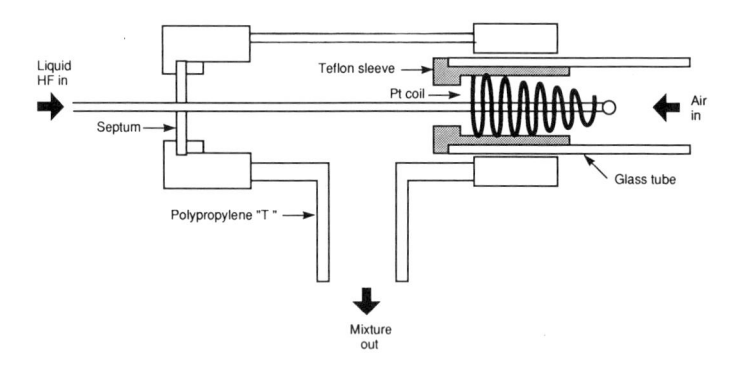

Figure 6.14 Sketch of an injection port for the introduction of a solution of hydrogen fluoride into a moving air stream.

The concentration produced from this injection technique is

$$C_{ppm} = \frac{22.4 \times 10^6 \left(\frac{T}{273}\right)\left(\frac{760}{P}\right) q_L \rho X}{q_D M} \quad (6.20)$$

where C is the concentration (ppm), T is the temperature (°K), P is the pressure (mmHg), q_L is the liquid injection rate (mL/min), ρ is the liquid density (g/mL),

Specialized Systems   205

X is the liquid concentration (g/mL), q_D is the diluent gas flow rate (L/min), and M is the molecular weight (g/mol).

Example 6.4. What is the injection rate of a 48.6% hydrofluoric acid solution needed to produce 70 ppm in air flowing at 64 L/min? Ambient conditions are 25°C and 760 mmHg, and the solution density is 1.14 g/mL.

Concentration in air	70 ppm
Flow rate	64 L/min
Molecular weight	20 g/mol
Solution concentration	48.6% or 0.486 g/mL
Solution density	1.14 g/mL
Pressure	760 mmHg
Temperature	25°C or 298°K

From Equation 6.20, the required injection rate is

$$q_L = \frac{(70)(20)(64)}{(22.4\text{E}+06)\left(\frac{298}{273}\right)\left(\frac{760}{760}\right)(1.14)(0.486)} = 0.00661 \text{ mL/min}$$

High concentrations above 100 ppm can be generated using the gas mixture technique shown in Figure 6.12. If the gas cylinder is placed in the oven with a mass flow controller and the temperature is raised to 40°C, a pressure of 15 psig can be obtained. This is more than sufficient for easy flow control. Mixtures can also be generated by dilution of calibration gas standards that are commercially available in concentrations of up to 0.5% in helium and nitrogen.

The output concentrations can be measured by a variety of chemical and instrumental techniques. The colorimetric method involves the collection of gas in 0.1 N sodium hydroxide and subsequent analysis using the method of Bellack and Schoubal.[115] Instrumental methods include ion-specific electrode and electrochemical analyzers.

6.6 FORMALDEHYDE

Formaldehyde is a most unusual material. When needed in industrial applications, it has a tendency to slowly react with its surroundings and to disappear. On the other hand, when it is least desired it seems to show up in the environment via processed wood products and insulation, and it has remarkable staying power. Formaldehyde cannot be obtained commercially in the gaseous state because of polymerization. Even if it is dissolved in water and stabilized with methanol, it still will form paraformaldehyde upon standing. Because of these reasons, the generation of gas mixtures requires several unusual approaches.

Because of its reactivity, formaldehyde mixing with any diluent gas is normally done in a dynamic fashion. Concentrations in the sub-parts per million range at flow rates near 1 L/min can be prepared using the permeation tubes discussed in Chapter 5.[116-119] The tubes normally contain paraformaldehyde or polyoxymethylene solid, but formaldehyde is given off as this material decomposes. Diffusion tubes containing paraformaldehyde have also been used.[120,121] Higher concentrations and flows using paraformaldehyde can be generated by passing air through a thermostatically controlled U-tube[122] or flask.[123]

Another approach for preparing long-term concentrations in the parts per million range involves the vapor-phase depolymerization of trioxane.[124] Figure 6.15 shows a schematic of this system. A diffusion tube is first filled with trioxane using a heat gun. Dry nitrogen sweeps the trioxane vapor from the diffusion chamber through the catalyst bed held in a tube furnace. The catalyst consists of carborundum coated with phosphoric acid. To prevent repolymerization of the formaldehyde, the mixture is immediately diluted with air. The generation rate can be checked chemically or approximated by weighing the diffusion vial. This technique has an output of about 750 µg/hr and takes about 1 week to stabilize.

Concentrations near 100 ppm at flows from 16 to 100 L/min can be produced using the vaporizer shown in Figure 5.13. First, paraformaldehyde is dissolved in warm water to obtain a 1 to 15% by weight solution. Formalin (37% by weight) can also be used but 10% methanol will be present. The solution is then injected at a known rate into an SV-20 vaporizer.[125] The dilution air can either be dry or humidified.[114] The output concentration is then monitored with an infrared analyzer operating at 3.58 µ. A variation of this technique was used by Kennedy and Hull[126] and by Andrawes.[127] Concentrations using this technique can be calculated from Equation 6.20.

Example 6.5. What is the injection rate of a 12.5% formaldehyde solution needed to produce 100 ppm in air flowing at 32 L/min? Ambient conditions are 25°C and 760 mmHg and the solution density is 1.04 g/mL.

Concentration in air	100 ppm
Flow rate	32 L/min
Molecular weight	30 g/mol
Solution concentration	12.5% or 0.125 g/mL
Solution density	1.04 g/mL
Pressure	760 mmHg
Temperature	25°C or 298°K

From Equation 6.21, the required injection rate is

$$q_L = \frac{(100)(30)(32)}{(22.4E+06)\left(\frac{298}{273}\right)\left(\frac{760}{760}\right)(1.04)(0.126)} = 0.0302 \text{ mL/min}$$

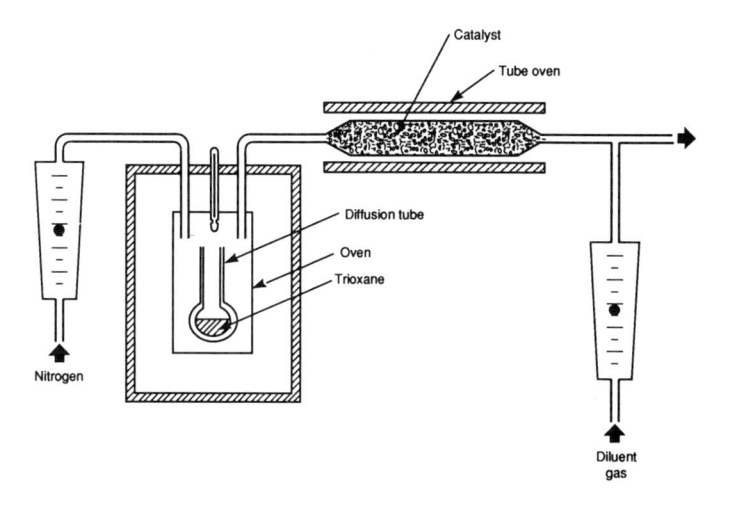

Figure 6.15 Sketch of an apparatus for producing formaldehyde from the depolymerization of trioxane.

A number of techniques are available to monitor formaldehyde. The most popular chemical methods involve chromotropic acid[128] and pararosaniline.[129] Instrumental methods include infrared, electrochemical, and automated wet chemical techniques.

6.7 CHLORINE DIOXIDE

Chlorine dioxide, like formaldehyde, is not available in the pure gaseous state because of its explosive reactivity. Even dilute mixtures of chlorine dioxide in inert gases are unavailable commercially because of its extreme reactivity and decomposition potential. Because of these limitations, chlorine dioxide is usually produced using dynamic systems.

A readily available source for chlorine dioxide is a pulp-mill bleach plant. This gas is used in large quantities to bleach the pulp before the final stages of paper manufacture. Sample streams from such facilities have been used to actually test air-purifying respirator cartridges.[130] A sample gas stream from the plant is diluted with air, cooled, and sent to the cartridges being evaluated. Using this technique, 500 ppm at 64 L/min can be produced.

If a pulp mill is not available, another more expedient approach exists. Chlorine gas, diluted with air, can be passed over sodium chlorite to produce the following reaction:[59]

$$Cl_2 + 2NaClO_2 = 2ClO_2 + NaCl \qquad (6.21)$$

Although this reaction is well known, the experimental details have only

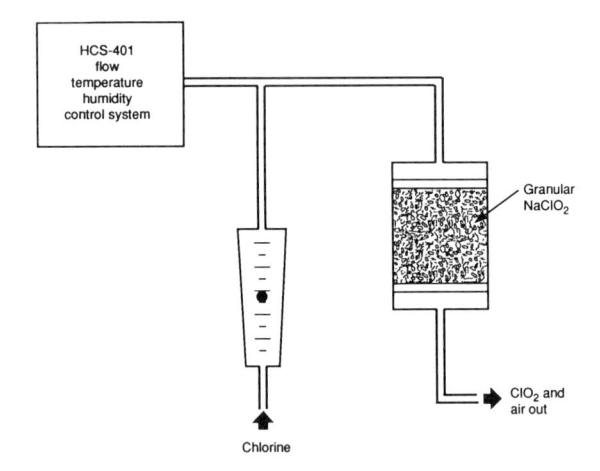

Figure 6.16 Sketch of an apparatus for producing chlorine dioxide from chlorine and sodium chlorite.

recently been brought to light by Frustaci.[131] His system is shown in Figure 6.16. Here, a flow-temperature-humidity control system generates air at 32 or 64 L/min.[114] Chlorine is then blended with the air at a known flow. It must be emphasized that the initial chlorine concentration should not exceed 1%, since concentrations above 5% are reported to be explosive.[59] The chlorine and air mixture then passes through a granular bed of sodium chlorite, which produces the chlorine dioxide. A 100-g bed is effective for about 6 hr at 64 L/min. The optimum humidity is between 40 and 60%. As the humidity approaches 85%, the granular bed begins to soften, building up the back pressure of the system.

Concentrations as high as 1700 ppm have been produced by Frustaci with no complications. However, the National Institute of Occupational Safety and Health suggests the lower explosive limit to be in the region of 3900 ppm. To maximize the efficiency at these high concentrations, gas lines should be inert and as short as possible. Also direct light on the experiment should be minimized to prevent decomposition.

Low concentrations (less than 1 ppm) can be produced statically for instrument calibration using the volume dilution technique. A 200 to 500 ppm air sample is collected in a 1-L glass bottle fitted with a silicone septum. A portion of this air is then injected into a 20-L Tedlar gas sampling bag that has previously been filled with a known quantity of air. This gas mixture can be used for 24 hr without noticeable gas loss.

The concentration can be estimated using the mass balance obtained gravimetrically. A more precise wet chemical technique involves bubbling the gas mixture through a potassium iodide solution and titrating with sodium thiosulfate to a colorless end point. Instrumental techniques include infrared spectrometry (9.1 μ) and electrochemical chlorine analyzers.

6.8 BROMINE AND IODINE

Halogen compounds comprise one of the most reactive families of chemical compounds because of their powerful oxidizing capabilities. This is especially true of bromine and iodine. Static mixtures involving these gases are generally unacceptable because of their reaction with, and sorption onto, the containing vessel's surface.

Both bromine and iodine can be produced in the parts per million range at 1 L/min using the permeation tubes discussed in Chapter 5. Higher concentrations of bromine can be achieved using diffusion tubes, since bromine is a liquid at room temperature (the boiling point is 59.5°C). Higher bromine concentrations have been generated dynamically by bubbling gas through a solution saturated with bromine, as shown in Figure 6.17.[59] The purified gas is preconditioned to the same temperature as the bromine solution by passing it through a heat-controlled condenser just before bubbling it through a saturated bromine solution containing excess bromine. If air is used as the carrier gas through a midget fritted tube, an output concentration of 13,000 ppm can be obtained at 26°C and kept constant for many hours. Lower concentrations can be obtained by further downstream dilution. Solubility data for bromine in water are given in Table 6.5. Bromine is infinitely soluble in alcohol, chloroform, ether, carbon tetrachloride, and carbon disulfide at room temperature, so saturated solutions cannot be obtained.

Iodine is difficult to produce in the parts per million range at flows above 10 L/min because of the crystalline nature of the compound. The passage of gas over or through the crystals often produces uneven concentrations, even with close temperature control. Reproducibility is often poor, since the rate of sublimation is very dependent on iodine grain size. Iodine, however, is partially soluble in a number of solvents, as shown in Table 6.6. As with bromine, the carrier gas can be bubbled through the saturated solution and further diluted downstream with additional gas.

Ludwich[132] has proposed a dynamic system for continuous iodine production, which is shown in Figure 6.18. Molecular iodine is generated by the continuous addition of a sodium nitrite solution, one drop at a time, to sodium iodide dissolved in 3 N sulfuric acid via the reaction

$$4H^+ + 2NO_2^- + 2I^- = I_2 + 2NO + 2H_2O \tag{6.22}$$

As iodine is generated, it is swept into the metered diluent air by bubbling helium into the reaction mixture. Since some of the iodine remains dissolved in the solution, the exact concentration must be determined by an alternate procedure, such as an ion-specific electrode.

6.9 HUMIDITY

Most of this book has addressed techniques to produce gases and vapors in the

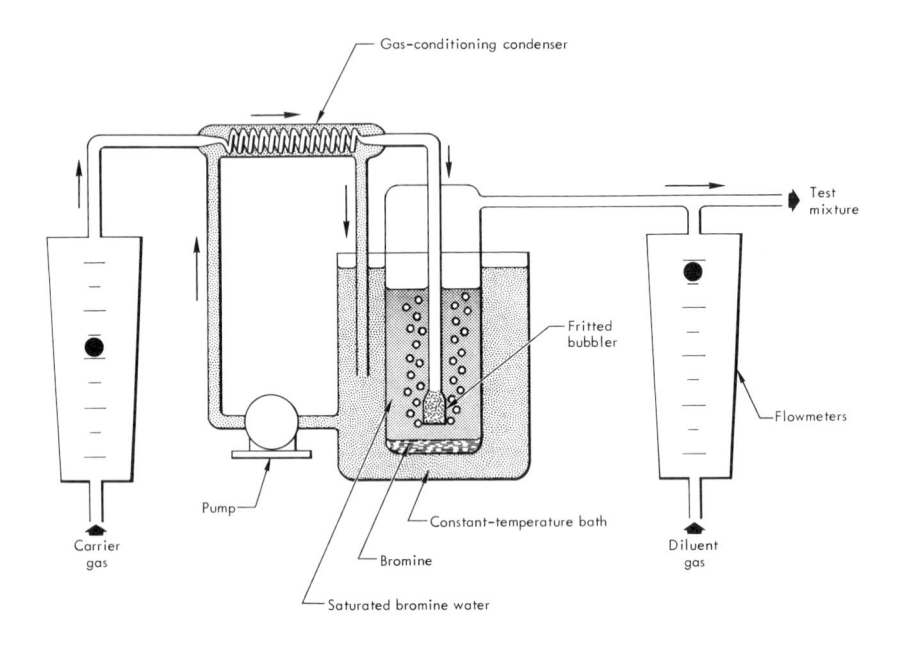

Figure 6.17 Sketch of an apparatus for dynamically generating bromine.

Table 6.5 Solubility of Bromine in Water[a]

Temperature (°C)	Solubility (wt %)
0	2.28
3	3.08
5.8	3.73
10	3.60
20	3.41
25	3.39
30.1	3.34
36	3.36
41	3.39
44.8	3.41
48.8	3.45
52.8	3.50
53.6	3.50

[a] *Source:* Data taken from Stephen and Stephen.[133]

parts per million range. The production of known humidities, however, moves the concentration up several orders of magnitude on the parts per million scale. For example, air at a relative humidity of 50% and 25°C contains 1.6% or 16,000 ppm of water vapor. Although humidity can be expressed in a number of different fashions, this section will focus primarily on relative humidity.

Table 6.6 Solubility of Iodine in Various Solvents[a]

Solvent	Temp (°C)	Solubility (wt %)
Benzene	25	14.1
	30	16.1
Carbon disulfide	10	10.5
	20	14.6
	30	19.3
Carbon tetrachloride	15	2.05
	18	2.25
	21	2.51
	25	2.91
Chloroform	15	3.26
	18	3.54
	21	3.95
	25	4.52
Ethanol	0	16.7
	25	21.3
	35	24.6
Ethyl ether	0	19.3
	25	25.2
	35	28.5
Heptane	25	1.70
	35	2.49
Hexane	25	0.456
Tetrachloroethane	25	0.416
Toluene	25	0.356
Tribromomethane	25	1.90
Water	10	0.200
	20	0.285
	25	0.335
	30	0.385

[a] *Source:* Data taken from Stephen and Stephen.[133]

6.9.1 Static Systems

Numerous temperature-controlled humidity chambers are commercially available from almost every scientific supply company. Typically, the relative humidity ranges are 20 to 95% over a 3 to 90°C range. However, specially designed

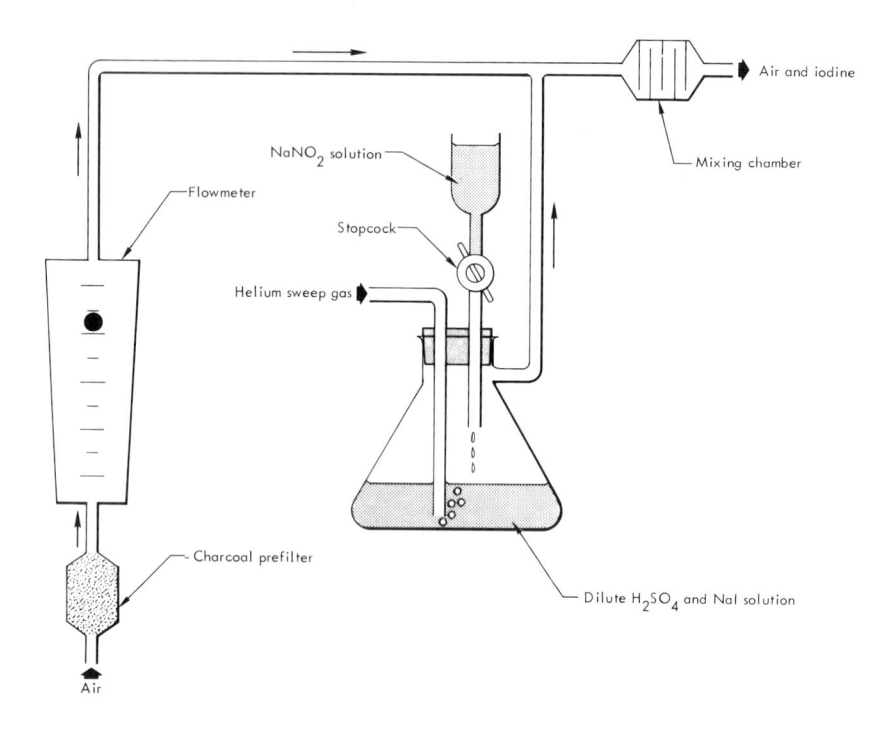

Figure 6.18 Sketch of an apparatus for continuously producing iodine in air.

systems such as the NBS two-pressure system that operate at a 3 to 98% relative humidity range can be obtained.[134]

Constant humidity can also be maintained in a static system by exposing certain saturated aqueous solutions to an enclosed air space at a known temperature.[135] These are tabulated for reference in Appendix N. Such solutions are available from 3 to 98% relative humidity at 25°C, depending on the saturated solution used. There is, however, some temperature dependence in these systems. For example, potassium nitrate yields a 60.4% humidity at 0°C and a 52.9% humidity at 25°C. Although this technique is not considered a primary standard, it is considered a transfer or secondary standard because the values are NBS traceable.[136] This particular technique is extremely useful for humidity probe calibration, and the accuracy for many solutions is ±1% if the temperature is known.

6.9.2 Dynamic Systems

Producing humidities of 5 to 95% can involve all the usual techniques of dynamic gas mixture production. Mechanical pump feed, evaporation, and saturation of air streams have all been used. Figure 6.19 summarizes some of the more traditional techniques involving partial and complete saturation. The top figure illustrates the partial saturation, split stream method. Here, air is metered to a T junction where the stream is split. One stream is partially saturated while the other

Specialized Systems

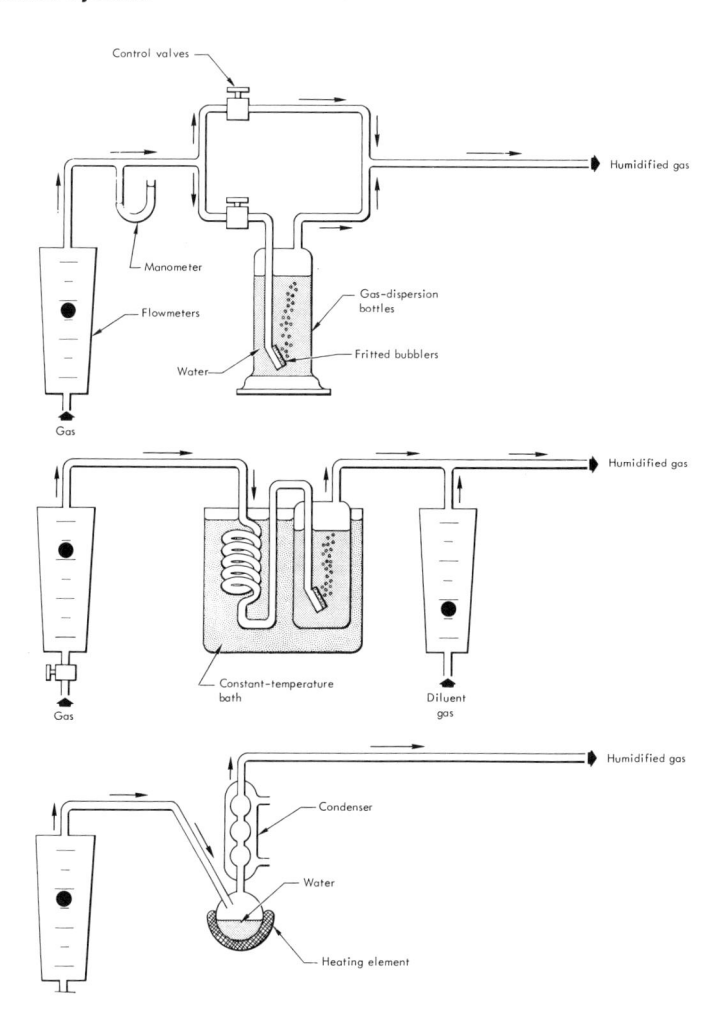

Figure 6.19 Sketch of three systems for humidifying a moving gas stream via partial saturation with a split stream (top), partial saturation with double dilution (middle), and complete saturation (bottom).

stream is used for a subsequent dilution.[137] This technique appears, at first glance, to be straightforward, but several problems can exist. First of all, humidities above 70% are difficult to achieve. The addition of more bubblers (the brute-force method) creates additional back pressure which can, under certain conditions, vent spectacularly and dampen everything within the laboratory. Also the back pressure will alter the flow rate calibration of many flow meter types.

What is often overlooked in many humidifier systems is the heat required to evaporate water. It takes heat to increase the vapor pressure of water, and it takes considerable heat to convert a mole of liquid to a mole of gas. If one attempts to use the heat transferred from the environment, high humidities are almost impossible to achieve.

The middle sketch in Figure 6.19 illustrates the partial saturation, double dilution technique. In this method, the gas flow rate through the partial saturation unit is adjusted before the flow meter. Again, a gas dispersion bottle is useful in assisting in the evaporation of the water. A constant-temperature bath surrounding the bottle and the incoming air line greatly enhance the heat transfer and the maximum relative humidity available. Since the back pressure remains essentially constant for a given flow, flow meter corrections are not usually needed. However, as the water is consumed and the level in the bubbler falls, the humidity also shows a corresponding decrease.

The bottom sketch in Figure 6.19 is one way to raise the humidity to the 90+% range.[138] Here, air is metered into a chamber of hot, refluxing, distilled water; cooled to the required temperature with a Graham, West, Allihn, or other suitable condenser; and is ready for further dilution or subsequent trace contaminant injection. This addition of water vapor will increase the total flow rate by 2 to 3% at room temperature. However, as the temperature increases, the saturation percent also increases and must be accounted for in the total flow assessment.

For relative humidities below 3% at room temperature, the technique shown in Figure 6.20 can be used. Here a syringe or infusion pump can inject a precise rate of water into a nebulizer assembly. A fine mist is produced and any large particles are trapped on extra coarse porous discs. The heat of vaporization is supplied using a heating tape.

Concentrations in the parts per million range can be supplied using the permeation or diffusion tubes discussed in Chapter 5. Another method involves flowing nitrogen saturated with water vapor into a chamber until a certain specified pressure is reached. Dilution is achieved by further pressurizing the holding tank with dry nitrogen passed through a liquid-nitrogen chiller.[139]

Most of the techniques just discussed have at least one major deficiency in their design — the lack of proper temperature control. When working with relative humidity, the temperature and relative humidity must be measured in close proximity to ensure reproducible results. Table 6.7 gives a comparison of conditions that have the same weight percent composition. Note the large effect of temperature on the relative humidity. Another major obstacle is the large pressure drop in many of the systems discussed. Nebulizers, valves, fritted bubblers, and discs all contribute to this problem. There is, however, another alternative. Figure 6.21 shows a schematic of a system that circumvents many of the problems previously encountered. Here, air controlled with a mass flow controller (calibrated at the median humidity) regulates the air flow into the water reservoir. A level switch controls the height of the water in the reservoir. As the air passes over the water, the humidity is maintained using a water heater. As the air exits the reservoir, the temperature is controlled using a cooling system working in conjunction with an air heater. As the humidified air exits the system, a sensor measures temperature and humidity and heaters with controllers maintain a feed-

Specialized Systems

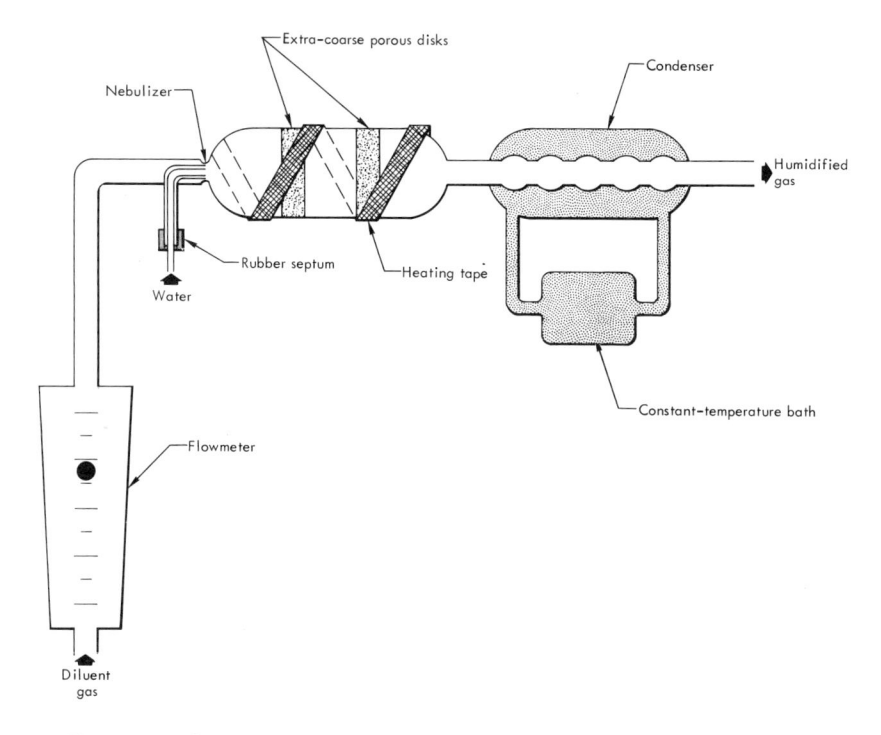

Figure 6.20 Sketch of an apparatus for humidifying a gas stream with injection.

back loop. In this way, both the parameters of interest — temperature and humidity — are precisely controlled at the same physical location. Pressure, air flow, and water level switches are provided to ensure that all the key supply lines are functioning properly. This technique is the basic design employing the commercially available Miller-Nelson Research Inc. flow-temperature-humidity control system.[114]

6.9.3 Calculations

The concentration in parts per million of gas mixture is calculated in the usual way using

$$C_{ppm} = \frac{10^6 p_w}{P} \quad (6.23)$$

Sometimes, however, the concentration is expressed in parts per million of dry gas. If this is the case, Equation 6.23 becomes

$$C_{ppm} = \frac{10^6 p_w}{P - p_w} \quad (6.24)$$

Table 6.7 Water Vapor Concentration as a Function of Humidity and Temperature

	Temperature (°C)	Humidity (%)	Water Concentration (g/m³)
Constant water concentration			
	22.0	101.0	19.6
	23.0	95.4	19.6
	24.0	90.0	19.6
	25.0	85.0	19.6
	26.0	80.4	19.6
	27.0	76.1	19.6
	28.0	72.0	19.6
	29.0	68.1	19.6
	30.0	64.5	19.6
Constant humidity			
	20.0	85.0	14.7
	21.0	85.0	15.6
	22.0	85.0	16.5
	23.0	85.0	17.5
	24.0	85.0	18.5
	25.0	85.0	19.6
	26.0	85.0	20.7
	27.0	85.0	21.9
	28.0	85.0	23.1
	29.0	85.0	24.4
	30.0	85.0	25.8
Constant temperature			
	25.0	50.0	11.5
	25.0	55.0	12.7
	25.0	60.0	13.8
	25.0	65.0	15.0
	25.0	70.0	16.1
	25.0	75.0	17.3
	25.0	80.0	18.4
	25.0	85.0	19.6
	25.0	90.0	20.7
	25.0	85.0	19.6
	25.0	95.0	21.9
	25.0	100.0	23.0

where C is the concentration (ppm), p_w is the vapor pressure of water (mmHg), and P is the total atmospheric pressure (mmHg).

Specialized Systems

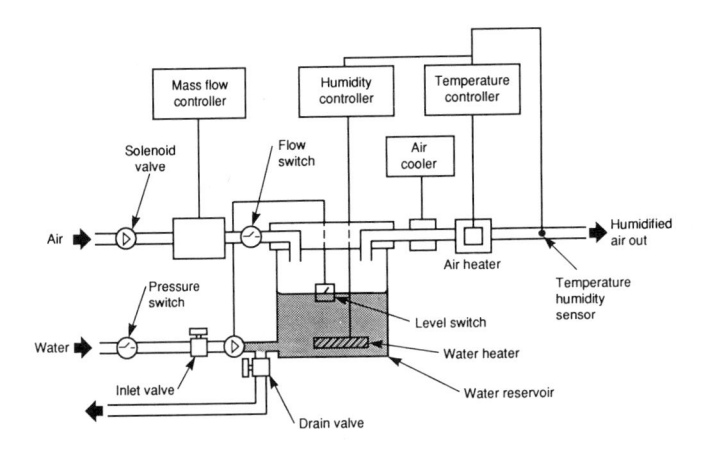

Figure 6.21 Sketch of an apparatus for continually humidifying an air stream using a feedback control system to air and water heaters.

Example 6.6. Calculate the vapor pressure of water at 85% relative humidity at 25°C.

From Appendix G, the Antoine constants are

A	8.1076
B	1750.3
C	235
Temperature	25°C
Humidity	85% or 0.85 decimal percent

From Equation 5.41, the vapor pressure is

$$\log p = 8.1076 - \left[\frac{1750.3}{(235+25)}\right] = 1.3757$$

$$p_s = 23.8 \text{ mmHg}$$

At 85% relative humidity, the vapor pressure is

$$p_w = (23.8)(0.85) = 20.2 \text{ mmHg}$$

Example 6.7. What is the concentration of water in pure air at 20°C at 25% relative humidity and 760 mmHg?

Vapor pressure	17.5 mmHg from Appendix H
Atmos. pressure	760 mmHg
Humidity	25% or 0.25 decimal percent

From Equation 6.24, the concentration is

$$C = \frac{(E+06)(17.5)(0.25)}{760-(0.25)(17.5)} = 5790 \text{ ppm}$$

Example 6.8. What is the water vapor concentration of air at 25°C and 85% relative humidity at 1 atm? Express the answer in g/m³.

Vapor pressure	23.8 mmHg
Atmos. pressure	1 atm or 760 mmHg
Humidity	85% RH or 0.85 decimal percent
Temperature	25°C or 298°K
Mol. wt. water	18 g/mol
L/m³	1000
mmHg/atm	760

From Equation 6.23, the water concentration of the mixture is

$$C = \frac{(23.8)(0.85)(E+06)}{760} = 26{,}618 \text{ ppm}$$

From Equation 4.5, the concentration of water in air is

$$\frac{W}{V} = \frac{(26{,}618)(18)}{(22.4)(E+06)\left(\frac{298}{273}\right)\left(\frac{760}{760}\right)} = 0.0196 \text{ g/L}$$

$$W/V = 19.6 \text{ g/m}^3$$

The concentration of water vapor is more commonly expressed as the relative humidity, which is the ratio of the quantity of water vapor present to the amount that would saturate the gas at the prevailing temperature. Expressed in terms of partial pressure, this would be

$$H_R = \frac{10^2 p_w}{P_s} \qquad (6.25)$$

where H_R is the relative humidity (%), p_w is the partial pressure of water at the temperature of interest (mmHg), and p_s is the pressure of water at complete saturation (mmHg). The vapor pressures at saturation are given in Appendix H. Appendix I gives the mass of water vapor in air at various temperatures.

Humidity, sometimes expressed in percent, is given by

Specialized Systems

$$H_P = \frac{10^2 P_w (P - P_s)}{P_s (P - P_w)} \quad (6.26)$$

and is related to relative humidity by

$$H_P = H_R \left(\frac{P - P_s}{P - P_w} \right) \quad (6.27)$$

where P is the total system pressure. At room temperature, the partial pressure of atmospheric moisture is so small compared with the total pressure that

$$\frac{P - P_s}{P - P_w} = 1 \quad (6.28)$$

Example 6.9. What is the concentration of water vapor in percent of wet air and parts per million of dry air at 90% relative humidity at 40°C and 1 atm? The vapor pressure at 40°C is 55.3 mmHg.

Vapor pressure	55.3 mmHg
Atmos. pressure	1 atm or 760 mmHg
Humidity	90% RH or 0.90 decimal percent
mmHg/atm	760

From Equation 6.24, the concentration in ppm is

$$C_{ppm} = \frac{(E+06)(55.3)(0.90)}{760 - (55.3)(0.90)} = 70{,}076 \text{ ppm}$$

From Equation 6.25, the concentration in percent is

$$C_\% = \frac{(100)(55.3)(0.90)}{760} = 6.55\%$$

Example 6.10. How many grams per minute of water are required to maintain a relative humidity of 70% in an air stream flowing at 64 L/min? Atmospheric temperature and pressure are 25°C and 760 mmHg.

Mass H$_2$O in air	23.0 g H$_2$O/m^3 (from Appendix I)
Flow rate	64 L/min
Vapor pressure	23.8 mmHg (from Appendix H)
Atmos. pressure	760 mmHg

Humidity	70% RH or 0.70 decimal percent
Temperature	25°C or 298°K
Mol. wt. water	18 g/mol
L/m³	1000

There are two approaches to this problem.

Approach 1 uses Appendix I:

$$q_{(H_2O)} = \frac{(23.0)(0.70)(64)}{1000} = 1.03 \text{ g/min}$$

Approach 2 uses Equations 6.23 and 5.14:

The vapor pressure of water at 70% relative humidity is

$$p_w = (23.8)(0.70) = 16.7 \text{ mmHg}$$

From Equation 6.23, the concentration of the mixture is

$$C = \frac{(16.7)(E+06)}{760} = 21,921 \text{ ppm}$$

From a variation of Equation 5.14, the injection rate is

$$q_w = \frac{(21,921)(18)}{(22.4)(E+06)\left(\frac{298}{273}\right)\left(\frac{760}{760}\right)} = 1.03 \text{ g/min}$$

Example 6.11. Determine the percent of saturation at 20 and 80°C for air at 50% relative humidity at 760 mmHg using Appendix H.

Vapor pressure at 20°C	17.5 mmHg
Vapor pressure at 80°C	355 mmHg
Atmospheric pressure	760 mmHg
Relative humidity	50% or 0.50 decimal percent

From Equation 6.26, the percent humidity at 20°C is

$$H_p = \frac{(0.50)(17.5)(760-17.5)(100)}{(17.5)[760-(0.50)(17.5)]} = 49.4\%$$

From Equation 6.26, the percent humidity at 80°C is

$$H_p = \frac{(0.50)(355)(760-355)(100)}{(355)[760-(0.50)(355)]} = 34.8\%$$

Example 6.12. Calculate the relative humidity when the temperature of a flowing gas stream at 65% relative humidity is raised from 15 to 75°C.

Vapor pressure at 15°C 12.8 mmHg
Vapor pressure at 75°C 289 mmHg
Relative humidity 65% or 0.65 decimal percent

From Equation 6.25, the relative humidity is

$$H_R = \frac{(0.65)(12.8)(100)}{289} = 2.88\%$$

and the percent of relative humidity and saturation are approximately equal. However, as the temperature is raised and the partial pressure becomes more significant for a given humidity, the numerical difference between saturation and relative humidity becomes more significant.[140]

6.9.4 Methods of Measurement

Traditionally, the method of continually monitoring a moving air stream for humidity is wet bulb psychrometry. In addition, there are numerous electronic methods that display the humidity, as well as the temperature, in several ways.

A wet bulb psychrometer is shown in Figure 6.22. It consists basically of two thermometers placed in close proximity. The bulb of one is housed in a wick saturated with water (the wet bulb), while the other (the dry bulb) is left completely open to the gas of interest. The gas, usually flowing past at a velocity of at least 1000 ft/min, cools the wet bulb thermometer to a new equilibrium temperature. The relative humidity can then be obtained by knowing the dry bulb temperature and the difference between the dry- and wet bulb temperatures; such a table is shown in Appendix J. In addition, the data can be obtained from other suitable psycrometric charts.[140]

Example 6.13. Using the table in Appendix J, determine the relative humidity if the dry-bulb reading is 21°C and the wet-bulb reading is 16.5°C.

Dry bulb temp. (t_D) = 21°C
Wet bulb temp. (t_W) = 16.5°C

From Appendix J:

$$t_D - t_W = 21 - 16.5 = 4.5°C$$

The relative humidity from the table is 64%.

Although this method seems simple enough, there are several disadvantages.

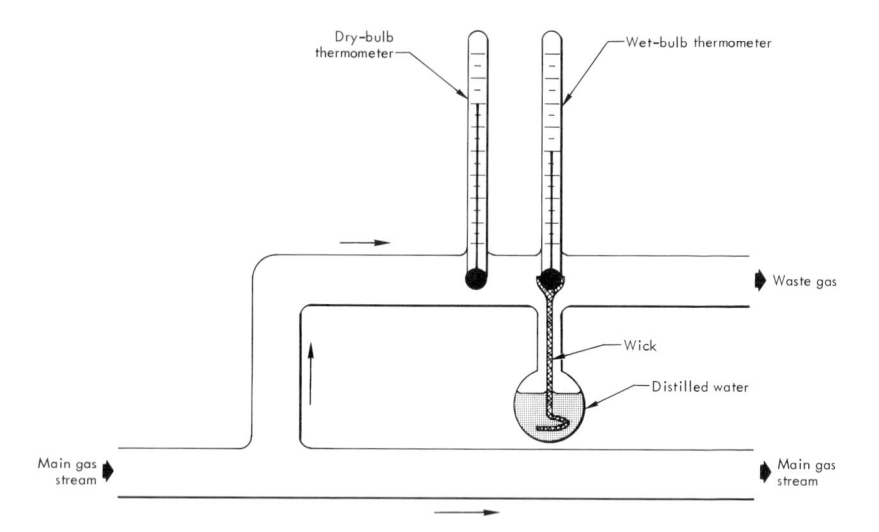

Figure 6.22 Sketch of an apparatus for continually measuring humidity using the wet-bulb psychrometric technique.

Table 6.8 Characteristics of Several Humidity Detection Devices[a]

Manufacturer	Model	Measurement Principle	Range (RH %)	Temp Range (°C)	Response Time (sec)
General Eastern	RH-2	Resistance	0–99	−40–54	
Hy-Cal Engineering	HS-3552-C	Strain gauge	0–100	−40–66	60
Modus	210	Capacitance	5–98	−10–70	1
Ondyne	Delta	Capacitance	5–98	−10–120	20
Panametrics	MC-2	Capacitance	5–95	−40–180	60
Rosemont Analytical	TR-8L	Electrolytic	10–95	−20–80	
Rotronic	H3V-200S	Capacitance	0–100	0–50	10
Science/Electronics	VH-H	Capacitance	0–100	−20–90	1
Solomat	155RH	Capacitance	0–98	−10–70	1
Testoterm	6100	Capacitance	2–98	−12–60	20
Vaisala	HMW-31YB	Capacitance	0–100	−5–55	15

[a] This represents only a few random samples from a very extensive field. The specifications were taken from the product bulletins and are, at times, somewhat optimistic.

In dynamic systems, water must be continually supplied to the wet bulb to maintain wick saturation. Also, the measuring unit must be placed in a separate gas stream, since water evaporating from the wick further adds to the moisture of the gas stream, especially at lower humidities. Additional errors occur from inadequate flow rate, radiant heating effects, contaminated wick, contaminated water, and an improperly fitted wick.[141-143] For these reasons, it is often desirable to use calibrated electronic measurement devices.

Electronic methods, a sampling of which are summarized in Table 6.8, have the advantage of direct measurement of both temperature and a variety of humidity outputs. Currently, relative humidity detectors often rely on bulk polymer resistance, thin-film capacitance,[144] and strain-gauge sensors.[145] The response time is usually less than 2 min, with an accuracy of better than ±5% full scale. The gas streams should be clean and free of dust, moisture droplets, and chemical vapors. Additional techniques have been discussed in a number of summaries.[146-151]

REFERENCES

1. Conner, W. D. and J. S. Nader. *Am. Ind. Hyg. Assoc. J.* 25:291 (1964).
2. Swearengen, P. Lawrence Livermore National Laboratory, Livermore, CA, private communication.
3. Pring, J. N. and G. M. Westrip. *Nature* 170:530 (1952).
4. Thienes, C. H., R. G. Skillen, A. Hoyt, and E. Bogen. *Am. Ind. Hyg. Assoc. J.* 26:255 (1965).
5. Deutsch, S. *J. Air Pollut. Control Assoc.* 18:78 (1968).
6. Altshuller, A. P. and A. F. Wartburg. *Int. J. Air Water Pollut.* 4:70 (1961).
7. Pierce, L. B., K. Nishikawa, and N. O. Fansah. "Validation of Calibration Techniques," paper presented at the 8th Conference on Methods in Air Pollution and Industrial Hygiene Studies, Oakland, CA (1967).
8. Adams, D. F. *J. Air Pollut. Control Assoc.* 13:88 (1963).
9. Horwood, J. H. *J. Air Pollut. Control Assoc.* 9:42 (1959).
10. Priante, S. J. Miller-Nelson Research Inc., Monterey, CA, private communication.
11. Regener, V. H. "Ozone Measuring Devices," Report No. AFCRL-64-212, University of New Mexico, Albuquerque, NM (1963).
12. Regener, V. H. *J. Geophys. Res.* 69:3795 (1964).
13. Hersch, P. and R. Deuringer. *Anal. Chem.* 35:897 (1963).
14. Lubke, M. *Staub-Reinhalt. Luft* 30:28 (1970).
15. McAdie, H. G. and F. J. Hopton. *Am. Lab.* 7:13 (1975).
16. McQuacker, N. R., H. Haboosheh, and W. Best. *Am. Lab.* 13:105 (1981).
17. Brewer, A. W. and J. R. Milford. *Proc. Roy. Soc. London* (Ser. A.) 256:470 (1960).
18. Menser, H. A. and H. E. Heggestad. *Crop. Sci.* 4:103 (1964).
19. Richards, B. L., O. C. Taylor, and G. F. Edmunds. *J. Air Pollut. Control Assoc.* 18:73 (1968).
20. Watanabe, I. and E. R. Stephens. *Anal. Chem.* 5: 313 (1979).
21. Brochure, Fischer America, Inc., Waukesha, WI (9/1989).
22. Lunt, R. W. "Mechanism of Ozone Formation in Electrical Discharges," in *Ozone Technology* (Washington, D.C.: American Chemical Society, 1956).
23. Hern, A. G. *Proc. Phys. Soc. (London)* 78:932 (1961).
24. Hendricks, R. H. and L. B. Larson. *Am. Ind. Hyg. Assoc. J.* 27:80 (1966).
25. Jaffe, L. S. *Am. Ind. Hyg. Assoc. J.* 28:267 (1967).
26. Nash, T. *Atmos. Environ.* 1:679 (1967).
27. Jellinek, H. H. G. and F. J. Kryman. *Environ. Sci. Technol.* 1:658 (1967).
28. Brochure, SKC Inc. Bulletin 881, Fullerton, CA (1988).
29. Byers, D. H., and B. E. Saltzman. *Am. Ind. Hyg. Assoc. J.* 19:251 (1958).
30. Hodgeson, J. A., K. J. Krost, A. E. O'Keeffe, and R. K. Stevens. *Anal. Chem.* 42:1795 (1970).

31. Kopczynski, S. L. and J. J. Bufalini. *Anal. Chem.* 43:1126 (1971).
32. DeMore, W. B., J. C. Romanovsky, M. Feldstein, W. J. Hamming, and P. K. Mueller, "Interagency Comparison of Iodometric Methods for Ozone Determination," ASTM Technical Publication 598, Philadelphia, PA (1975).
33. McKelvey, J. M. and H. E. Hoelscher. *Anal. Chem.* 29:123 (1957).
34. Lodge, J. P. and H. A. Bravo. *Anal. Chem.* 36:671 (1964).
35. Flamm, D. L. *Environ. Sci. Technol.* 11:978 (1977).
36. Layton, R. F., D. Spath, and J. O. Pierce. *Am. Ind. Hyg. Assoc. J.* 31:738 (1970).
37. Saltzman, B. E. and N. Gilbert. *Am. Ind. Hyg. Assoc. J.* 20:379 (1959).
38. Cherniack, I. and R. J. Brian. *J. Air Pollut. Control Assoc.* 15:351 (1965).
39. Brochure, Mast Development Co., 724-5, Reno, NV.
40. Potter, L. and S. Duckworth. *J. Air Pollut. Control Assoc.* 15:207 (1965).
41. Regener, V. H. "The Preparation of Chemiluminescent Substance for the Measurement of Atmospheric Ozone," Report No. AFCRL-66-246, University of New Mexico, Albuquerque, NM (1966).
42. Nederbragt, G. W., A. Vander Horst, and J. Van Duijn. *Nature* 206:87 (1965).
43. Gregory, G. L., C. H. Hudgins, and R. A. Edahl. *Environ. Sci. Technol.* 17:100 (1983).
44. Johnston, A. R., J. F. Dyrud, and Y. T. Shih. *Am. Ind. Hyg. Assoc. J.* 50:451 (1989).
45. Stedman, D. H., E. E. Daby, F. Stuhl, and H. Niki. *J. Air Pollut. Control Assoc.* 22:260 (1972).
46. Hodgeson, J. A., R. E. Baumgardner, B. E. Martin, and K. A. Rehme. *Anal. Chem.* 43:1123 (1971).
47. Fontijn, A., A. J. Sabadell, and R. J. Ronco. *Anal. Chem.* 42:575 (1970).
48. Paur, R. J. "Accuracy of Ozone Calibration Methods," in NBS Publication 464, *Methods and Standards for Environmental Measurement* (1977).
49. Bass, A. M., A. E. Ledford, and J. K. Whittaker. in NBS Publication 464, *Methods and Standards for Environmental Measurement* (1977).
50. Chleck, Z. *Int. J. Appl. Radiat. Isotopes* 7:141 (1959).
51. Hommel, C. O., D. Chleck, and F. J. Brousaides. *Nucleonics* 19:94 (1961).
52. Zafonte, L., W. Long, and J. N. Pitts. *Anal. Chem.* 46:1872 (1974).
53. Verhoek, F. H. and F. Daniels. *J. Am. Chem. Soc.* 53:1250 (1931).
54. Coon, E. D. *J. Am. Chem. Soc.* 59:1910 (1937).
55. Giauque, W. F. and J. D. Kemp. *J. Chem. Phys.* 6:40 (1938).
56. Atwood, K. and G. K. Rollefson. *J. Chem. Phys.* 9:506 (1941).
57. Adley, F. E. and C. P. Skillern. *Am. Ind. Hyg. Assoc. J.* 19:233 (1958).
58. Buck, M. and H. Stratmann. *Staub* 27:11 (1967).
59. Saltzman, B. E. *Anal. Chem.* 33:1100 (1961).
60. Gill, W. E. *Am. Ind. Hyg. Assoc. J.* 21:87 (1960).
61. Kinosian, J. R. and B. R. Hubbard. *Am. Ind. Hyg. Assoc. J.* 19:453 (1958).
62. Moore, W. J. *Physical Chemistry,* 2nd ed. (Englewood Cliffs, NJ: Prentice-Hall, Inc., 1960).
63. Elkins, H. B. *J. Ind. Hyg. Toxicol.* 28:37 (1946).
64. Wade, H. A., H. B. Elkins, and B. P. W. Ruotolo. *Arch, Ind. Hyg. Occup. Med.* 1:8 (1940).
65. Stratmann, H. and M. Buck. *Int. J. Air Water Pollut.* 10:313 (1966).
66. Hughes, E. E., H. L. Rook, and E. R. Deardorff. *Anal. Chem.* 49:1823 (1977).
67. Fahey, D. W., C. S. Eubank, G. Hubler, and F. C. Fehsenfeld. *Atmos. Environ.* 19:1883 (1985).
68. Rehme, K. A., B. E. Martin, and J. A. Hodgeson. "Tentative Method for the

Calibration of Nitric Oxide, Nitrogen Dioxide, and Ozone Analyzers by Gas Phase Titration," Report No. EPA-R2-73-246, (Research Triangle Park, NC: National Environmental Research Center, March, 1974).
69. Palmes E. D., A. F. Gunnison, J. DiMattio, and C. Tomczyk, *Am. Ind. Hyg. Assoc. J.* 37:570 (1976).
70. Jones, W. and T. A. Ridgik. *Am. Ind. Hyg. Assoc. J.* 41:433 (1980).
71. Huygen, C. *Atmos. Environ.* 5:55 (1971).
72. Hemenway, D. R. and G. J. Jakab. *Appl. Ind. Hyg.* 2:18 (1987).
73. Higuchi, J. E., A. Hsu, and R. D. MacPhee. *J. Air Pollut. Control Assoc.* 26:136 (1976).
74. Hama, G. M. *Am. Ind. Hyg. Assoc. J.* 19:477 (1958).
75. Serat, W. F., J. Kyono, and P. K. Mueller. *Atmos. Environ.* 3:303 (1969).
76. Catalog 88, Matheson Gas Products, Newark, CA (1987).
77. Saltzman, B. E. and N. Gilbert. *Am. Ind. Hyg. Assoc. J.* 20:379 (1959).
78. Saltzman, B. E. and A. F. Wartburg. *Anal. Chem.* 37:1261 (1965).
79. Kusnetz, H. L., B. E. Saltzman, and M. E. Lanier. *Am. Ind. Hyg. Assoc. J.* 21:361 (1960).
80. Saltzman, B. E. *Anal. Chem.* 26:1949 (1954).
81. Shaw, J. T. *Atmos. Environ.* 1:81 (1967).
82. Singh, T., R. F. Sawyer, E. S. Starkman, and L. S. Caretto. *J. Air Pollut. Control Assoc.* 18:102 (1968).
83. Hersch, P. A. *J. Air Pollut. Control Assoc.* 19:164 (1969).
84. Nelson, G. O. "Sulfur Dioxide and Nitrogen Dioxide Cartridge Testing System Operating Manual," M-041, (Livermore, CA: Lawrence Livermore National Laboratory, 1973).
85. Fisher, G. E. and T. A. Huis. *J. Air Pollut. Control Assoc.* 20:666 (1970).
86. Crecelius, H. J. and W. Forwerg. *Staub Reinhalt. Luft* 30:23 (1970).
87. Purdue, L. J., J. E. Dudley, J. B. Clements, and R. J. Thompson. "Studies of Air Sampling for Nitrogen Dioxide." I. A Reinvestigation of the Jacobs-Hochheiser Procedure, U. S. Dept. Health, Education, and Welfare, DAQED, LSB (1970).
88. Hartkamp, H. *Staub Reinhalt. Luft* 29:6 (1969).
89. Rigdon, L. P. and R. W. Crawford. "An Experimental System for Determining Nitrogen Oxides in Air," UCRL-51057, (Livermore, CA: Lawrence Livermore National Laboratory, 1971).
90. Groth, R. H. and D. S. Calabro. *J. Air Pollut. Control Assoc.* 19:884 (1969).
91. Stevens, R. K. and J. A. Hodgeson. *Anal. Chem.* 45:443A (1973).
92. Di Martini, R. *Anal. Chem.* 42:1102 (1970).
93. Lefers, J. B. and P. J. van den Berg. *Anal. Chem.* 52 (1982).
94. Scheide, E. P., E. E. Hughes, and J. K. Taylor. *Am. Ind. Hyg. Assoc. J.* 40:180 (1979).
95. Trujillo, P. E. and E. E. Campbell. Anal. Chem. 47:1629 (1975).
96. McCammon, C. S. and J. W. Woodfin. *Am. Ind. Hyg. Assoc. J.* 38:378 (1977).
97. Shepherd, M. and S. Schuhmann. *J. Res. Natl. Bur. Std.* 26:357 (1941).
98. Nelson, G. O. *Rev. Sci. Instrum.* 41:776 (1970).
99. Corte, G. L., G. Dowd, and L. DuBois. *Am. Ind. Hyg. Assoc. J.* 37:873 (1975).
100. Brochure, Model 411, Arizona Instrument Corp, Jerome, AZ.
101. Van Sandt, W. Lawrence Livermore National Laboratory, Livermore, CA, private communication.
102. Hashimoto, Y. and S. Tanaka. *Environ. Sci. Technol.* 14:413, (1980).
103. Monahan, M. Calgon Carbon Corp., Pittsburgh, PA, private communication.

104. Aldridge, W. J. *Analyst* 69:262 (1944).
105. Silverman, L. "Experimental Test Methods," in *Air Pollution Handbook* P. L. Magill, F. R. Holden, and C. Ackley, Eds. (New York: McGraw-Hill Book Company, Inc., 1956).
106. Pella, P. A., E. E. Hughes, and J. K. Taylor. *Am. Ind. Hyg. Assoc. J.* 36:755 (1975).
107. Young, M. S. and J. P. Monat. *Am. Ind. Hyg. Assoc. J.* 43:890 (1982).
108. Elfers, L. A. and C. E. Decker. *Anal. Chem.* 40:1658 (1958).
109. Ryan, R. and P. W. West. *Am. Ind. Hyg. Assoc.* 43:640 (1982).
110. Box, W. D. *Am. Ind. Hyg. Assoc. J.* 24:618 (1963).
111. Cassinelli, M. E. *Am. Ind. Hyg. Assoc. J.* 47:219 (1986).
112. Hill, A. C., L. G. Transtrum, M. R. Pack, and A. Holloman. *J. Air Pollut. Control Assoc.* 9:22 (1959).
113. Thompson, C. R. and J. O. Ivie. *Int. J. Air Water Pollut.* 9:799 (1965).
114. Brochure, HCS-401 Control System, Miller-Nelson Research Inc., Monterey, CA.
115. Bellack, E. and P. J. Schoubal. *Anal. Chem.* 30:2032 (1958).
116. Coyne, L. B., R. E. Cook, J. R. Mann, S. Bouyoucos, O. F. McDonnald, and C. L. Baldwin. *Am. Ind. Hyg. Assoc. J.* 46:609 (1985).
117. Bisgaard, R., L. Molhave, B. Rietz, and P. Wilhardt. *Am. Ind. Hyg. Assoc. J.* 45:425 (1984).
118. Silberstein, S. *Am. Ind. Hyg. Assoc. J.* 51:102 (1990).
119. Muller, R. E. and U. Schurath. *Anal. Chem.* 55:1440 (1983).
120. Balmat, J. L. *Am. Ind. Hyg. Assoc. J.* 46:690 (1985).
121. Henry, N. W. *Am. Ind. Hyg. Assoc. J.* 42:853 (1981).
122. Levin, J. O., R. Lindahl, and K. Andersson. *Environ. Sci. Technol.* 20:1273 (1986).
123. Green, D. J. and T. J. Kulle. *Am. Ind. Hyg. Assoc. J.* 47:50 (1986).
124. Geisling, K. L. and R. R. Miksch. *Anal. Chem.* 54:140 (1982).
125. Brochure, Miller-Nelson Research Inc, SV-20, Monterey, CA.
126. Kennedy, E. R. and R. D. Hull. *Am. Ind. Hyg. Assoc. J.* 47:94 (1986).
127. Andrawes, F. F. *J. Chromatogr. Sci.* 22:506 (1984).
128. Balmat, J. L. and G. W. Meadows. *Am. Ind. Hyg. Assoc. J.* 46:578 (1985).
129. Miksch, R. R., D. W. Anthon, L. Z. Fanning, C. D. Hollowell, K. Revzan, and J. Glanville. *Anal. Chem.* 53:2118 (1981).
130. Simon, C. G., R. P. Fisher, and J. D. Davison. *Am. Ind. Hyg. Assoc. J.* 48:1 (1987).
131. Frustaci, D. "Chlorine Dioxide Test System — Design and Performance," paper presented at the American Industrial Hygiene Conference, St. Louis, MO (May, 1989).
132. Ludwich, J. D. *Anal. Chem.* 41:1907 (1969).
133. Stephen, H. and T. Stephen. *Solubilities of Inorganic and Organic Compounds* (New York: Macmillan Company, 1963).
134. Huang, P. H. "Humidity Sensing, Measurements, and Calibration Standards," Sensors Expo West Proceedings, 303a-1 (1989).
135. Greenspan, L. *J. Res. Natl. Bur. Stds.* 81A:89 (1977).
136. Schoen, O. W. *Instr. Control Systems* 45:61 (1972).
137. Harris, F. E. and L. K. Nash. *Anal. Chem.* 23:736 (1951).
138. Henry, N. W. *Am. Ind. Hyg. Assoc. J.* 40:1017 (1979).
139. Gelizunas, V. L. *Anal Chem.* 41:1400 (1969).
140. Zimmerman, O. T. and I. Lavine. *Psychrometric Tables and Charts*, 2nd ed. (Dover, NH: Industrial Research Service, Inc., 1964).
141. Humphreys, C. M. and K. J. Kronoveter. *Am. Ind. Hyg. Assoc. J.* 31:609 (1970).

142. Technical Bulletin No. 1, Hydrodynamics, Inc., Rev. 1263, Silver Spring, MD.
143. Handbook, "Selecting Humidity Sensors for Industrial Processes," (Watertown, MA: General Eastern, March, 1982).
144. Lafarie, J. P. "Humidity Sensors Based on Dielectric Polymers," Sensors Expo West Proceedings, 303c-1 (1989).
145. Brownawell, M. *Sensors* 23:12 (1989).
146. Hollander, L. E., D. S. Mills, and T. A Perls. *ISA J.* 7:50 (1960).
147. Cole, K. M. and J. A. Reger. *Instr. Control Systems* 43:77 (1970).
148. Staff, *Instr. Control Systems* 45:75 (1972).
149. Lafarie, J. P. *Sensors* 21:33 (1987).
150. Quinn, F. C. *Heating, Piping & Air Condit.* 15:107 (1966).
151. Wiederhold, P. R. *Instr. Control Systems* 51:31 (1978).
152. Ganzer, K. M. *Am. Lab.* 19:40 (1987).

Appendix A

Conversion Factors

Length

cm	m	mm	in.	ft
1	0.01000	10.00	0.3937	0.03281
100	1	1000	39.37	3.281
0.100	0.001000	1	0.03937	0.003281
2.540	0.02540	25.40	1	0.08333
30.48	0.3048	304.8	12.00	1

Mass

g	kg	lb	oz
1	0.001000	0.002205	0.03528
1000	1	2.205	35.28
453.6	0.4536	1	16.00
28.35	0.02835	0.06251	1

Volume

cm^3	L	$in.^3$	ft^3	gal
1	0.001000	0.06103	3.532E-05	2.642E-04
1000	1	61.03	0.03532	0.2642
16.39	0.01639	1	5.787E-04	4.329E-03
28320	28.32	1728	1	7.481
3785	3.785	231.0	0.1337	1.000

Conversion Factors (Continued)

Density

g/cm^3	g/L	lb/in.^{3}	lb/ft^3
1	1000	0.03613	62.43
0.001000	1	3.613E-05	0.06243
27.68	27680	1	1728
0.01602	16.02	5.788E-04	1

Appendix A

Pressure

lb/in.²	atm	mmHg or Torr at 0°C	in. Hg at 0°C	mm H₂O at 4°C	in. H₂O at 4°C	bar	mbar	Pascal N/m²	kg/cm²	dynes/cm²	lb/ft²
1	0.06805	51.71	2.036	703.0	27.68	0.06894	68.94	6894	0.0703	68950	144.0
14.70	1	760	29.92	10330	406.7	1.013	1013	101300	1.034	1.014E+06	2117
0.01934	0.001316	1	0.03937	13.59	0.5352	0.001333	1.333	133.3	0.001360	1333	2.785
0.4912	0.03342	25.40	1	345.3	13.60	0.03386	33.86	3386	0.03454	33870	70.73
0.001421	9.669E-05	0.07349	0.002893	1	0.03937	9.806E-05	0.09806	9.806	0.000100	97.98	0.2046
0.03610	0.002456	1.867	0.07350	25.40	1	0.002491	2.491	249.1	0.002538	2489	5.198
14.50	0.9869	750.1	29.53	10197	401.4	1	1000	100000	1.020	1.000E+06	2089
0.01450	9.869E-04	0.7501	0.02953	10.20	0.4015	0.001000	1	100	0.001020	1000	2.089
1.450E-04	9.869E-06	0.007501	2.953E-04	0.1020	0.004014	1.00E-05	0.01000	1	1.020E-05	10.00	0.02089
14.22	0.9676	735.4	28.95	10000	393.7	0.9806	980.6	98060	1	9.805E+05	2048
1.450E-05	9.867E-07	7.499E-04	2.952E-05	0.01019	4.013E-04	1.00E-06	1.00E-03	0.1000	1.019E-06	1	0.002088
0.006946	4.726E-04	0.3592	0.01414	4.883	0.1923	4.789E-04	0.4789	47.89	4.884E-04	478.9	1

Appendix B

Atomic Weights and Numbers

Name	Symbol	Atomic Number	Atomic Weight	Name	Symbol	Atomic Number	Atomic Weight
Actinium	Ac	89		Gadolinium	Gd	64	157.25
Aluminum	Al	13	26.89	Gallium	Ga	31	69.72
Americium	Am	95		Germanium	Ge	32	72.57
Antimony	Sb	51	121.75	Gold	Au	79	196.97
Argon	Ar	18	39.95	Hafnium	Hf	72	178.49
Arsenic	As	33	74.92	Helium	He	2	4.00
Astatine	At	85		Holmium	Ho	67	164.93
Barium	Ba	56	137.34	Hydrogen	H	1	1.008
Berkelium	Bk	97		Indium	I	53	126.90
Beryllium	Be	4	9.01	Iron	Fe	26	55.45
Bismuth	Bi	83	208.98	Krypton	Kr	36	83.80
Boron	B	5	10.81	Lanthanum	La	57	138.91
Bromine	Br	35	79.90	Lawrencium	Lr	103	
Cadmium	Cd	48	112.40	Lead	Pb	82	207.19
Cesium	Cs	55	132.91	Lithium	Li	3	6.94
Calcium	Cq	20	40.08	Lutetium	Lu	71	174.97
Californium	Cf	98		Magnesium	Mg	12	24.31
Carbon	C	6	12.01	Manganese	Mn	25	54.94
Cerium	Ce	58	140.12	Mendelevium	Md	101	
Chlorine	Cl	17	35.45	Mercury	Hg	80	200.59
Chromium	Cr	24	51.00	Molybdenum	Mo	42	95.94
Cobalt	Co	27	58.93	Neodymium	Nd	60	144.24
Copper	Cu	29	63.55	Neon	Ne	10	20.18
Curium	Cm	96		Neptunium	Np	93	
Dysprosium	Dy	66	162.50	Nickel	Ni	38	58.71
Einsteinium	Es	99		Niobium	Nb	41	92.91
Erbium	Er	68	167.26	Nitrogen	N	7	14.01
Europium	Eu	63	151.96	Nobelium	No	102	
Fermium	Fm	100		Osmium	Os	75	190.20
Fluorine	F	9	18.00	Oxygen	O	8	16.00
Francium	Fr	87		Palladium	Pd	46	106.40

Atomic Weights and Numbers (Continued)

Name	Symbol	Atomic Number	Atomic Weight	Name	Symbol	Atomic Number	Atomic Weight
Phosphorus	P	15	30.97	Sulfur	S	16	32.06
Platinum	Pt	78	195.09	Tantalum	Ta	73	180.95
Plutonium	Pu	94		Technetium	Tc	43	
Polonium	Po	84		Tellurium	Te	52	127.60
Potassium	K	19	39.10	Terbium	Tb	65	158.92
Praseodymium	Pr	59	140.91	Thallium	Tl	81	204.37
Promethium	Pm	61		Thorium	Ti	90	232.04
Protactinium	Pa	91		Thullium	Tm	59	168.93
Radium	Ra	88		Tin	Sn	50	118.69
Rhodium	Rh	45	102.91	Titanium	Th	22	47.90
Rubidium	Rb	37	84.57	Tungsten	W	74	183.85
Ruthenium	Ru	44	101.07	Uranium	U	92	238.03
Samarium	Sm	62	150.35	Vanadium	V	23	50.94
Scandium	Sc	21	44.96	Xenon	Xe	54	131.30
Selenium	Se	34	78.96	Ytterbium	Yb	70	173.04
Silicon	Si	14	28.09	Yttrium	Y	39	88.91
Silver	Ag	47	107.87	Zinc	Zn	30	65.37
Sodium	Na	11	22.99	Zirconium	Zr	40	91.22
Strontium	Sr	38	87.62				

Appendix C

Values for the Molar Gas Constant

Volume	Temp	mol	atm	psi	mmHg	in. Hg	in. H₂O	ft H₂O
ft³	°K	g	0.002897	0.04257	2.202	0.08668	1.178	0.09821
ft³	°K	lb	1.314	19.31	998.7	39.32	534.5	44.55
ft³	°R	g	0.001609	0.02365	1.223	0.04816	0.6547	0.05456
ft³	°R	lb	0.7300	10.73	554.8	21.84	297.0	24.75
cm³	°K	g	82.06	1206	62360	2455	33380	2782
cm³	°K	lb	37210	546800	2.828E+07	1.113E+06	1.514E+07	1.261E+06
cm³	°R	g	45.59	669.9	34650	1364	18540	1545
cm³	°R	lb	20680	303900	1.572E+07	6.188E+05	8.412E+06	7.011E+05
L	°K	g	0.08206	1.206	62.36	2.455	33.38	2.782
L	°K	lb	37.22	547.0	28290	1114	15140	1262
L	°R	g	0.04559	0.6699	34.65	1.364	18.54	1.545
L	°R	lb	20.68	303.9	15720	618.7	8412	701.0

Appendix D

Density of Dry Air

Temp °C	\multicolumn{7}{c}{Density in g/L at Stated Pressure in mmHg}
	700	710	720	730	740	750	760
0	1.191	1.208	1.225	1.242	1.259	1.276	1.293
5	1.170	1.186	1.203	1.220	1.237	1.253	1.270
10	1.149	1.165	1.181	1.198	1.214	1.231	1.247
15	1.129	1.145	1.161	1.178	1.194	1.210	1.226
20	1.110	1.126	1.142	1.157	1.173	1.189	1.205
25	1.091	1.106	1.122	1.137	1.153	1.168	1.184
30	1.073	1.088	1.104	1.119	1.134	1.150	1.165
35	1.056	1.071	1.086	1.101	1.116	1.131	1.146
40	1.039	1.054	1.069	1.083	1.098	1.113	1.128
45	1.022	1.037	1.052	1.066	1.081	1.095	1.110
50	1.007	1.021	1.035	1.050	1.064	1.079	1.093

Appendix E

Densities of Common Gases and their Deviation from the Perfect Gas Law

			Vapor density at 0°C and 1 atm			
Compound	Molecular Weight (g/mol)	Boiling Point (°C)	Actual ρA (g/L)	Ideal ρI (g/L)	Density Ratios at 0°C and 1 atm (ρA/ρI)	(ρA/ρair)
Acetylene	26.04	−84	1.1709	1.162	1.008	0.906
Air	28.96	−183	1.2929	1.292	1.000	1.000
Allene	40.07	−34	1.787	1.788	0.999	1.382
Ammonia	17.03	−33	0.771	0.760	1.015	0.596
Argon	39.94	−185	1.784	1.782	1.001	1.380
Arsine	77.94	−62	3.485	3.478	1.002	2.695
Boron trichloride	117.17	−12.5	5.326	5.228	1.019	4.119
Boron trifluoride	67.81	−100	3.077	3.026	1.017	2.380
Bromotrifluoroethylene	160.92	−2	7.249	7.181	1.010	5.607
Bromotrifluoromethane	148.93	−58	6.870	6.646	1.034	5.314
Butadiene 1,3-	54.09	−4	2.428	2.414	1.006	1.878
Butane, n-	58.12	−0.5	2.730	2.593	1.053	2.112
Butene, 1-	56.11	−6	2.583	2.504	1.032	1.998
Butene, cis-2-	56.11	3.7	2.582	2.504	1.031	1.997
Butene, trans-2-	56.11	1	2.582	2.504	1.031	1.997
Butyne	54.09	8	2.542	2.414	1.053	1.966
Carbon dioxide	44.01	−79	1.977	1.964	1.007	1.529
Carbon monoxide	28.01	−192	1.251	1.250	1.001	0.968
Carbonyl fluoride	66.01	−85	2.966	2.946	1.007	2.294
Carbonyl sulfide	60.07	−52	2.711	2.680	1.011	2.097
Chlorine	70.91	−34	3.198	3.164	1.011	2.474
Chlorine dioxide	67.46	10	3.21	3.010	1.07	2.48
Chlorine trifluoride	92.45	11.8	4.145	4.125	1.005	3.206
Chlorodifluoromethane	86.47	−41	4.01	3.859	1.04	3.10
Chloropentafluoroethane	154.47	−39	7.17	6.893	1.04	5.55
Chlorotrifluoroethylene	116.47	−28	5.339	5.197	1.027	4.129
Chlorotrifluoromethane	104.46	−81	4.666	4.661	1.001	3.609
Cyanogen	52.04	−21	2.335	2.322	1.006	1.806
Cyanogen chloride	61.47	13	2.793	2.743	1.018	2.160
Cyclobutane	56.11	12.5	2.581	2.504	1.031	1.996

Densities of Common Gases and their Deviation from the Perfect Gas Law (Continued)

Compound	Molecular Weight (g/mol)	Boiling Point (°C)	Actual ρA (g/L)	Ideal ρI (g/L)	Density Ratios at 0°C and 1 atm (ρA/ρI)	(ρA/ρair)
Cyclopropane	42.08	−33	1.879	1.878	1.001	1.453
Deuterium	4.03	−250	0.1800	0.180	1.000	0.139
Diborane	27.67	−93	1.247	1.235	1.010	0.964
Dibromodifluoromethane	209.82	24	9.372	9.363	1.001	7.249
Dichlorodifluoromethane	120.91	−30	5.425	5.395	1.005	4.196
Dichlorofluoromethane	102.92	8.9	4.621	4.593	1.006	3.574
Dichlorosilane	101.01	8.2	4.551	4.507	1.010	3.520
Dichlorotetrafluoroethane, 1,2-	170.92	3.6	7.673	7.627	1.006	5.935
Difluoroethane, 1,1-	66.05	−25	2.950	2.947	1.001	2.282
Difluoroethylene, 1,1-	64.04	−86	2.860	2.858	1.001	2.212
Difluoro-1-chloroethane, 1,1-	100.50	−10	4.509	4.485	1.005	3.488
Dimethyl ether	46.07	−25	2.091	2.056	1.017	1.617
Dimethylamine	45.08	6.9	2.014	2.012	1.001	1.558
Dimethylpropane, 2,2-	72.15	9.5	3.390	3.220	1.053	2.622
Ethane	30.07	−89	1.356	1.342	1.011	1.049
Ethyl chloride	64.52	12.4	2.885	2.879	1.002	2.231
Ethylene	28.05	−104	1.262	1.252	1.008	0.976
Ethylene oxide	44.05	10.7	1.926	1.966	0.980	1.490
Fluorine	38.00	−188	1.696	1.696	1.000	1.312
Fluoroethane	48.06	−38	2.16	2.145	1.01	1.67
Fluoroform	70.01	−84	3.12	3.124	1.00	2.41
Germane	76.66	−90	3.420	3.421	1.000	2.645
Helium	4.003	−269	0.1785	0.179	0.999	0.138
Hexafluoroacetone	166.02	−28	7.449	7.408	1.005	5.761
Hexafluoroethane	138.01	−78	6.236	6.158	1.013	4.823
Hydrogen	2.016	−253	0.08986	0.090	0.999	0.070
Hydrogen bromide	80.91	−66	3.636	3.610	1.007	2.812
Hydrogen chloride	36.46	−85	1.639	1.627	1.007	1.268
Hydrogen cyanide	27.03	25.7	1.224	1.206	1.015	0.947
Hydrogen fluoride	20.01	19.5	2.403	0.893	2.691	1.859
Hydrogen iodide	127.91	−35.5	5.766	5.708	1.010	4.460
Hydrogen selenide	80.98	−41	3.617	3.614	1.001	2.798
Hydrogen sulfide	34.08	−60	1.538	1.521	1.011	1.190
Isobutane	58.12	−12	2.669	2.593	1.029	2.064
Isobutene	56.11	−7	2.582	2.504	1.031	1.997
Krypton	83.80	−153	3.749	3.739	1.003	2.900
Methane	16.04	−162	0.7168	0.716	1.001	0.554
Methyl bromide	94.95	3.5	4.265	4.237	1.007	3.299
Methyl chloride	50.49	−24	2.307	2.253	1.024	1.784
Methyl fluoride	34.03	−78	1.545	1.519	1.017	1.195
Methyl mercaptan	48.10	6	2.146	2.146	1.000	1.660
Methyl vinyl ether	58.08	5.5	2.570	2.592	0.992	1.988
Methylacetylene	40.07	−23	1.826	1.788	1.021	1.412
Monoethylamine	45.09	16.6	2.088	2.012	1.038	1.615
Monomethylamine	31.06	−6.5	1.397	1.386	1.008	1.081

Appendix E

Neon	20.18	−246	0.9002	0.900	1.000	0.696
Nickel carbonyl	170.75	43	7.727	7.619	1.014	5.976
Nitric oxide	30.01	−152	1.341	1.339	1.001	1.037
Nitrogen	28.01	−196	1.2506	1.250	1.000	0.967
Nitrogen dioxide	46.01	21.3	—[a]	2.053	—	—
Nitrogen trifluoride	71.01	−129	3.176	3.169	1.002	2.456
Nitrogen trioxide	76.02	3.5	—[a]	3.392	—	—
Nitrosyl chloride	65.47	−6	2.993	2.921	1.024	2.315
Nitrous oxide	44.02	−89.5	1.978	1.964	1.007	1.530
Oxygen	32.00	−183	1.429	1.428	1.001	1.105
Oxygen difluoride	54.00	−145	2.424	2.410	1.006	1.875
Ozone	48.00	−182	2.144	2.142	1.001	1.658
Perchloryl fluoride	102.45	−47	4.712	4.572	1.031	3.645
Perfluorobutane	238.03	−2	10.849	10.622	1.021	8.391
Perfluorocyclobutane	200.03	−6	9.329	8.926	1.045	7.216
Perfluoropropane	188.02	−37	8.588	8.390	1.024	6.642
Perfluoro−2−butene	200.03	1.2	9.096	8.926	1.019	7.035
Phosgene	98.92	8	4.498	4.414	1.019	3.479
Phosphine	34.00	−88	1.482	1.517	0.977	1.146
Phosphorous pentafluoride	125.97	−85	5.805	5.621	1.033	4.490
Phosphorous trifluoride	87.97	−101	3.920	3.925	0.999	3.032
Propane	44.10	−42	2.015	1.968	1.024	1.559
Propylene	42.08	−48	1.937	1.878	1.032	1.498
Silane	32.12	−112	1.440	1.433	1.005	1.114
Silicon tetrafluoride	104.08	−65	4.689	4.644	1.010	3.627
Stibine	124.77	−17	5.61	5.568	1.01	4.34
Sulfur dioxide	64.07	−10	2.927	2.859	1.024	2.264
Sulfur hexafluoride	146.05	−64	6.612	6.517	1.015	5.114
Sulfur tetrafluoride	108.06	−40	4.891	4.822	1.014	3.783
Sulfuryl fluoride	102.06	−55	4.813	4.554	1.057	3.723
Tetrafluoroethylene	100.02	−76	4.566	4.463	1.023	3.532
Tetrafluorohydrazine	104.01	−73	4.727	4.641	1.018	3.656
Tetrafluoromethane	88.01	−128	3.946	3.927	1.005	3.052
Trichlorofluoromethane	137.37	23.8	6.36	6.130	1.04	4.92
Trimethylamine	59.11	3.5	2.795	2.638	1.060	2.162
Vinyl bromide	106.96	16	4.898	4.773	1.026	3.788
Vinyl chloride	62.50	−14	2.861	2.789	1.026	2.213
Vinyl fluoride	46.04	−51	2.064	2.054	1.005	1.596
Vinyl methyl ether	58.08	7	2.6	2.592	1.00	2.01
Xenon	131.30	−108	5.890	5.859	1.005	4.556

[a] Equilibrium mixture

Appendix F

Diffusion Coefficients at 25°C and 760 mmHg in Air

Compound	Diffusion coefficient (cm^2/sec)
Acetic acid	0.1235
Acetone	0.1049
Acrylonitrile	0.1059
Allyl alcohol	0.1021
Allyl chloride	0.0975
Amyl acetate	0.0610
Amyl alcohol	0.0716
Amyl alcohol, sec-	0.0728
Amyl butyrate	0.0486
Amyl formate	0.0663
Amyl formate, iso-	0.0675
Amyl propionate, n-	0.0559
Amyl-isobutyrate	0.0496
Aniline	0.0735
Benzene	0.0932
Benzonitrile	0.0710
Benzyl acetate	0.0600
Benzyl alcohol	0.0712
Benzyl chloride	0.0713
Bromine	0.1064
Bromochloromethane	0.0953
Bromoform	0.0767
Butyl acetate	0.0672
Butyl acetate, iso-	0.0690
Butyl alcohol	0.0861
Butyl alcohol, iso-	0.0880
Butyl alcohol, sec-	0.0891
Butyl alcohol, tert-	0.0873
Butyl amine	0.0872
Butyl amine, iso-	0.0900
Butyl butyrate, iso-	0.0559

Diffusion Coefficients at 25°C and 760 mmHg in Air (Continued)

Compound	Diffusion coefficient (cm²/sec)
Butyl ether	0.0536
Butyl formate, iso-	0.0784
Butyl propionate	0.0608
Butyl propionate, iso-	0.0611
Butyl valerate, iso-	0.0494
Butyl-iso-butyrate, iso-	0.0551
Butyric acid	0.0775
Butyric acid, iso-	0.0785
Caproic acid	0.0602
Caproic acid, iso-	0.0596
Carbon disulfide	0.1045
Carbon tetrachloride	0.0828
Chlorobenzene	0.0747
Chloroform	0.0888
Chloropicrin	0.0811
Chlorotoluene, m-	0.0645
Chlorotoluene, o-	0.0688
Chlorotoluene, p-	0.0621
Cymene, p-	0.0630
Diacetone alcohol	0.0647
Dibromoethane, 1,2-	0.0826
Dibromo-3-chloropropane, 1,2-	0.0686
Dibutyl phthalate	0.0421
Dichloroethane, 1,1-	0.0919
Dichloroethane, 1,2-	0.0907
Dichloroethyl ether, sym-	0.0694
Dichloromethane	0.1037
Dichloropropane, 1,2-	0.0794
Diethyl amine	0.0993
Diethyl phthalate	0.0497
Diethylene glycol	0.0730
Diethylene glycol monoethyl ether	0.0610
Dimethyl formamide	0.0973
Dioxane	0.0922
Di-2-ethyl hexyl phosphate	0.0394
Di-iso-octyl phthalate	0.0337
Ethyl acetate	0.0861
Ethyl alcohol	0.1181
Ethyl bromide	0.0989
Ethyl butyrate	0.0669
Ethyl cyanoacetate	0.0710
Ethyl ether	0.0918
Ethyl formate	0.0976
Ethyl propionate	0.0766
Ethyl valerate	0.0603
Ethylbenzene	0.0755
Ethylene chlorohydrin	0.0964
Ethylene diamine	0.1009

Appendix F

Ethylene glycol	0.1005
Ethylene glycol monoethyl ether	0.0788
Ethylene glycol monomethyl ether	0.0884
Ethyl-1-butanol, 2-	0.0656
Ethyl-iso-butyrate	0.0675
Formic acid	0.1530
Heptyl alcohol, n-	0.0544
Hexane, n-	0.0732
Hexyl alcohol	0.0621
Mercury	0.1423
Mesityl oxide	0.0760
Mesitylene	0.0663
Methyl acetate	0.0978
Methyl alcohol	0.1520
Methyl butyrate	0.0745
Methyl ethyl ketone	0.0903
Methyl formate	0.1090
Methyl n-caproate	0.0610
Methyl propionate	0.0862
Methyl propyl ketone	0.0793
Methyl valerate	0.0665
Methyl-iso-butyrate	0.0748
Nitrobenzene	0.0721
Octane	0.0616
Octyl alcohol	0.0506
Pentachloroethane	0.0673
Pentane	0.0842
Propionic acid	0.0952
Propyl acetate	0.0768
Propyl acetate, iso-	0.0770
Propyl alcohol	0.0993
Propyl alcohol, iso-	0.1013
Propyl bromide	0.0875
Propyl bromide, iso-	0.0914
Propyl butyrate	0.0610
Propyl ether, iso-	0.0683
Propyl formate	0.0831
Propyl iodide	0.0868
Propyl iodide, iso-	0.0878
Propyl valerate	0.0556
Propylbenzene	0.0669
Propylbenzene, iso-	0.0677
Propylene glycol	0.0879
Propyl-iso-butyrate	0.0622
Propyl-iso-butyrate, iso-	0.0638
Pseudocumene	0.0642
Stryene	0.0701
Tetrachloroethane, 1,1,2,2-	0.0722
Tetrachloroethylene	0.0797
Tetraethyl pyrophosphate	0.0475
Toluene	0.0849
Toluene di-iso-cyanate	0.0583
Tributyl phosphate	0.0432
Trichloroethane, 1,1,1-	0.0794

Diffusion Coefficients at 25°C and 760 mmHg in Air (Continued)

Compound	Diffusion coefficient (cm^2/sec)
Trichloroethane, 1,1,2-	0.0792
Trichloroethylene	0.0875
Triethyl amine	0.0754
Triethyl phosphate	0.0552
Triethylene glycol	0.0590
Valeric acid, iso-	0.0653
Xylene, m-	0.0670
Xylene, o-	0.0727
Xylene, p-	0.0670

Note: Data taken from Lugg, G.A., *Anal. Chem.* 40:1072 (1969) with permission.

Appendix G

Constants for Vapor Pressure Calculations

Compound	Temperature Range (°C)	A	B	C
Acetaldehyde	-45–70	6.8109	992.0	230
Acetic acid	0–36	7.803	1651	225
Acetone	—	7.0245	1161	224
Acetonitrile	—	7.120	1314	230
Acrylonitrile	-20–140	7.039	1233	222.5
Ammonia	-83–60	7.555	1003	247.9
Amylbenzene	15–104	7.3517	1858.4	212.0
Aniline	—	7.2418	1675.3	200
Benzene	—	6.9056	1211.0	220.79
Benzonitrile	—	6.7463	1436.7	181.0
Benzotrifluoride	-20–180	7.0071	1331.3	220.58
Benzyl alcohol	20–113	7.8184	1950.3	194.4
Bromine	—	6.8330	1133	228
Bromobenzene	—	6.8834	1440	204
Bromocyclohexane	0–68	7.3414	1778.8	235
Bromotoluene, p-	10–85	7.2284	1743.7	218
Butadiene, 1,3-	-80–65	6.8594	935.5	239.5
Butane, n-	—	6.8303	945.9	240
Butane, iso-	—	6.7481	882.8	240
Butene, 1-	—	6.8429	926.1	240
Butyl alcohol, tert-	—	8.1360	1582.4	218.9
Butylchloride, n-	—	6.7520	1125.8	212
Carbon disulfide	-10–160	6.8514	1122.5	236.5
Carbon tetrachloride	—	6.9339	1242.4	230
Chlorine	—	6.8677	821.1	240
Chlorobenzene	0–42	7.1069	1500	224
Chloroethylbenzene, o-	—	6.9817	1556	201
Chloroform	-30–150	6.90328	1163.0	227.4
Chlorophenol, o-	15–80	7.2420	1668	210
Chlorotoluene, o-	0–65	7.3680	1735.8	230
Cresol, p-	—	7.0059	1493	160
Cyclohexane	-50–200	6.845	1203.5	222.9
Cyclohexene	—	6.8862	1230	224

Constants for Vapor Pressure Calculations (Continued)

Compound	Temperature Range (°C)	A	B	C
Cyclopentane	—	6.8868	1124.2	231.4
Cyclopentene	—	6.9207	1121.8	233.4
Decane	10–80	7.3151	1705.6	212.6
Dibromopropane, 1,3-	0–71	7.5498	1890.56	240
Dichlorobenzene, o-	—	6.9240	1538.3	200
Diethylamine	-30–100	6.8319	1057.2	212.0
Diethylbenzene, o-	—	6.9902	1577.9	200.6
Diethyl ether	—	6.7857	994.2	220.0
Diethyl ketone	—	6.8579	1216.3	204
Diethyl sulfide	0–150	6.9284	1257.8	218.7
Dimethyl ether	—	6.7367	791.18	230.0
Dimethyl formamide	15–60	7.3438	1624.7	216.2
Dipropyl sulfide	0–53	7.2831	1599	222.2
Ethane	—	6.8027	656.4	256
Ethyl acetate	-20–150	7.0981	1238.7	217
Ethyl alcohol	—	8.0449	1554.3	222.7
Ethyl benzene	—	6.9572	1424.3	213.2
Ethyl bromide	-50–130	6.8929	1083.8	231.7
Ethyl chloride	-65–70	6.8027	949.6	230
Ethylene	—	6.7476	585.0	255.0
Ethylene bromide	—	7.0624	1469.7	220.1
Ethylene chloride	—	7.1843	1358.5	232.2
Ethylene oxide	-70–100	7.4078	1181.3	250.6
Ethyl formate	-30–235	7.1170	1176.6	223
Ethyl mercaptan	-40–100	6.9521	1084.5	231.4
Ethyl toluene, p-	—	6.9980	1527.1	209
Fluorobenzene	-40–180	6.9367	1736.4	220.0
Formic acid	—	6.9446	1295.3	218
Furan	-35–90	6.9752	1060.9	227.7
Heptane, n-	—	6.9024	1268.1	216.9
Heptene, 1-	—	6.9007	1257.5	219.2
Hexachloropropene	20–109	7.2664	1863.7	213
Hexane, n-	—	6.8778	1171.5	224.4
Hexene, 1-	—	6.8657	1153.0	225.85
Hydrazine	-10–39	8.2623	1881.6	238.0
Hydrogen cyanide	-40–70	7.2976	1206.8	247.5
Hydrogen fluoride	-55–105	8.3804	1952.6	335.5
Iodine	—	7.2630	1697.9	204
Iodobenzene	—	6.8951	1562.87	201
Methyl acetate	—	7.2021	1232.8	228
Methyl alcohol	-20–140	7.8786	1473.1	230
Methyl amine	-45–50	6.9121	838.1	214.2
Methyl aniline	—	7.2258	1728.2	202
Methylene chlorobromide	-10–155	6.9278	1166.0	220
Methyl ethyl ketone	—	6.9742	1209.6	216
Methyl formate	—	7.1362	1111	229.2
Morpholine	0–44	7.71813	1745.8	235
Naphthalene	—	6.8458	1606.5	187.2
Nonane	-10–60	7.2643	1607.1	217.5
Octane	-20–40	7.3720	1587.8	230.1
Ozone	—	6.7260	566.9	260
Pentane	—	6.8522	1064.6	232
Phenol	—	7.1362	1518.1	175
Phosgene	-68–68	6.8430	941.3	230

Appendix G

Phosphine	—	6.7010	643.7	256
Phosphorous pentachloride	—	9.4274	2422.2	208
Propadiene	-100–40	5.6457	441	194
Propane	—	6.8297	813.2	248
Proprionic acid	0–60	7.7156	1690	210
Propyl acetate	0–170	7.0677	1304.1	210
Propyl alcohol, n-	—	7.9973	1569.7	209.5
Propyl alcohol, iso-	0–113	6.6604	813.0	132.9
Propyl benzene	—	6.9514	1491.3	207.1
Propylene	—	6.8196	785	247
Propylene oxide, 1,2-	-35–130	7.0649	1113.6	232
Selenium dioxide	—	6.5778	1879.8	179
Styrene	—	6.9241	1420	206
Sulfur	—	6.6954	2285.4	155
Sulfur dioxide	—	7.3278	1022.8	240
Tetrachloroethylene	—	7.0200	1415.5	221
Thiophene	-10–180	6.9593	1246	221
Toluene	—	6.9546	1344.8	219.5
Tribromoethane, 1,1,2-	20–60	7.337	1789	215
Trichlorobenzene, 1,2,4-	20–109	7.555	2064	230.1
Trichloroethane, 1,1,2-	—	6.852	1263	205.2
Trichloroethene, 1,1,2-	—	7.028	1315	230
Triethylamine	0–130	6.826	1161	205
Trimethylamine	-50–50	6.816	937.5	235
Trimethylbenzene, 1,2,3-	—	7.041	1594	207.1
Trimethylbenzene, 1,3,5-	—	7.07436	1569.62	209.58
Undecane	15–100	7.3685	1803.90	208.32
Undecene, 1-	—	6.9666	1562.5	189.74
Vinyl chloride	-100–50	6.4971	7.834	230.0
Water	0–60	8.1076	1750.3	235.0
Water	60–150	7.9668	1668.2	228.0
Xylene, m-	—	7.0091	1462.3	215.1

Note: Data taken from Lange, N.A., Ed. *Handbook of Chemistry,* 10th ed. rev. (New York: McGraw-Hill Book Company, Inc., 1967) with permission.

Appendix H

Vapor Pressures of Water at Various Temperatures

Temperature (°F)	(°C)	Vapor Pressure (mmHg)	Temperature (°F)	(°C)	Vapor Pressure (mmHg)	Temperature (°F)	(°C)	Vapor Pressure (mmHg)
32.0	0.0	4.56	93.2	34.0	39.9	154.4	68.0	214.3
33.8	1.0	4.91	95.0	35.0	42.2	156.2	69.0	223.9
35.6	2.0	5.28	96.8	36.0	44.5	158.0	70.0	233.8
37.4	3.0	5.67	98.6	37.0	47.1	159.8	71.0	244.1
39.2	4.0	6.08	100.4	38.0	49.7	161.6	72.0	254.8
41.0	5.0	6.53	102.2	39.0	52.4	163.4	73.0	265.8
42.8	6.0	7.00	104.0	40.0	55.3	165.2	74.0	277.3
44.6	7.0	7.50	105.8	41.0	58.3	167.0	75.0	289.2
46.4	8.0	8.03	107.6	42.0	61.5	168.8	76.0	301.5
48.2	9.0	8.59	109.4	43.0	64.8	170.6	77.0	314.3
50.0	10.0	9.19	111.2	44.0	68.2	172.4	78.0	327.5
51.8	11.0	9.83	113.0	45.0	71.9	174.2	79.0	341.1
53.6	12.0	10.50	114.8	46.0	75.6	176.0	80.0	355.3
55.4	13.0	11.22	116.6	47.0	79.6	177.8	81.0	369.9
57.2	14.0	11.97	118.4	48.0	83.7	179.6	82.0	385.0
59.0	15.0	12.77	120.2	49.0	88.0	181.4	83.0	400.7
60.8	16.0	13.62	122.0	50.0	92.5	183.2	84.0	416.9
62.6	17.0	14.52	123.8	51.0	97.2	185.0	85.0	433.6
64.4	18.0	15.46	125.6	52.0	102.1	186.8	86.0	450.9
66.2	19.0	16.46	127.4	53.0	107.2	188.6	87.0	468.7
68.0	20.0	17.52	129.2	54.0	112.5	190.4	88.0	487.2
69.8	21.0	18.64	131.0	55.0	118.0	192.2	89.0	506.2
71.6	22.0	19.82	132.8	56.0	123.8	194.0	90.0	525.9
73.4	23.0	21.06	134.6	57.0	129.8	195.8	91.0	546.2
75.2	24.0	22.37	136.4	58.0	136.1	197.6	92.0	567.1
77.0	25.0	23.75	138.2	59.0	142.6	199.4	93.0	588.7
78.8	26.0	25.20	140.0	60.0	149.4	201.2	94.0	611.0
80.6	27.0	26.73	141.8	61.0	156.5	203.0	95.0	634.0
82.4	28.0	28.34	143.6	62.0	163.8	204.8	96.0	657.7
84.2	29.0	30.03	145.4	63.0	171.5	206.6	97.0	682.1
86.0	30.0	31.81	147.2	64.0	179.4	208.4	98.0	707.3
87.8	31.0	33.68	149.0	65.0	187.6	210.2	99.0	733.3
89.6	32.0	35.65	150.8	66.0	196.2	212.0	100.0	760.0
91.4	33.0	37.72	152.6	67.0	205.1	213.8	101.0	787.6

Appendix I

Mass of Water Vapor in Saturated Air at 1 atm Pressure

Temperature (°F)	(°C)	Mass (g H$_2$O/ m^3 air)	Temperature (°F)	(°C)	Mass (g H$_2$O/ m^3 air)	Temperature (°F)	(°C)	Mass (g H$_2$O/ m^3 air)
32.0	0.0	4.83	93.2	34.0	37.5	154.4	68.0	181
33.8	1.0	5.17	95.0	35.0	39.5	156.2	69.0	189
35.6	2.0	5.54	96.8	36.0	41.6	158.0	70.0	197
37.4	3.0	5.93	98.6	37.0	43.8	159.8	71.0	205
39.2	4.0	6.34	100.4	38.0	46.1	161.6	72.0	213
41.0	5.0	6.78	102.2	39.0	48.5	163.4	73.0	222
42.8	6.0	7.24	104.0	40.0	51.0	165.2	74.0	231
44.6	7.0	7.73	105.8	41.0	53.6	167.0	75.0	240
46.4	8.0	8.25	107.6	42.0	56.3	168.8	76.0	249
48.2	9.0	8.80	109.4	43.0	59.2	170.6	77.0	259
50.0	10.0	9.38	111.2	44.0	62.1	172.4	78.0	269
51.8	11.0	9.99	113.0	45.0	65.2	174.2	79.0	280
53.6	12.0	10.6	114.8	46.0	68.4	176.0	80.0	291
55.4	13.0	11.3	116.6	47.0	71.8	177.8	81.0	302
57.2	14.0	12.0	118.4	48.0	75.3	179.6	82.0	313
59.0	15.0	12.8	120.2	49.0	78.9	181.4	83.0	325
60.8	16.0	13.6	122.0	50.0	82.7	183.2	84.0	337
62.6	17.0	14.4	123.8	51.0	86.6	185.0	85.0	350
64.4	18.0	15.3	125.6	52.0	90.7	186.8	86.0	363
66.2	19.0	16.3	127.4	53.0	94.9	188.6	87.0	376
68.0	20.0	17.3	129.2	54.0	99.3	190.4	88.0	390
69.8	21.0	18.3	131.0	55.0	103.9	192.2	89.0	404
71.6	22.0	19.4	132.8	56.0	108.6	194.0	90.0	418
73.4	23.0	20.5	134.6	57.0	113.5	195.8	91.0	433
75.2	24.0	21.7	136.4	58.0	118.7	197.6	92.0	448
77.0	25.0	23.0	138.2	59.0	124.0	199.4	93.0	464
78.8	26.0	24.3	140.0	60.0	129.5	201.2	94.0	481
80.6	27.0	25.7	141.8	61.0	135.2	203.0	95.0	497
82.4	28.0	27.2	143.6	62.0	141.2	204.8	96.0	514
84.2	29.0	28.7	145.4	63.0	147.3	206.6	97.0	532
86.0	30.0	30.3	147.2	64.0	153.6	208.4	98.0	550
87.8	31.0	32.0	149.0	65.0	160.2	210.2	99.0	569
89.6	32.0	33.7	150.8	66.0	167.0	212.0	100.0	588
91.4	33.0	35.6	152.6	67.0	174.1	213.8	101.0	608

Appendix J

Relative Humidity from Wet- and Dry-Bulb Thermometer Readings

tD−tW tD	0.2	0.4	0.6	0.8	1.0	1.2	1.4	1.6	1.8	2.0	2.2	2.4	2.6	2.8	3.0	3.2	3.4	3.6
−10	93	87	80	74	67	61	54	48	41	35	28	22	16	9	—	—	—	—
−9	94	88	81	75	69	63	57	51	45	39	33	27	21	15	9	—	—	—
−8	94	88	83	77	71	65	60	54	48	43	37	32	26	20	15	10	—	—
−7	95	89	84	78	73	67	62	57	52	46	41	36	31	25	20	15	10	5
−6	95	90	85	79	74	69	64	59	54	49	45	40	35	30	25	20	15	11
−5	95	90	86	81	76	71	66	62	57	52	48	43	39	34	29	25	20	16
−4	95	91	86	82	77	73	68	64	59	55	51	46	42	38	33	29	25	21
−3	96	91	87	82	78	74	70	66	62	57	53	49	45	41	37	33	29	25
−2	96	92	88	84	79	75	71	68	64	60	56	52	48	44	40	37	33	29
−1	96	92	88	84	81	77	73	69	66	62	58	54	51	47	43	40	36	33
0	96	93	89	85	81	78	74	71	67	64	60	57	53	50	46	43	40	36
1	97	93	90	86	83	80	76	73	70	66	63	59	56	53	49	46	43	40
2	97	93	90	87	84	81	78	74	71	68	65	62	59	55	52	49	46	43
3	97	94	91	88	84	82	78	76	73	70	67	64	61	58	55	52	49	46
4	97	94	91	88	85	82	79	77	74	71	68	65	62	60	57	54	51	48
5	97	94	91	88	86	83	80	77	75	72	69	67	64	61	58	56	53	51
6	97	94	92	89	86	84	81	78	76	73	70	68	65	63	60	58	55	53
7	97	95	92	89	87	84	82	79	77	74	72	69	67	64	62	59	57	54
8	97	95	92	90	87	85	82	80	77	75	73	70	68	65	63	61	58	56
9	98	95	93	90	88	85	83	81	78	76	74	71	69	67	64	62	60	58
10	98	95	93	90	88	86	83	81	79	77	74	72	70	68	66	63	61	59
11	98	96	93	91	89	86	84	82	80	78	75	73	71	69	67	65	62	60
12	98	96	93	91	89	87	85	82	80	78	76	74	72	70	68	66	64	62
13	98	96	93	91	89	87	85	83	81	79	77	75	73	71	69	67	65	63
14	98	96	94	92	90	88	86	84	82	79	78	76	74	72	70	68	66	64
15	98	96	94	92	90	88	86	85	82	80	78	76	74	73	71	69	67	65

Gas Mixtures: Preparation and Control

Relative Humidity from Wet- and Dry-Bulb Thermometer Readings (Continued)

tD−tW / tD	0.5	1.0	1.5	2.0	2.5	3.0	3.5	4.0	4.5	5.0	5.5	6.0	6.5	7.0	7.5	8.0	8.5	9.0	9.5
16	95	90	85	81	76	71	67	63	58	54	50	46	42	38	34	30	26	23	19
17	95	90	86	81	76	72	68	64	60	55	51	47	43	40	36	32	28	25	21
18	95	91	86	82	77	73	69	65	61	57	53	49	45	41	38	34	30	27	23
19	95	91	87	82	78	74	70	65	62	58	54	50	46	43	39	36	32	29	26
20	96	91	87	83	78	74	70	66	63	59	55	51	48	44	41	37	34	31	28
21	96	91	87	83	79	75	71	67	64	60	56	53	49	46	42	39	36	32	29
22	96	92	87	83	80	76	72	68	64	61	57	54	50	47	44	40	37	34	31
23	96	92	88	84	80	76	72	69	65	62	58	55	52	48	45	42	39	36	33
24	96	92	88	84	80	77	73	69	66	62	59	56	53	49	46	43	40	37	34
25	96	92	88	84	81	77	74	70	67	63	60	57	54	50	47	44	41	39	36
26	96	92	88	85	81	78	74	71	67	64	61	58	54	51	49	46	43	40	37
27	96	93	89	85	82	78	75	71	68	65	62	58	56	52	50	47	44	41	38
28	96	93	89	85	82	78	75	72	69	65	62	59	56	53	51	48	45	42	40
29	96	93	89	86	82	79	76	72	69	66	63	60	57	54	52	49	46	43	41
30	96	93	89	86	83	79	76	73	70	67	64	61	58	55	52	50	47	44	42
31	96	93	90	86	83	80	77	73	70	67	64	61	59	56	53	51	48	45	43
32	96	93	90	86	83	80	77	74	71	68	65	62	60	57	54	51	49	46	44
33	96	93	90	87	83	80	77	74	71	68	66	63	60	57	55	52	50	47	45
34	97	93	90	87	84	81	78	75	72	69	66	63	61	58	56	53	51	48	46
35	97	94	90	87	84	81	78	75	72	69	67	64	61	59	56	54	51	49	47
36	97	94	90	87	84	81	78	75	73	70	67	64	62	59	57	54	52	50	48
37	97	94	91	87	84	82	79	76	73	70	68	65	63	60	58	55	53	51	48
58	97	94	91	88	84	82	79	76	74	71	68	66	63	61	58	56	54	51	49
39	97	94	91	88	85	82	79	77	74	71	69	66	64	61	59	57	54	52	50
40	97	94	91	88	85	82	80	77	74	72	69	67	64	62	59	57	54	53	51

Appendix J

tD-tW \ tD	3.8	4.0	4.5	5.0	5.5	6.0	6.5	7.0	7.5	8.0	8.5	9.0	9.5	10.0	10.5	11.0
−10	—	—	—	—	—	—	—	—	—	—	—	—	—	—	—	—
−9	—	—	—	—	—	—	—	—	—	—	—	—	—	—	—	—
−8	—	—	—	—	—	—	—	—	—	—	—	—	—	—	—	—
−7	—	—	—	—	—	—	—	—	—	—	—	—	—	—	—	—
−6	6	7	—	—	—	—	—	—	—	—	—	—	—	—	—	—
−5	11	12	—	—	—	—	—	—	—	—	—	—	—	—	—	—
−4	17	17	8	—	—	—	—	—	—	—	—	—	—	—	—	—
−3	21	22	12	8	—	—	—	—	—	—	—	—	—	—	—	—
−2	25	26	17	13	5	—	—	—	—	—	—	—	—	—	—	—
−1	29	29	21	17	10	7	—	—	—	—	—	—	—	—	—	—
0	33	33	25	22	14	12	5	—	—	—	—	—	—	—	—	—
1	36	37	29	26	19	16	9	7	—	—	—	—	—	—	—	—
2	40	40	33	29	22	20	13	11	5	—	—	—	—	—	—	—
3	43	43	36	33	26	24	17	15	10	8	—	—	—	—	—	—
4	46	45	39	35	29	26	21	19	14	12	7	—	—	—	—	—
5	48	48	41	38	32	29	24	22	17	15	10	6	—	—	—	—
6	50	50	44	40	35	32	27	24	20	18	13	9	5	—	—	—
7	52	51	46	42	37	34	27	27	22	21	16	12	8	—	—	—
8	54	53	48	44	39	36	32	29	25	23	19	15	11	7	—	—
9	55	55	50	46	41	39	34	32	28	26	22	18	14	10	6	—
10	57	56	51	48	43	41	36	34	30	27	24	20	16	13	9	6
11	58	58	53	50	45	42	38	36	32	—	—	—	—	—	—	—
12	60	59	54	51	47	44	40	—	—	—	—	—	—	—	—	—
13	61	60	56	53	48	—	—	—	—	—	—	—	—	—	—	—
14	62	61	57	—	—	—	—	—	—	—	—	—	—	—	—	—
15	63	—	—	—	—	—	—	—	—	—	—	—	—	—	—	—

Relative Humidity from Wet— and Dry—Bulb Thermometer Readings (Continued)

tD–tW / tD	10.0	10.5	11.0	11.5	12.0	12.5	13.0	13.5	14.0	14.5	15.0	16.0	17.0	18.0	19.0	20.0
16	15	12	8	5	—	—	—	—	—	—	—	—	—	—	—	—
17	18	14	11	8	—	—	—	—	—	—	—	—	—	—	—	—
18	20	17	14	10	7	—	—	—	—	—	—	—	—	—	—	—
19	22	19	16	13	10	7	—	—	—	—	—	—	—	—	—	—
20	24	21	18	15	12	9	6	—	—	—	—	—	—	—	—	—
21	26	23	20	17	14	12	9	6	—	—	—	—	—	—	—	—
22	28	25	22	19	17	14	11	8	6	—	—	—	—	—	—	—
23	30	27	24	21	19	16	13	11	8	6	—	—	—	—	—	—
24	31	29	26	23	20	18	15	13	10	8	5	—	—	—	—	—
25	33	30	28	25	22	20	17	15	12	10	8	—	—	—	—	—
26	34	32	29	26	24	21	19	17	14	12	10	5	—	—	—	—
27	36	33	31	28	26	23	21	18	16	14	12	7	—	—	—	—
28	37	34	32	29	27	25	22	20	18	16	13	9	5	—	—	—
29	38	36	33	31	28	26	24	22	19	17	15	11	7	—	—	—
30	39	37	35	32	30	28	25	23	21	19	17	13	9	5	—	—
31	40	38	36	33	31	29	27	25	22	20	18	14	11	7	5	—
32	41	39	37	35	32	30	28	26	24	22	20	16	12	9	7	5
33	42	40	38	36	33	31	29	27	25	23	21	17	14	10	8	7
34	43	41	39	37	35	32	30	28	26	24	23	19	15	12	10	8
35	44	42	40	38	36	34	32	30	28	26	24	20	17	13	11	10
36	45	43	41	39	37	35	33	31	29	27	25	21	18	15	13	11
37	46	44	42	40	38	36	34	32	30	28	26	23	19	16	14	12
38	47	45	43	41	39	37	35	33	31	29	27	24	20	17	15	13
39	48	46	43	42	39	38	36	34	32	30	28	25	22	18	15	12
40	48	46	44	42	40	38	36	35	33	31	29	26	23	20	16	14

Note: Data taken from U.S. Weather Bureau 1971. With permission.

[a] This table gives the approximate relative humidity directly from a reading of the dry-bulb temperature, tD, and wet-bulb temperature, tW (both in °C) at a pressure of 743 mmHg.

Appendix K

Gas Flow Conversion Factors for Mass Flow Controllers

Gas	Test Gas	Conversion Factor Relative to Test Gas	Conversion Factor Relative to Nitrogen	Specific Heat, Cp cal/g °C
Acetylene	C_2H_4	0.97	0.58	0.4036
Air	N_2	1.00	1.00	0.240
Allene (propadiene)	$CHCLF_2$	0.95	0.43	0.352
Ammonia	N_2O	1.03	0.74	0.492
Argon	Ar	1.00	1.42	0.1244
Arsine	N_2O	0.95	0.67	0.1167
Boron tribromide	CCL_2F_2	1.07	0.38	0.0674
Boron trichloride	$CHCLF_2$	0.89	0.41	0.1279
Boron trichloride	$CHCLF_2$	1.11	0.51	0.1778
Bromine	N_2O	1.14	0.81	0.0539
Bromine pentafluoride	CCL_2F_2	0.72	0.26	0.1369
Bromine trifluoride	CCL_2F_2	1.09	0.39	0.1161
Bromotrifluoromethane	CCL_2F_2	1.04	0.37	0.1113
Butadiene, 1,3-	CCL_2F_2	0.91	0.32	0.3514
Butane	CCL_2F_2	0.74	0.26	0.4007
Butene, 1-	CCL_2F_2	0.85	0.30	0.3648
Butene, cis-2-	CCL_2F_2	0.92	0.33	0.336
Butene, trans-2-	CCL_2F_2	0.82	0.29	0.374
Carbon dioxide	N_2O	1.04	0.74	0.2016
Carbon disulfide	C_2H_4	1.00	0.60	0.1428
Carbon monoxide	N_2	1.00	1.00	0.2488
Carbon tetrachloride	CCL_2F_2	0.88	0.31	0.128
Carbon tetrafluoride	$CHCLF_2$	0.92	0.42	0.1654
Carbonyl fluoride	C_2H_4	0.91	0.55	0.171
Carbonyl sulfide	N_2O	0.93	0.66	0.1651
Chlorine	N_2	0.86	0.86	0.1144
Chlorine trifluoride	$CHCLF_2$	0.87	0.40	0.165
Chlorodifluoromethane	$CHCLF_2$	1.00	0.46	0.1544
Chloroform	CCL_2F_2	1.11	0.39	0.1309
Chloropentafluoroethane	CCL_2F_2	0.68	0.24	0.164

Gas Flow Conversion Factors for Mass Flow Controllers (Continued)

Gas	Test Gas	Conversion Factor Relative to Test Gas	Conversion Factor Relative to Nitrogen	Specific Heat, Cp cal/g °C
Chlorotrifluoromethane	CCl$_2$F$_2$	1.08	0.38	0.153
Cyanogen	C$_2$H$_4$	0.98	0.45	0.2613
Cyanogen chloride	C$_2$H$_4$	1.02	0.61	0.1739
Cyclopropane	CHClF$_2$	1.00	0.46	0.3177
Deuterium	N$_2$	1.00	1.00	1.722
Diborane	CHClF$_2$	0.95	0.44	0.508
Dibromodifluoromethane	CCL$_2$F$_2$	0.55	0.19	0.150
Dibromomethane	CHClF$_2$	1.02	0.47	0.075
Dichlorodifluoromethane	CCL$_2$F$_2$	1.00	0.35	0.1432
Dichlorofluoromethane	CHCL$_2$F	0.93	0.42	0.140
Dichloromethylsilane	CCL$_2$F$_2$	0.71	0.25	0.1882
Dichlorosilane	CHClF$_2$	0.88	0.40	0.150
Dichlorotetrafluoroethane, 1,2-	CCL$_2$F$_2$	0.63	0.22	0.160
Difluoroethylene, 1,1-	CHClF$_2$	0.93	0.43	0.224
Dimethyl ether	CCl$_2$F$_2$	1.10	0.39	0.3414
Dimethylamine	CCl$_2$F$_2$	1.05	0.37	0.366
Dimethylpropane, 2,2-	CCl$_2$F$_2$	0.61	0.22	0.3914
Disilane	CCl$_2$F$_2$	0.89	0.32	0.310
Ethane	CHClF$_2$	1.08	0.50	0.4097
Ethanol	CCl$_2$F$_2$	1.11	0.39	0.3395
Ethyl acetylene	CCl$_2$F$_2$	0.91	0.32	0.3513
Ethyl chloride	CCl$_2$F$_2$	1.10	0.39	0.244
Ethylene	C$_2$H$_4$	1.00	6.00	0.365
Ethylene oxide	CHClF$_2$	1.13	0.52	0.268
Fluorine	N$_2$	0.98	0.98	0.1873
Fluoroform	CHClF$_2$	1.08	0.50	0.176
Freon 113	CCl$_2$F$_2$	0.57	0.20	0.161
Freon 1132A	CHClF$_2$	0.93	0.43	0.224
Freon 114	CCL$_2$F$_2$	0.63	0.22	0.160
Freon 115	CCL$_2$F$_2$	0.68	0.24	0.164
Freon 116	CCL$_2$F$_2$	0.68	0.24	0.1843
Freon 12	CCL$_2$F$_2$	1.00	0.35	0.1432
Freon 13	CCL$_2$F$_2$	1.08	0.38	0.153
Freon 13B1	CCL$_2$F$_2$	1.04	0.37	0.1113
Freon 14	CHClF$_2$	0.92	0.42	0.1654
Freon 21	CHCL$_2$F	0.93	0.42	0.140
Freon 22	CHClF$_2$	1.00	0.46	0.1544
Freon 23	CHClF$_2$	1.08	0.50	0.176
Freon C318	CCL$_2$F$_2$	0.47	0.17	0.185
Germane	C$_2$H$_4$	0.95	0.57	0.1404
Germanium tetrachloride	CCL$_2$F$_2$	0.75	0.27	0.1071
Helium	He	1.00	1.42	1.241
Hexafluoroethane	CCL$_2$F$_2$	0.68	0.24	0.1843
Hexane	CHClF$_2$	0.51	0.18	0.3968
Hydrogen	H$_2$	1.00	1.01	3.419
Hydrogen bromide	N$_2$	1.00	1.00	0.0861

Appendix K

Hydrogen chloride	N_2	1.00	1.00	0.1912
Hydrogen cyanide	N_2O	1.07	0.76	0.3171
Hydrogen fluoride	H_2	1.00	1.00	0.3479
Hydrogen iodide	N_2	1.00	1.00	0.0545
Hydrogen selenide	N_2O	1.11	0.79	0.1025
Hydrogen sulfide	N_2O	1.13	0.80	0.2397
Iodine pentafluoride	CCL_2F_2	0.70	0.25	0.1108
Isobutane	CCL_2F_2	0.56	0.20	0.3872
Isobutylene	CCL_2F_2	0.83	0.30	0.3701
Krypton	Ar	1.00	1.41	0.0593
Methane	N_2O	1.01	0.72	0.5328
Methanol	C_2H_4	0.98	0.58	0.3274
Methyl acetylene	$CHCLF_2$	0.94	0.43	0.3547
Methyl bromide	C_2H_4	0.97	0.58	0.1106
Methyl chloride	C_2H_4	1.05	0.63	0.1926
Methyl fluoride	C_2H_4	0.93	0.56	0.3221
Methyl mercaptan	$CHCLF_2$	1.13	0.52	0.2459
Methyl trichlorosilane	CCL_2F_2	0.71	0.25	0.164
Molybdenum hexafluoride	CCL_2F_2	0.60	0.21	0.1373
Monoethylamine	CCL_2F_2	0.99	0.35	0.387
Monomethylamine	$CHCLF_2$	0.99	0.45	0.4343
Neon	Ar	1.00	1.42	0.246
Nitric oxide	N_2	1.00	1.00	0.2328
Nitrogen	N_2	1.00	1.00	0.2485
Nitrogen dioxide	N_2O	1.03	0.74	0.1933
Nitrogen trifluoride	$CHCLF_2$	1.05	0.48	0.1797
Nitrosyl chloride	C_2H_4	1.02	0.61	0.1632
Nitrous oxide	N_2O	1.00	0.71	0.2088
Octafluorocyclobutane	CCL_2F_2	0.47	0.17	0.185
Oxygen	N_2	1.00	1.00	0.2193
Oxygen difluoride	C_2H_4	1.06	0.63	0.1917
Pentaborane	CCL_2F_2	0.72	0.26	0.380
Pentane	CCL_2F_2	0.60	0.21	0.398
Perchloryl fluoride	CCL_2F_2	1.16	0.39	0.1514
Perfluoropropane	CCL_2F_2	0.47	0.17	0.194
Phosgene	$CHCLF_2$	0.97	0.44	0.1394
Phosphine	N_2O	1.07	0.76	0.2374
Phosphorous oxychloride	CCL_2F_2	0.85	0.30	0.1324
Phosphorous pentafluoride	CCL_2F_2	0.85	0.30	0.161
Phosphorous trichloride	CCL_2F_2	1.01	0.36	0.125
Propane	CCL_2F_2	1.01	0.36	0.3885
Propylene	$CHCLF_2$	0.90	0.41	0.3541
Silane	C_2H_4	1.00	0.60	0.3189
Silicon tetrachloride	CCL_2F_2	0.80	0.28	0.127
Silicon tetrafluoride	CCL_2F_2	0.98	0.35	0.1691
Sulfur dioxide	N_2O	0.96	0.69	0.1488
Sulfur hexafluoride	CCL_2F_2	0.74	0.26	0.1592
Sulfur tetrafluoride	CCL_2F_2	1.00	0.36	0.1593
Sulfuryl fluoride	CCL_2F_2	1.10	0.39	0.1543
Tetrafluorohydrazine	CCL_2F_2	0.91	0.32	0.182
Titanium tetrachloride	CCL_2F_2	0.76	0.27	0.120
Trichloroethane	CCL_2F_2	0.78	0.28	0.1654
Trichloroethylene	CCL_2F_2	0.91	0.32	0.1459
Trichlorofluoromethane	CCL_2F_2	0.93	0.33	0.1357

Gas Flow Conversion Factors for Mass Flow Controllers (Continued)

Gas	Test Gas	Conversion Factor Relative to Test Gas	Conversion Factor Relative to Nitrogen	Specific Heat, Cp cal/g °C
Trichlorosilane	CCL_2F_2	0.93	0.33	0.138
Trichlorotrifluoroethane	CCL_2F_2	0.57	0.20	0.161
Triisobutyl aluminum	CCL_2F_2	0.17	0.06	0.508
Trimethylamine	CCL_2F_2	0.79	0.28	0.371
Tungsten hexafluoride	CCL_2F_2	0.54	0.19	0.1079
Uranium hexafluoride	CCL_2F_2	0.55	0.20	0.0888
Vinyl bromide	$CHCLF_2$	1.01	0.46	0.1241
Vinyl chloride	$CHCLF_2$	1.04	0.48	0.2054
Water				0.445
Xenon	Ar	1.00	1.42	0.0378

Appendix L

Table of Atomic Volumes Used for Diffusion Coefficient Calculation

$$D = 0.0043\ T^{1.5}\ (A/B)/P$$
$$A = [(1/Ma) + (1/Mb)]^{1/2}$$
$$B = (Va^{1/3} + Vb^{1/3})^2$$

Sample calculation for nitrobenzene, $C_6H_5NO_2$.

Fill in the number of each element and calculate the molecular weight and atomic volume.

Atm press (P)	1 atm	
MW air (Ma)	28.9 g/mol	
MW vapor (Mb)	123.0 g/mol	
Temperature	25°C	

At vol air (Va)	29.9 mL/g-mol
At vol vap (Vb)	124.5 mL/g-mol
A = 0.2066	B = 65.5630
298°K	

Calculated diffusion coefficient 0.0697 cm²/sec

Number	Element	Atomic Volume per Atom (mL/g-mol)	Total Atomic Volume (mL/g-mol)	Atomic Weight (g/mol)	Total Molecular Weight (g/mol)
	Air	29.9	0.0	29.0	0.0
	Ar	30.5	0.0	74.9	0.0
	Bi	48.0	0.0	209.0	0.0
	Br	79.9	0.0	79.9	0.0
6	C	14.8	88.8	12.0	72.0
	Cl Terminal R-Cl	21.6	0.0	35.5	0.0
	Cl Medial RCH-Cl-R	24.6	0.0	35.5	0.0
	Cr	27.4	0.0	52.0	0.0
	F	8.7	0.0	19.0	0.0
	Ge	34.5	0.0	72.6	0.0
5	H	3.7	18.5	1.0	5.0
	I	37.0	0.0	126.9	0.0
1	N	15.6	15.6	14.0	14.0
	N Primary amines	10.5	0.0	14.0	0.0

Table of Atomic Volumes Used for Diffusion Coefficient Calculation (Continued)

Number	Element	Atomic Volume per Atom (mL/g-mol)	Total Atomic Volume (mL/g-mol)	Atomic Weight (g/mol)	Total Molecular Weight (g/mol)
	N Secondary amines	12.0	0.0	14.0	0.0
	O Double bonded	7.4	0.0	16.0	0.0
Oxygen coupled to two other elements					
	O R-O-H, alcohols	7.4	0.0	16.0	0.0
	O R-CHO, R-CO-R	7.4	0.0	16.0	0.0
	O CH$_3$COOR Me esters	9.1	0.0	16.0	0.0
	O CH$_3$-O-R Me ethers	9.9	0.0	16.0	0.0
	O Higher esters	11.0	0.0	16.0	0.0
	O Higher ethers	11.0	0.0	16.0	0.0
	O RCOOH acids	12.0	0.0	16.0	0.0
2	O In union with S,P,N	8.3	16.6	16.0	32.0
	P	27.0	0.0	31.0	0.0
	Sn	42.3	0.0	118.7	0.0
	S	25.6	0.0	32.1	0.0
	Sb	34.2	0.0	121.8	0.0
	Si	32.0	0.0	28.1	0.0
	Ti	35.7	0.0	47.9	0.0
	V	32.0	0.0	50.9	0.0
	Zn	20.4	0.0	65.4	0.0
	Water	18.8	0.0	18.0	0.0
	3-Membered ring	-3.0	0.0	0.0	0.0
	4-Membered ring	-8.5	0.0	0.0	0.0
	5-Membered ring	-11.5	0.0	0.0	0.0
1	6-Membered ring	-15.0	-15.0	0.0	0.0
	Naphthalene ring	-30.0	0.0	0.0	0.0
	Anthracene ring	-47.5	0.0	0.0	0.0
	Total		124.5 mL/g-mol		123.0 g/mol

Note: Data taken from LeBas, *The Molecular Volumes of Liquid Chemical Compounds*, (London: Longmans, 1915).

Appendix M

Density of Dry Air as a Function of Altitude at 20°C

Altitude		Pressure			Density
ft	m	Atm	mmHg	mbar	g/L
0	0	1.00	760	1013	1.205
100	30	0.996	757	1009	1.200
200	61	0.993	755	1006	1.197
300	91	0.989	752	1002	1.192
400	122	0.986	749	999	1.188
500	152	0.982	746	995	1.183
600	183	0.979	744	992	1.180
700	213	0.975	741	988	1.175
800	244	0.971	738	984	1.170
900	274	0.968	736	981	1.166
1000	305	0.964	733	977	1.162
1100	335	0.961	730	973	1.158
1200	366	0.957	727	969	1.153
1300	396	0.954	725	966	1.150
1400	427	0.950	722	962	1.145
1500	457	0.947	720	959	1.141
1600	488	0.944	717	956	1.138
1700	518	0.940	714	952	1.133
1800	549	0.937	712	949	1.129
1900	579	0.933	709	945	1.124
2000	610	0.930	707	942	1.121
2100	640	0.926	704	938	1.116
2200	671	0.923	701	935	1.112
2300	701	0.920	699	932	1.109
2400	732	0.916	696	928	1.104
2500	762	0.913	694	925	1.100
2600	792	0.909	691	921	1.095
2700	823	0.906	689	918	1.092
2800	853	0.903	686	915	1.088
2900	884	0.899	683	911	1.083
3000	914	0.896	681	908	1.080
3200	975	0.890	676	902	1.072

Density of Dry Air as a Function of Altitude at 20°C

Altitude		Pressure			Density
ft	m	Atm	mmHg	mbar	g/L
3400	1036	0.883	671	894	1.064
3600	1097	0.877	667	888	1.057
3800	1158	0.870	661	881	1.048
4000	1219	0.864	657	875	1.041
4200	1280	0.857	651	868	1.033
4400	1341	0.851	647	862	1.025
4600	1402	0.845	642	856	1.018
4800	1463	0.838	637	849	1.010
5000	1524	0.832	632	843	1.003
5200	1585	0.826	628	837	0.995
5400	1646	0.820	623	831	0.988
5600	1707	0.814	619	825	0.981
5800	1768	0.807	613	817	0.972
6000	1829	0.801	609	811	0.965
6500	1981	0.786	597	796	0.947
7000	2134	0.772	587	782	0.930
7500	2286	0.757	575	767	0.912
8000	2438	0.743	565	753	0.895
8500	2591	0.729	554	738	0.878
9000	2743	0.715	543	724	0.862
9500	2896	0.701	533	710	0.845
10000	3048	0.688	523	697	0.829
15000	4572	0.564	429	571	0.680
20000	6096	0.460	350	466	0.554
25000	7620	0.371	282	376	0.447
30000	9144	0.297	226	301	0.358
35000	10668	0.235	179	238	0.283
40000	12192	0.185	141	187	0.223

Note: Data taken from Caplan, K. J. *Am. Ind. Hyg. Assoc. J.* 46:B-10 (1985) with permission.

Appendix N

Equilibrium Relative Humidity of Selected Saturated Salt Solutions

Temperature (°C)	0	5	10	15	20	25	30	35	40	45	50
Saturated salt											
Cesium fluoride		5.52	4.89	4.33	3.83	3.39	3.01	2.69	2.44	2.24	2.11
Lithium bromide	7.75	7.43	7.14	6.86	6.61	6.37	6.16	5.97	5.80	5.65	5.53
Zinc bromide		8.86	8.49	8.19	7.94	7.75	7.62	7.55	7.54	7.59	7.70
Potassium hydroxide		14.34	12.34	10.68	9.32	8.23	7.38	6.73	6.26	5.94	5.72
Sodium hydroxide				9.57	8.91	8.24	7.58	6.92	6.26	5.60	4.94
Lithium chloride	11.23	11.26	11.29	11.30	11.31	11.30	11.28	11.25	11.21	11.16	11.10
Calcium bromide			21.62	20.20	18.50	16.50					
Lithium iodide		21.68	20.61	19.57	18.56	17.56	16.57	15.57	14.55	13.49	12.38
Potassium acetate						22.51	21.61				
Potassium fluoride			23.38	23.40	23.11						
Magnesium chloride	33.66	33.60	33.47	33.30	33.07	32.78	32.44	32.05	31.60	31.10	30.54
Sodium iodide		42.42	41.83	40.88	39.65	38.17	36.15	34.73	32.88	31.02	29.21
Potassium carbonate	43.13	43.13	43.14	43.15	43.16	43.16	43.17				
Magnesium nitrate	60.35	58.86	57.36	55.87	54.38	52.89	51.40	49.91	48.42	46.93	45.44
Sodium bromide		63.51	62.15	60.68	59.14	57.57	56.03	54.55	53.17	51.95	50.93
Cobalt chloride						64.92	61.83	58.63	55.48	52.56	50.01
Potassium iodide		73.30	72.11	70.98	69.90	68.86	67.89	66.96	66.09	65.26	64.49
Strontium chloride		77.13	75.66	74.13	72.52	70.85	69.12				
Sodium nitrate		78.57	77.53	76.46	75.36	74.25	73.14	72.06	71.00	69.99	69.04
Sodium chloride	75.51	75.65	75.67	75.61	75.47	75.29	75.09	74.87	74.68	74.52	74.43
Ammonium chloride			80.55	79.89	79.23	78.57	77.90				
Potassium bromide		85.09	83.75	82.62	81.67	80.89	80.27	79.78	79.43	79.18	79.02
Ammonium sulfate	82.27	82.42	82.06	81.70	81.34	80.99	80.63	80.27	79.91	79.56	79.20

Equilibrium Relative Humidity of Selected Saturated Salt Solutions (Continued)

Temperature (°C)	0	5	10	15	20	25	30	35	40	45	50
Potassium chloride	88.61	87.67	86.77	85.92	85.11	84.34	83.62	82.95	82.32	81.74	81.20
Strontium nitrate		92.38	90.55	88.72	86.89	85.06					
Potassium nitrate	96.33	96.27	95.96	95.41	94.62	93.58	92.31	90.79	89.03	87.03	84.78
Potassium sulfate	98.77	98.48	98.18	97.89	97.59	97.30	97.00	96.71	96.41	96.12	95.82
Potassium chromate						97.88	97.08	96.42	95.89	95.50	95.25

Note: Data taken from Greenspan, L. J. *Resear. Nat. Bur. Stds.*, 81A:89 (1977) with permission.

Appendix O

Characteristics of Commercially Available Permeation Sources

Permeating Material	Rate (ng/min/cm)	Temp (°C)	K (25°C)	Life (mo)	Mol Wt (g/mol)	Source
Acetaldehyde	33	30	0.555		44.1	Metronics
Acetaldehyde	45	30	0.555	24	44.1	Kin-Tek
Acetaldehyde	235	30	0.555	11	44.1	AID
Acetaldehyde	280	30	0.555	8	44.1	Kin-Tek
Acetaldehyde	360	30	0.555		44.1	Metronics
Acetic acid	53	30	0.407	67	60.1	AID
Acetic acid	340	60	0.407	9	60.1	Kin-Tek
Acetic acid	3450	100	0.407		60.1	Metronics
Acetic anhydride	239	70	0.239	15	102.1	AID
Acetone	270	70	0.421	10	58.1	AID
Acetone	330	50	0.421	9	58.1	Kin-Tek
Acetone	330	50	0.421		58.1	Metronics
Acetonitrile	74	30	0.595	24	41.1	Kin-Tek
Acrolein	26	30	0.436	6	56.1	Kin-Tek
Acrolein	160	30	0.436	6	56.1	Kin-Tek
Acrolein	164	30	0.436	17	56.1	AID
Acrylonitrile	50	50	0.460		53.1	Metronics
Acrylonitrile	90	30	0.460	30	53.1	AID
Acrylonitrile	130	90	0.460		53.1	Metronics
Acrylonitrile	166	40	0.460	6	53.1	Kin-Tek
Acrylonitrile	550	60	0.460	6	53.1	Kin-Tek
Allyl alcohol	130	90	0.421		58.1	Metronics
Allyl chloride	690	50	0.320		76.5	Metronics
Allyl chloride	1710	90	0.320		76.5	Metronics
Allyl sulfide	48	70	0.214	62	114.2	AID
Ammonia	135	30	1.436	24	17.0	Kin-Tek
Ammonia	210	30	1.436	4.7	17.0	AID
Ammonia	290	30	1.436		17.0	Metronics
Ammonia	300	35	1.436	3.3	17.0	AID
Ammonia	350	30	1.436	3	17.0	Kin-Tek
Ammonia	470	40	1.436	2.1	17.0	AID

Characteristics of Commercially Available Permeation Sources (Continued)

Permeating Material	Rate (ng/min/cm)	Temp (°C)	K (25°C)	Life (mo)	Mol Wt (g/mol)	Source
Aniline	35	70	0.263	99	93.1	AID
Benzaldehyde	240	80	0.230		106.1	Metronics
Benzene	34	30	0.313	88	78.1	AID
Benzene	240	50	0.313		78.1	Metronics
Benzene	260	70	0.313	11	78.1	AID
Benzene	371	60	0.313	7	78.1	Kin-Tek
Benzene	1297	80	0.313	2	78.1	Kin-Tek
Bromine	37	30	0.154		158.8	Metronics
Bromine	40	30	0.154	60	158.8	Kin-Tek
Bromine	240	30	0.154	42	158.8	AID
Bromine	300	30	0.154	36	158.8	Kin-Tek
Bromine	400	30	0.154		158.8	Metronics
Bromoform	127	70	0.097	77	252.8	AID
Bromomethane	225	30	0.258	26	94.9	AID
Butadiene	177	30	0.452	12	54.1	AID
Butane	24	30	0.421	82	58.1	AID
Butane	106	50	0.421	7	58.1	Kin-Tek
Butane	346	30	0.421	5	58.1	Kin-Tek
Butane	400	30	0.421		58.1	Metronics
Butanol, iso-	50	80	0.330	10	74.1	Kin-Tek
Butanol, iso-	240	100	0.330	5	74.1	Kin-Tek
Butanol, n-	94	80	0.330	27	74.1	Kin-Tek
Butanol, n-	340	100	0.330	7	74.1	Kin-Tek
Butene, 1-	87	30	0.436	23	56.1	AID
Butene, cis-	57	30	0.436	37	56.1	AID
Butene, trans-	140	30	0.436	14.6	56.1	AID
Butyl acetate	453	80	0.210	4	116.2	Kin-Tek
Butyl acetate	1365	100	0.210	2	116.2	Kin-Tek
Butyl acetate, iso-	314	80	0.210	8	116.2	Kin-Tek
Butyl amine	326	70	0.334	7.7	73.1	AID
Butyl glycidyl ether	230	90	0.188		130.2	Metronics
Butyl mercaptan, n-	75	50	0.271		90.2	Metronics
Butyl mercaptan, n-	490	80	0.271	5	90.2	Kin-Tek
Butyl mercaptan, sec-	35	80	0.271		90.2	Metronics
Butyl mercaptan, tert-	50	70	0.271	54	90.2	AID
Butyl mercaptan, t-	170	80	0.271		90.2	Metronics
Butyraldehyde	165	60	0.339	12	72.1	Kin-Tek
Carbon disulfide	65	30	0.321		76.1	Metronics
Carbon disulfide	75	30	0.321	12	76.1	Kin-Tek
Carbon disulfide	140	30	0.321	30.6	76.1	AID
Carbon disulfide	500	30	0.321	8.6	76.1	AID
Carbon disulfide	680	30	0.321		76.1	Metronics
Carbon disulfide	735	30	0.321	5	76.1	Kin-Tek
Carbon tetrachloride	12	50	0.159		153.8	Metronics
Carbon tetrachloride	27	30	0.159	50	153.8	Kin-Tek
Carbon tetrachloride	190	50	0.159		153.8	Metronics
Carbon tetrachloride	220	70	0.159	24.5	153.8	AID
Carbon tetrachloride	225	30	0.159	23	153.8	AID
Carbon tetrachloride	339	60	0.159	12	153.8	Kin-Tek

Appendix O

Carbonyl sulfide	300	30	0.407		60.1	Metronics
Carbonyl sulfide	1200	30	0.407		60.1	Metronics
Carbonyl sulfide	1500	30	0.407	15	60.1	AID
Chlorine	320	30	0.345	72	70.9	AID
Chlorine	825	30	0.345	5	70.9	Kin-Tek
Chlorine	940	30	0.345		70.9	Metronics
Chlorine	1300	30	0.345	1	70.9	Kin-Tek
Chlorine	1400	30	0.345	1.4	70.9	AID
Chlorine	1500	30	0.345	14	70.9	AID
Chlorine	2000	30	0.345		70.9	Metronics
Chlorine	7000	30	0.345	3.0	70.9	AID
Chloroacetophenone, a-	36	80	0.158		154.6	Metronics
Chlorobenzene	150	50	0.217		112.6	Metronics
Chlorobenzene	280	60	0.217	12	112.6	Kin-Tek
Chlorobenzene	780	80	0.217	4	112.6	Kin-Tek
Chloroethane	30	30	0.379		64.5	Metronics
Chloroethane	56	30	0.379	54	64.5	AID
Chloroform	95	30	0.205	53	119.4	AID
Chloroform	125	30	0.205	12	119.4	Kin-Tak
Chloroform	640	50	0.205		119.4	Metronics
Chloroform	713	70	0.205	7.1	119.4	AID
Chloroform	1085	60	0.205	4	119.4	Kin-Tek
Chloromethane	200	30	0.484	16.9	50.5	AID
Chloromethane	330	30	0.484		50.5	Metronics
Chloromethane	607	30	0.484	5.6	50.5	AID
Chloromethane	3900	30	0.484		50.5	Metronics
Chloroprene	40	70	0.276	81	88.5	AID
Chlorotoluene, o-	47	50	0.193		126.6	Metronics
Cyclohexane	20	70	0.290	132	84.2	AID
Cyclohexane	20	50	0.290		84.2	Metronics
Cyclohexane	260	80	0.290		84.2	Metronics
Cyclopropane	130	30	0.581		42.1	Metronics
Decane, n-	58	70	0.172		142.3	Metronics
Diaminoethane, 1,2-	448	70	0.407	10	60.1	AID
Dibromoethane, 1,2-	115	70	0.130	64	187.9	AID
Dichlorobenzene, m-	1080	90	0.166		147.0	Metronics
Dichlorobenzene, o-	270	80	0.166	12	147.0	Kin-Tek
Dichlorobenzene, o-	640	90	0.166		147.0	Metronics
Dichlorobenzene, o-	860	100	0.166	4	147.0	Kin-Tek
Dichlorobenzene, p-	1370	90	0.166		147.0	Metronics
Dichlorodifluoromethane	1000	30	0.202	24	120.9	AID
Dichloroethane, 1,1-	265	30	0.247	15	99.0	AID
Dichloroethane, 1,1-	270	50	0.247		99.0	Metronics
Dichloroethane, 1,2-	69	30	0.247	60	99.0	AID
Dichloroethane, 1,2-	260	70	0.247	16	99.0	AID
Dichloroethane, 1,2-	320	50	0.247		99.0	Metronics
Dichloroethylene, 1,1-	53	30	0.252		96.9	Metronics
Dichloroethylene, 1,1-	79	30	0.252	52	96.9	AID
Dichloroethylene, 1,1-	880	30	0.252		96.9	Metronics
Dichloroethylene, 1,2-cis-	270	30	0.252	29	96.9	AID
Dichloroethylene, 1,2-trans	145	30	0.252	29	96.9	AID
Dichloroethylene, 1,2-trans	1310	30	0.252		96.9	Metronics
Dichlorofluoromethane	900	30	0.238	5.1	102.9	AID
Dichloromethane	60	30	0.288	75	84.9	AID

Characteristics of Commercially Available Permeation Sources (Continued)

Permeating Material	Rate (ng/min/cm)	Temp (°C)	K (25°C)	Life (mo)	Mol Wt (g/mol)	Source
Dichloromethane	300	30	0.288	15	84.9	AID
Dichloromethane	390	30	0.288	10	84.9	Kin-Tek
Dichloromethane	420	30	0.288		84.9	Metronics
Dichloromethane	1600	70	0.288	2.8	84.9	AID
Dichloromethane	2350	60	0.288	1	84.9	Kin-Tek
Dichloropropane, 1,2-	1150	90	0.216		113.0	Metronics
Dichloropropene, trans-1,3-	200	50	0.220		111.0	Metronics
Dichloropropene, trans-1,3-	3080	90	0.220		111.0	Metronics
Dichlorotetrafluoroethane, 1,2-	670	30	0.143	2.6	170.9	AID
Diethyl amine	82	30	0.334		73.1	Metronics
Diethyl disulfide	28	60	0.200	36	122.2	Kin-Tek
Diethyl disulfide	58	70	0.200	58	122.2	AID
Diethyl disulfide	370	100	0.200	8	122.2	Kin-Tek
Diethyl mercury	40	35	0.0945		258.7	Metronics
Diethyl sulfide	14	30	0.271	202	90.2	AID
Diethyl sulfide	51	70	0.271	56	90.2	AID
Diethyl sulfide	170	60	0.271	12	90.2	Kin-Tek
Diethyl sulfide	190	60	0.271		90.2	Metronics
Diethyl sulfide	560	80	0.271	4	90.2	Kin-Tek
Diisopropyl amine	157	70	0.242	30	101.2	AID
Dimethyl acetamide, N,N-	48	70	0.281	66	87.1	AID
Dimethyl amine	22	30	0.542		45.1	Metronics
Dimethyl amine	240	30	0.542	9.6	45.1	AID
Dimethyl amine	275	30	0.542		45.1	Metronics
Dimethyl disulfide	9	30	0.260		94.2	Metronics
Dimethyl disulfide	47	50	0.260		94.2	Metronics
Dimethyl disulfide	75	60	0.260	36	94.2	Kin-Tek
Dimethyl disulfide	130	70	0.260	28	94.2	AID
Dimethyl disulfide	275	80	0.260	12	94.2	Kin-Tek
Dimethyl ether	190	30	0.530		46.1	Metronics
Dimethyl formamide, N-N-	77	70	0.334	41	73.1	AID
Dimethyl formamide, N-N	190	80	0.334		73.1	Metronics
Dimethyl hydrazine, 1,1-	320	70	0.407	6.4	60.1	AID
Dimethyl hydrazine, 1,1-	550	30	0.407	8.6	60.1	AID
Dimethyl sulfide	9	30	0.394		62.1	Metronics
Dimethyl sulfide	70	30	0.394	41	62.1	AID
Dimethyl sulfide	86	30	0.394	24	62.1	Kin-Tek
Dimethyl sulfide	95	30	0.394		62.1	Metronics
Dimethyl sulfide	300	70	0.394	9.5	62.1	AID
Dimethyl sulfide	637	60	0.394	4	62.1	Kin-Tek
Dimethyl sulfide	1100	70	0.394	2.6	62.1	AID
Dimethylamine	370	30	0.542	5	45.1	Kin-Tek
Dimethylhydrazine, uns-	28	40	0.407	12	60.1	Kin-Tek
Dimethylhydrazine, uns-	120	60	0.407	6	60.1	Kin-Tek
Dinitrotoluene, 2,4-	20	80	0.134		182.1	Metronics
Dioxane	320	80	0.278	10	88.1	Kin-Tek
Dodecane	160	100	0.144		170.3	Metronics
Epichlorohydrin	390	80	0.264	9	92.5	Kin-Tek

Appendix O

Epichlorohydrin	1050	90	0.264		92.5	Metronics
Epichlorohydrin	1228	100	0.264	3	92.5	Kin-Tek
Ethanol	30	50	0.530	60	46.1	Kin-Tek
Ethanol	49	70	0.530	54	46.1	AID
Ethanol	290	80	0.530	8	46.1	Kin-Tek
Ethanol	1500	100	0.530		46.1	Metronics
Ethyl acetate	296	70	0.278	10	88.1	AID
Ethyl acetate	320	50	0.278	6	88.1	Kin-Tek
Ethyl acetate	420	50	0.278		88.1	Metronics
Ethyl acetate	1725	80	0.278	1	88.1	Kin-Tek
Ethyl amine	77	30	0.542	31	45.1	AID
Ethyl amine	140	30	0.542		45.1	Metronics
Ethyl benzene	122	60	0.230	24	106.2	Kin-Tek
Ethyl benzene	210	70	0.230		106.2	Metronics
Ethyl benzene	446	80	0.230	6	106.2	Kin-Tek
Ethyl chloride	39	30	0.379	36	64.5	Kin-Tek
Ethyl chloride	400	30	0.379	7	64.5	Kin-Tek
Ethyl mercaptan	37	50	0.394		62.1	Metronics
Ethyl mercaptan	64	30	0.394	44	62.1	AID
Ethyl mercaptan	67	30	0.394	36	62.1	Kin-Tek
Ethyl mercaptan	88	30	0.394		62.1	Metronics
Ethyl mercaptan	505	60	0.394	5	62.1	Kin-Tek
Ethyl methyl sulfide	290	60	0.321	9	76.2	Kin-Tek
Ethyl methyl sulfide	1029	80	0.321	2	76.2	Kin-Tek
Ethylene dichloride	64	30	0.247	24	99.0	Kin-Tek
Ethylene dichloride	434	60	0.247	9	99.0	Kin-Tek
Ethylene glycol dimethylether	82	30	0.271		90.1	Metronics
Ethylene glycol dimethylether	860	70	0.271		90.1	Metronics
Ethylene oxide	82	30	0.555		44.1	Metronics
Ethylene oxide	100	30	0.555	6	44.1	Kin-Tek
Ethylene oxide	128	30	0.555	23	44.1	AID
Ethylene oxide	560	30	0.555	5.3	44.1	AID
Ethylene oxide	600	30	0.555	4	44.1	Kin-Tek
Ethylene oxide	750	30	0.555		44.1	Metronics
Ethyleneimine	46	50	0.568	61	43.1	AID
Formaldehyde	66	80	0.815	24	30.0	Kin-Tek
Formaldehyde	350	1006	0.815	24	30.0	Kin-Tek
Formaldehyde	510	100	0.815		30.0	Metronics
Formic acid	26	50	0.532		46.0	Metronics
Furan	180	50	0.359		68.1	Metronics
Holothane	625	30	0.124	10	197.0	AID
Hexane	160	70	0.284	14	86.2	AID
Hexane	316	60	0.284	6	86.2	Kin-Tek
Hexane, n-	50	30	0.284		86.2	Metronics
Hexane, n-	230	50	0.284		86.2	Metronics
Hydrazine	66	60	0.763	12	32.1	Kin-Tek
Hydrazine	140	70	0.763	24	32.1	AID
Hydrazine	226	80	0.763	8	32.1	Kin-Tek
Hydrogen chloride	300	30	0.671		36.5	Metronics
Hydrogen cyanide	48	30	0.906		27.0	Metronics
Hydrogen cyanide	57	30	0.906	5	27.0	Kin-Tek
Hydrogen cyanide	80	30	0.906	54	27.0	AID
Hydrogen cyanide	270	30	0.906	6	27.0	Kin-Tek

Characteristics of Commercially Available Permeation Sources (Continued)

Permeating Material	Rate (ng/min/cm)	Temp (°C)	K (25°C)	Life (mo)	Mol Wt (g/mol)	Source
Hydrogen cyanide	330	30	0.906	13	27.0	AID
Hydrogen cyanide	350	30	0.906		27.0	Metronics
Hydrogen fluoride	120	30	1.223	12	20.0	AID
Hydrogen fluoride	190	30	1.223		20.0	Metronics
Hydrogen fluoride	400	45	1.223	3.7	20.0	AID
Hydrogen fluoride	490	40	1.223	6	20.0	Kin-Tek
Hydrogen sulfide	50	30	0.717		34.1	Metronics
Hydrogen sulfide	240	30	0.717	6.0	34.1	AID
Hydrogen sulfide	330	35	0.717	4.4	34.1	AID
Hydrogen sulfide	410	30	0.717		34.1	Metronics
Hydrogen sulfide	575	30	0.717	2	34.1	Kin-Tek
Hydrogen sulfide	1205	50	0.717	1	34.1	Kin-Tek
Indene	20	50	0.210		116.2	Metronics
Iodine	310	100	0.0963		253.8	Metronics
Isopropyl mercaptan	74	70	0.321	37	76.2	AID
Isopropyl mercaptan	300	70	0.321	9.2	76.2	AID
Isopropylamine	550	70	0.414	4.3	59.1	AID
Mercury	356	100	0.122		200.6	Metronics
Mesitylene	170	80	0.203	12	120.2	Kin-Tek
Methanol	2.16	70	0.764	12	32.0	AID
Methanol	140	50	0.764	12	32.0	Kin-Tek
Methanol	175	50	0.764		32.0	Metronics
Methanol	310	60	0.764	8	32.0	Kin-Tek
Methanol	3800	100	0.764		32.0	Metronics
Methyl amine	65	30	0.787		31.1	Metronics
Methyl amine	136	30	0.787	17	31.1	AID
Methyl amine	650	30	0.787		31.1	Metronics
Methyl bromide	134	30	0.258		94.9	Metronics
Methyl bromide	225	30	0.258	26	94.9	AID
Methyl chloride	485	30	0.484	2	50.5	Kin-Tek
Methyl chloroform	144	60	0.183	24	133.4	Kin-Tek
Methyl ethyl ketone	34	30	0.339	80	72.1	AID
Methyl ethyl ketone	100	70	0.339	27	72.1	AID
Methyl ethyl ketone	205	50	0.339		72.1	Metronics
Methyl ethyl ketone	305	60	0.339	8	72.1	Kin-Tek
Methyl ethyl ketone	880	80	0.339	2	72.1	Kin-Tek
Methyl hydrazine	24	40	0.532	12	46.0	Kin-Tek
Methyl hydrazine	125	60	0.532	8	46.0	Kin-Tek
Methyl hydrazine	198	70	0.532	14.8	46.0	AID
Methyl iodide	33	30	0.172	234	141.9	AID
Methyl iodide	200	30	0.172	39	141.9	AID
Methyl iodide	890	50	0.172		141.9	Metronics
Methyl isocyanate	125	30	0.428	11.3	57.1	AID
Methyl isocyanate	210	30	0.428		57.1	Metronics
Methyl isocyanate	1780	30	0.428	1.8	57.1	AID
Methyl mercaptan	39	30	0.508		48.1	Metronics
Methyl mercaptan	65	30	0.508	46.7	48.1	AID
Methyl mercaptan	270	30	0.508	11.2	48.1	AID

Appendix O

Methyl mercaptan	376	30	0.508	7	48.1	Kin-Tek
Methyl mercaptan	410	30	0.508		48.1	Metronics
Methyl mercaptan	580	40	0.508	4	48.1	Kin-Tek
Methyl vinyl ketone	50	30	0.349		70.1	Metronics
Methylethyl sulfide	130	70	0.321	21.8	76.2	AID
Methylethyl sulfide	208	80	0.321		76.2	Metronics
Methylethyl sulfide	370	60	0.321		76.2	Metronics
Methylisopropyl sulfide	100	60	0.271		90.2	Metronics
Naphthalene	440	100	0.191		128.2	Metronics
Nitric acid	209	40	0.388	5	63.0	Kin-Tek
Nitric acid	1028	80	0.388	1	63.0	Kin-Tek
Nitrobenzene	29	50	0.199		123.1	Metronics
Nitrobenzene	230	80	0.199		123.1	Metronics
Nitrogen dioxide	300	30	0.532	91	46.0	AID
Nitrogen dioxide	770	30	0.532		46.0	Metronics
Nitrogen dioxide	987	30	0.532	2	46.0	Kin-Tek
Nitrogen dioxide	1150	30	0.532	1.8	46.0	AID
Nitrogen dioxide	1200	30	0.532	22	46.0	AID
Nitrogen dioxide	1715	30	0.532		46.0	Metronics
Nitrogen dioxide	5800	30	0.532	5.0	46.0	AID
Octanethiol	210	100	0.167		146.3	Metronics
Pentane, n-	125	30	0.339		72.2	Metronics
Pentane, n-	445	50	0.339	4	72.2	Kin-Tek
Perchloro methyl mercaptan	125	70	0.132	46	185.9	AID
Phosgene	400	30	0.247	5	98.9	AID
Phosgene	530	30	0.247		98.9	Metronics
Phosgene	1250	30	0.247	3.8	98.9	AID
Pinene	2.5	50	0.180		136.2	Metronics
Propane	37	30	0.554	53	44.1	AID
Propane	65	30	0.554		44.1	Metronics
Propane	100	30	0.554	20	44.1	AID
Propane	121	30	0.554	6	44.1	Kin-Tek
Propanol	17	70	0.407	160	60.1	AID
Propanol, 1-	166	80	0.407	12	60.1	Kin-Tek
Propanol, 1-	1590	100	0.407	4	60.1	Kin-Tek
Propanol, iso-	135	80	0.407	12	60.1	Kin-Tek
Propionaldehyde	455	50	0.421		58.1	Metronics
Propionaldehyde	690	60	0.421	3	58.1	Kin-Tek
Propionitrile	165	60	0.444	12	55.1	Kin-Tek
Propyl acetate, n-	829	80	0.239	3	102.1	Kin-Tek
Propyl mercaptan	170	80	0.322		76.0	Metronics
Propyl mercaptan	239	60	0.322	11	76.0	Kin-Tek
Propylbenzene, iso-	620	100	0.203		120.2	Metronics
Propylbenzene, n-	659	100	0.203	4	120.2	Kin-Tek
Propylene	245	30	0.581		42.1	Metronics
Propylene oxide	186	30	0.421	15	58.1	AID
Propylmercaptan, iso-	35	40	0.321		76.2	Metronics
Propylsulfide, di-n-	220	80	0.207		118.2	Metronics
Pyridene	10	50	0.309		79.1	Metronics
Pyridene	95	50	0.309		79.1	Metronics
Styrene	72	70	0.234	43	104.4	AID
Styrene	150	60	0.234	4	104.4	Kin-Tek
Sulfur dioxide	30	30	0.382	89	64.1	AID

Characteristics of Commercially Available Permeation Sources (Continued)

Permeating Material	Rate (ng/min/cm)	Temp (°C)	K (25°C)	Life (mo)	Mol Wt (g/mol)	Source
Sulfur dioxide	50	35	0.382	5.0	64.1	AID
Sulfur dioxide	175	30	0.382	24	64.1	Kin-Tek
Sulfur dioxide	215	30	0.382		64.1	Metronics
Sulfur dioxide	265	30	0.382	480	64.1	AID
Sulfur dioxide	410	30	0.382		64.1	Metronics
Sulfur dioxide	420	30	0.382	6.6	64.1	AID
Sulfur dioxide	528	30	0.382	3	64.1	Kin-Tek
Sulfur dioxide	730	30	0.382	7.9	64.1	AID
Sulfur dioxide	1300	30	0.382	20	64.1	AID
Sulfur dioxide	3750	30	0.382		64.1	Metronics
Sulfur hexafluoride	110	30	0.167		146.1	Metronics
Tetrachloroethylene	72	50	0.147		165.9	Metronics
Tetrachloroethylene	300	70	0.147	18	165.9	AID
Tetrachloroethylene	370	50	0.147		165.9	Metronics
Tetrachloroethylene	650	60	0.147	7	165.9	Kin-Tek
Tetrachloroethylene	3400	100	0.147	1.5	165.9	Kin-Tek
Tetrachlorosilane	12800	100	0.144		170.1	Metronics
Tetrahydrofuran	290	60	0.339	3	72.1	Kin-Tek
Thiophane	50	70	0.277		88.2	Metronics
Thiophene	143	70	0.291	25	84.1	AID
Thiophene	205	50	0.291		84.1	Metronics
Toluene	21	30	0.265	60	92.1	Kin-Tek
Toluene	22	30	0.265	133	92.1	AID
Toluene	120	70	0.265	24	92.1	AID
Toluene	150	50	0.265		92.1	Metronics
Toluene	786	80	0.265	3	92.1	Kin-Tek
Toluene	1900	90	0.265		92.1	Metronics
Toluene diisocyanate	5070	30	0.140		174.2	Metronics
Tolylene-2,4-diisocyanate	4	30	0.140	8	174.2	Kin-Tek
Tolylene-2,4-diisocyanate	201	100	0.140	1	174.2	Kin-Tek
Trichloroethane, 1,1,1-	113	70	0.183	40	133.4	AID
Trichloroethane, 1,1,1-	1900	90	0.183		133.4	Metronics
Trichloroethane, 1,1,2-	75	50	0.183		133.4	Metronics
Trichloroethane, 1,1,2-	274	70	0.183	18	133.4	AID
Trichloroethane, 1,1,2-	2320	100	0.183		133.4	Metronics
Trichloroethylene	55	40	0.186	36	131.4	Kin-Tek
Trichloroethylene	204	30	0.186	24	131.4	Kin-Tek
Trichloroethylene	260	30	0.186		131.4	Metronics
Trichloroethylene	1060	70	0.186	4.7	131.4	AID
Trichlorofluoromethane	92	30	0.178	24	137.4	Kin-Tek
Trichlorofluoromethane	180	30	0.178	27	137.4	AID
Trichlorofluoromethane	1623	30	0.178	3	137.4	Kin-Tek
Trichlorofluoromethane	1750	30	0.178		137.4	Metronics
Trichlorofluoromethane	2880	30	0.178		137.4	Metronics
Trichlorotrifluoromethane	1150	30	0.130		187.4	Metronics
Triethylamine	4	50	0.242		101.2	Metronics
Trimethyl amine	150	30	0.414		59.1	Metronics
Vinyl acetate	700	70	0.284	4.5	86.1	AID

Appendix O

Vinyl chloride	120	30	0.391	11	62.5	AID
Vinyl chloride	400	30	0.391	7.8	62.5	AID
Water	650	30	1.358		18.0	Metronics
Water	700	30	1.358	4.8	18.0	AID
Water	1040	35	1.358	3.0	18.0	AID
Water	1950	100	1.358		18.0	Metronics
Xylene, m-	44	70	0.230	67	106.2	AID
Xylene, m-	270	70	0.230		106.2	Metronics
Xylene, o-	40	70	0.230	76	106.2	AID

Note: Data taken from product information from Metronics, Santa Clara, CA, Kin-Tek, Texas City, TX, and Analytical Instrument Development (AID), Avondale, PA, with permission.

Appendix P

Cylinder Size and Capacities

Cylinder Designation	Approximate Dimensions Diameter × Length in. × in.	cm × cm	Tare Weight lb	kg	Internal Volume ft²	L	Approximate Capacity lb	ft²
1 A	9 × 51	23 × 130	122	55	1.55	43.8	11–130	175–286
1 B	12 × 38	30 × 97	177	80	2.15	60.9		330
1 C	15 × 54	38 × 137	195	88	4.52	128.0	170	
1 E	27 × 44	69 × 112	260	118				
1 F	15 × 43	38 × 109	72	33	3.87	109.6	100–300	322–389
1 H	9 × 51	23 × 130	188	85	1.55	43.8		
1 J	10 × 47	25 × 119	50	23	1.97	55.8	100–150	
1 K	15 × 52	38 × 132	158	72	4.46	126.3	150	
1 L	10 × 35	25 × 89	138	63	1.73	49.0		
1 M	12 × 50	30 × 127	114	52	3.00	85.0	100	
1 N	10 × 55	25 × 140	138	63	1.73	49.0	90	
1 O	12 × 30	30 × 76	75	34	1.36	38.4		
1 P	9 × 51	23 × 130	122	55	1.55	43.8	2.2–120	56
1 Q	9 × 51	23 × 130	122	55	1.55	43.8	20–100	
1 R	8 × 48	20 × 122	49	22	1.04	29.5	34–70	
1 S	8 × 51	20 × 130	77	35	1.37	38.8	35–100	
1 T	20 × 83	51 × 211	1300	590	15.72	445.0	600	
1 U	10 × 51	25 × 130	312	141	1.50	42.4		505–585
1 V	10 × 49	25 × 124	87	39	1.93	54.5		
1 X	9 × 51	23 × 130	122	55	1.55	43.8	100–100	
1 Y	9 × 51	23 × 130	122	55	1.55	43.8	2.6	
1 Z	9 × 56	23 × 142	140	63	1.72	48.7	26.4	

279

Cylinder Size and Capacities

Cylinder Designation	Approximate Dimensions Diameter × Length in. × in.	Approximate Dimensions Diameter × Length cm × cm	Tare Weight lb	Tare Weight kg	Internal Volume ft²	Internal Volume L	Approximate Capacity lb	Approximate Capacity ft³
2	9 × 26	23 × 66	64	29	0.59	16.7	6–40	35.3–100
2 A	8 × 38	20 × 97	63	29	0.84	23.9		
2 B	6 × 19	15 × 48	29	13	0.30	8.6		40
2 F	9 × 26	23 × 66	64	29	0.59	16.7		
2 L	8 × 22	20 × 56	36	16	0.52	14.8	14–50	
2 M	9 × 21	23 × 53	27	12	0.55	15.7		
2 P	9 × 26	23 × 66	64	29	0.59	16.7	39	20
2 R	7 × 33	18 × 84	32	15	0.55	15.7	18–21	
3	6 × 19	15 × 48	29	13	0.24	6.9	0.22–15	18–40
3 D	6 × 19	15 × 48	29	13	0.24	6.9	0.22	
3 F	6 × 19	15 × 48	30	14	0.24	6.7	0.5	
3 L	5 × 23	13 × 58	21	10	0.21	6.0	5–15	
3 M	9 × 12	23 × 30	13	6	0.19	5.5		
3 P	6 × 19	15 × 48	29	13	0.24	6.9	5–10	8.0
3 Q	6 × 19	15 × 48	29	13	0.24	6.9	3–15	
3 R	7 × 16	18 × 41	16	7	0.21	5.9	8–13	
3 S	4 × 25	10 × 64	18	8	0.13	3.8	8.25–10	
4	4 × 13	10 × 33	12	5	0.081	2.3	1–7.5	3.5–12
4 A	4 × 13	10 × 33	12	5	0.081	2.3		
4 B	4 × 12	10 × 30	11	5	0.070	2.0		
4 P	4 × 13	10 × 33	12	5	0.081	2.3	2–5	
4 X	4 × 13	10 × 33	12	5	0.081	2.3	0.19	
6	4 × 8	10 × 20	4.5	2	0.032	0.93	0.22–0.88	3.5
6 B	4 × 8	10 × 20	4.5	2	0.032	0.93		3.5
6 P	4 × 8	10 × 20	4.5	2	0.032	0.93	0.5–1.0	1.2

Appendix P

7	2	×	13	5 × 33	5	2	0.015	0.44	0.015–0.44	0.25
7 A	2	×	13	5 × 33	5	2	0.015	0.44	0.055	0.88
7 B	2	×	13	5 × 33	5	2	0.015	0.44		1.77
7 C	2	×	13	5 × 33	5	2	0.015	0.44	0.11	3.53
7 N	2	×	13	5 × 33	5	2	0.015	0.44		
7 Q	2	×	13	5 × 33	5	2	0.015	0.44	0.04–0.04	
7 X	2	×	13	5 × 33	5	2	0.015	0.44	0.015–0.12	
LB	2	×	12	5 × 30	4	2	0.015	0.44	0.056–1.5	1.75–2
LBA	2	×	12	5 × 30	4	2	0.015	0.44		1.77
LBB	2	×	12	5 × 30	4	2	0.015	0.44		0.71
8 X	2	×	9.5	5 × 24	2.3	1	0.010	0.3	0.055	
8 Y	3.5	×	10.2	9 × 26	8	4	0.035	1	0.220	

Note: Data taken from Matheson Gas Products, Catalog No 88, August 1987, with permission.

Appendix Q

Viscosity of Gases and Vapors

Gas	Temperature (°C)	Viscosity (μP)
Acetone	18	78.0
Air	20	180.8
Ammonia	20	108.0
Argon	23	221.0
Arsine	0	145.8
Benzene	16.8	75.9
Bromine	12.8	151.1
Butane, n-	16	84.1
Butane, iso-	23	75.5
Butyl acetate, iso-	16.1	76.4
Butyl formate, iso-	17.7	83.0
Carbon dioxide	20	160.0
Carbon disulfide	14.2	96.4
Carbon monoxide	20	184.0
Carbon oxysulfide	15	119.0
Chlorine	20	147.0
Chloroform	14.2	98.9
Cyanogen	20	107.0
Ethane	17.2	90.0
Ethyl alcohol	16.8	88.5
Ethyl chloride	20	105.0
Ethyl ether	18.9	73.5
Ethylene	20	109.0
Helium	21.4	199.4
Hydrogen	20	93.1
Hydrogen bromide	18.7	181.9
Hydrogen chloride	20	156.0
Hydrogen iodide	20.6	185.7
Hydrogen sulfide	17	124.1
Krypton	16.3	245.9
Methane	20	121.1
Methyl chloride	20	116.0

Viscosity of Gases and Vapors

Gas	Temperature (°C)	Viscosity (μP)
Methyl ether	20	102.0
Methyl formate	20	92.3
Neon	13.8	308.0
Nitric oxide	20	186.0
Nitrogen	20	184.0
Nitrous oxide	25	149.8
Oxygen	20	206.0
Phosphene	15	112.0
Propane	17.9	79.5
Sulfur dioxide	18	124.2
Xenon	20	226.0

Note: Data taken from Dean, J. A., Ed. *Lange's Handbook of Chemistry*, 11th ed. (New York: McGraw-Hill Book Co., 1973) with permission.

Appendix R

Miran 1A Operating Instructions

1. Plug in and turn on power.
2. Set slit to 1.
3. Turn IN (IN-OUT) handle to open position. Handle should be perpendicular to gas cell.
4. Turn OUT handle to open position. Handle should be in a horizontal position parallel to the table.
5. Attach sampling pump to the OUT port and turn on.
6. Set wavelength (microns) to desired position. Approach wavelength setting from a position that is lower than the desired setting.
7. Set path length adjustment dial setting to desired position. Again approach desired setting from a lower setting. Path length should be set at an absorbance minimum or transmittance maximum. Remember the dial setting is not the same as the actual path length. Remember the dial setting is not the same as the actual path length. See approximate relationship on the end of the 20-m gas cell.
8. Set absorbance range to desired position.
9. Hook up recorder if needed. The absorbance range selected is full scale on a 1-V recorder setting.
10. Set response time to 1 position and zero the instrument using ZERO (GAIN) control. Either ×10 or ×1 (HIGH or LOW) settings may be used. Set response time to 10 position for moderate signal damping.
11. Commence calibration operation using closed-loop pump or sampling procedure using air sampling pump.

INDEX

A

Absolute pressure calibration method, 152
Absorption, 11
Acetic acid, 76, 138
  diffusion coefficient, actual vs. calculated, 146
  evaporation methods, 159
Acetone, 76, 138
Acetylene, 19, 148
Acid gases
  and bubble meters, 32
  laboratory compressed-air system, 22
  removal of, 20
Acid-resistant plastics, 88
Aclar, 87
Activated alumina, see Alumina, activated
Activated charcoal, 19
Adsorption
  dessicants, 11
  in static systems, 83
Air
  chemical reactions, 92
  composition of, 12
  diffusion coefficients in, 243–246
  dry, density of, 237, 265, 266
  laboratory compressed-air system, 22
  saturated, mass of water vapor in at 1 atm, 253
Air purification, 11–22
  compressed-air system model, 21, 22
  miscellaneous contaminant removal, 19–22
  organic vapor removal, 19
  particulate removal, 16, 19–20
  water vapor removal, 11–18, see also Water vapor removal
    cooling, 16
    liquid dessicants, 16–18
    solid dessicants, 11–17
Alkali-resistant plastics, 88
Altitude
  density of dry air as function of, 265, 266
  and rotameter reading, 47–49

Alumina, activated, 11, 15, 16
  comparative efficiency, 14
  dessicant capacity vs. relative humidity, 13
Aluminosilicates, see Molecular sieves
Aluminum foil containers, 86, 87
Aluminum oxide, 13
Ammonia
  and bubble meters, 32
  injection techniques, 70
  porous plug meters, 59
Anemometer, heated-wire, 111
Anhydrocel, see Calcium sulfate
Anhydrone, 14, see also Magnesium perchlorate, anhydrous
Aniline, 137, 146
Antoine relationship, 164, 197
Asbestos plug flow meter
  features of, 59–62
  range of, 111
Ascarite, 14, 19, 102
Aspirator bottles, 34
Atmospheres (units), 10
Atmospheric pressure
  gas laws, nonideal gases, 8
  static systems at, 67–88, see also Static systems
    nonrigid chambers, 85–88
    rigid systems, in series, 83–85
    rigid systems, single, 67–83
Atomic volumes for diffusion coefficient calculation, 263, 264
Atomic weights and numbers, 233, 234
Azeotropic mixtures, 162

B

Barium oxide, 13, 14
Barium perchlorate, 13, 14
Bell provers, 25–29
Benzene, 76, 137, 138, 148
Benzyl chloride, 146
Blender, gas, 72

Boyle's law, 4
Bromine, 208, 210
Bubble meters, 31–33
Burets, 69, 71, 72
Butane, 149
Butanol, 146
Butyl acetate, 76, 138
Bypass, calibrated, 72, 73

C

Calcium carbonate, 20
Calcium chloride, 11, 13, 14, 18
Calcium hydroxide, 20
Calcium oxide, 13, 14
Calcium sulfate, 11, 15
    capacity of, 12
    comparative efficiency, 13, 14
    dessicant capacity vs. relative humidity, 13
Calculations
    humidity, 215–221
    injection methods, 133–138
    permeation methods, 152–157
    vapor pressure, constants for, 247–250
    volume dilution in single rigid chambers, 78–82
Calibrated bypass, 72, 73
Calibration
    permeation methods, 150–152
    volume measurement, see specific flow and volume measurement devices
Capacity, dessicant, 12–14
Capillary condensation, 11
Carbon, activated, 19
Carbon dioxide, 9, 19
    correction for purity, 96
    five-gas mixture specifications, 93
    generation of, 169, 172, 173
    removal of, 102
Carbon disulfide, 146, 159
Carbon monoxide, 19, 22, 148
    chemical reactions, 92
    five-gas mixtue specifications, 93
    generation of, 172, 173
    stability, cylinder material and, 95
Carbon tetrachloride, 146, 159
Carbonyl sulfide, chemical reactions, 92
Cellophane, 87, 88
Cellulose nitrate, 88
Cellulose triacetate, 88
Charles's law, 4
Chemical methods, dynamic systems, 171–176
Chemical reactions, 72
    dessicants, 11
    reactive gases, 92
Chlorine, 50

and mercury-sealed piston, 32–33
porous plug meters, 59
pulse diluters, 126
removal of, 20
Chlorine dioxide, 59, 172, 207–208
Chlorobenzene, 76, 138
Chloroform, 76, 138
Chlorotoluene, 146
Cobalt hydrocarbonyl, 122
Compressed-air dryer, 16
Compressed-air system, laboratory, 21, 22
Compressibility, 8, 9
Concentration
    characteristics of methods for producing gases and vapors, 2–3
    gas laws, ideal gases, 5–8
    static systems at atmospheric pressure, 73–78
Condensation methods, 161
Conductometry, 162
Controlled leak devices, 62, 63, 111
Conversion factors
    gas flow, for mass flow controllers, 259–262
    measuring units, 229–231
Conversion reactions, 172, 173
Cooling, water vapor removal, 16
Copper, 19, 20
Copper sulfate, 13
Correction factors
    measuring devices, see specific flow rate and volume measuring devices
    molecular, 53
Critical orifice devices, 56–59, 111
Cubic foot bottles, 26
Cyanates, 159
Cyclohexane, 76, 138
Cylinder sizes and capacities, 279–281

D

Dead volume determination, 104
Decane, 76, 138
Decay, sample, 87, 88
Dehydration, see Water vapor, removal of
Dehydrite, see Magnesium perchlorate
Density
    common gases, 239–241
    dry air, 237, 265, 266
    gas laws, ideal gases, 4, 5
Dessicants
    liquid, 16–18
    solid, 11–17
Dibromoethane, 76, 138
Dichromate solution, organic vapor removal, 19
Diethylamine, 146

Index

Diethylene glycol, 18
Diethyl ether, 76, 138
Diffusion coefficients
   in air, 243–246
   calculation of, atomic volumes for, 263, 264
Diffusion methods, dynamic systems, 137, 139–147
Dilution
   gas stream mixing, 110–116
   pulse diluters, 126, 127
Dimethylcarbamoyl chloride, 137
Dimethyl hydrazine, 159
Dioxane, 146
Dispersion containers, 159, 160
Double dilution gas stream mixing, 114–116
Drierite, see Calcium sulfate
Dry-bulb thermometers, relative humidity from, 255–258
Dry gas meters, 38–40, 111
Drying, see Water vapor, removal of
Dynamic systems, 109–176
   characteristics of, 2, 3
   chemical methods, 171–176
   diffusion methods, 137, 139–147
   electrolytic methods, 166–171
   evaporation methods, 158–165
   gas stream mixing, 110–116
      double dilution, 114–116
      single dilution, 110–114
   humidity in, 212–215
   hydrogen cyanide, 201, 202
   injection methods, 116–137
      calculations, 133–138
      electrolytic methods, 123, 125
      gravity feed methods, 125, 126
      liquid and gas pistons, 126, 127
      liquid pumps, 122–124
      liquid reservoirs, 116–117
      ports, 126–128
      pulse diluters, 126, 127
      syringe drive systems, 118–122
      ultimate system, 131–133
      vaporization techniques, 128–131
   nitrogen dioxide, 194–195
   permeation methods, 146–158
      calculations, 152–157
      calibration, 150–152
      device construction, 148–150

E

Efficiency, dessicant, 13
Electrobalance methods, permeation device calibration, 151
Electrolytic methods
   dynamic systems, 166–171
   injection, 123, 125
English units, 10
Equilibrium relative humidity of selected saturated salt solutions, 267, 268
Ethane, 19, 96, 148, 169, 172
Ethanol, 122
Ethyl cellulose, 88
Ethylene, 9, 19, 90, 148
Ethylene chlorhydrin, 146
Ethylene oxide, 137
   injection techniques, 70
   and mercury-sealed piston, 33
   porous plug meters, 59
Ethyl ether, 146
Evaporation methods, 158–165
Explosion hazards, static systems, 83
Exponential dilution, 82

F

Faraday's law, 168
Filters
   glass wool, 19
   HEPA, 22
   laboratory compressed-air system, 22
   miscellaneous contaminant removal, 19–22
   particulate removal, 16, 19, 20
Flame ionization, 162
Flow meters, see Meters
Flow rate
   conversion factors for mass flow controllers, 259–262
   gas density calculations, 4, 5
Flow rate and volume measurements, 25–63
   intermediate standards, 34–40
      dry gas meters, 38–40
      wet test meters, 34–38
   particle filtration, 19, 20
   primary standards, 25–34
      aspirator bottles, 34
      pistons, frictionless, 31–33
      pitot tubes, 29–31
      spirometers, 25–29
      syringe drive systems, 33–34
   secondary standards, 40–63
      controlled leaks, 62, 63
      critical orifices, 56–59
      mass flow meters, 50–54
      miscellaneous devices, 62, 63
      orifice meters, 54–56
      porous plugs, 59–62
      rotameters, 40–50
Fock's law, 152
Formaldehyde, 148, 162, 205–207
Frictionless pistons, 31–33

G

Gas blenders, 72
Gas constant, molar, 235
Gases
    characteristics of methods for producing, 2, 3
    density, see Density
    mixing, 72, 92, 109–116
    viscosity of, 283, 284
Gas laws, 1–9
    deviation of common gases from, 239–241
    ideal gases, 4–8
    nonideal gases, 8, 9
Gas meters, see Meters
Gas pistons, 126, 127
Gas stream mixing, 110–116
Glass wool filters, 19
Glass wool plugs, 129
Glycerol, 18
Gravimetric calibration devices, 151
Gravimetric methods, 92–94
Gravimetry, 162
Gravity feed systems, 125, 126

H

Halocarbons, 146
Heated-wire anemometer, 111
Heater characteristics, 130
Helium
    correction for purity, 94, 96
    five-gas mixture specifications, 93
HEPA (high-efficiency particulate air) filter, 22
Hexane, 76, 138, 146
High-efficiency particulate air (HEPA) filter, 22
Homogeneity of gas mixtures, 92
Hopcalite, 20
Humidity, see also Relative humidity; Water vapor
    calculations, 215–221
    dynamic systems, 212–215
    equilibrium relative humidity of selected salt solutions, 267, 268
    measurement methods, 221–223
    relative humidity from wet- and dry-bulb thermometers, 255–258
    static systems, 211, 212
    vapor pressure calculations, constants for, 247–250
    vapor pressure of water at various temperatures, 251
    water vapor mass in saturated air at 1 atm, 253

Hydrazine, 161
Hydrocarbons, 22
    partially evacuated systems, 102, 103
    plastics resistant to, 88
Hydrofluoric acid, 162
Hydrogen, 94, 96
Hydrogen chloride, 50
    and bubble meters, 32
    and mercury-sealed piston, 32, 33
    removal of, 20
Hydrogen cyanide
    dynamic systems, 201, 202
    generation of, 72
    removal of, 20
    static systems, 198–200
Hydrogen fluoride, 83, 201–205
Hydrogen sulfide, 92, 148

I

Icicle electrode, 166, 167
Ideal gas laws, 4–8
Implosion, 83
Injection systems
    calculations, 133–138
    characteristics of, 2, 3, 119–121
    dead volume determination, 104
    electrolytic methods, 123, 125
    gravity feed methods, 125, 126
    liquid and gas pistons, 126, 127
    liquid pumps, 122–124
    liquid reservoirs, 116–117
    ports, 126–128
    pulse diluters, 126, 127
    single rigid chambers, 68–70
    syringe drive systems, 118–122
    ultimate system, 131–133
    vaporization techniques, 128–131
Inlet, low flow, 111
Intermediate standards, meters, 34–40
International System of Units, 10
Iodine, 173, 208, 211

K

Kel-F, 86, 87
Kynar, 87

L

Laboratory air supplies
    compressed-air system, 21, 22
    particle filtration, 19
Leaks, controlled, 62, 63
Liquid dessicants, 16–18

Index

Liquid nitrogen, 16
Liquid pistons, 126, 127
Liquid pumps, 122–124
Liquid reservoirs, 116–117
Lithium chloride, 18
Lithium perchlorate, 14
Low-flow inlet, 110, 111

M

Magnesium oxide, 14
Magnesium perchlorate, 11, 15
    capacity of, 12
    comparative efficiency, 13, 14
Manganese, 20
Manometric techniques, 94–100
Mass flow controllers
    gas flow conversion factors for, 259–262
    range of, 111
Mass flow conversion factors, 259–262
Mass flow meters, 50–54, 111
Measurement units, 10
    conversion factors, 229–231
    gas laws, 5–8
Mekohbite, 14
Membrane filters, 19
Mercury, 196–198
    gas mixing methods, 72, 73
    vapor pressure, 197
Mercury pistons, 126, 127
Mercury-sealed pistons, 31–33
Mesitylene, 146
Metal aluminosilicates, see Molecular sieves
Meters
    dry gas, 38–40
    mass flow, 50–54
    orifice, 54–56
    rotameters, 40–50
    wet test, 34–38
Methane, 19, 22, 148
    correction for purity, 94, 96
    five-gas mixture specifications, 93
Methanol, 19, 76, 138
Methyl chloride, 19
Methyl ethyl ketone, 76, 138
Methyl pentane, 90
Metric system, 10
Microbalance calibration methods, 151
Microgasometric device, 152
Microm buret, 69
Micropipets, 69
Miller-Nelson Research, Inc., 131, 132
Miran 1A infrared analyzer, 132, 133, 285
Mixing techniques, 72, 92, 110–116
mmHG units, 10

Moisture, see Water vapor, removal of
Molar gas constant values, 235
Molecular correction factor, 53
Molecular sieves
    characteristics of, 17
    comparative efficiency, 14
    water vapor removal, 16, 17
Motor-driven syringes, 122
Mullite tube, 19
Multiple dilution techniques, 114–116
Multiple syringe injectors, 118
Mylar, 86, 87, 90, 91

N

Needle specifications, 129
Nickel carbonyl, 122
Nitric oxide, 59
Nitrogen
    correction for purity, 94, 96
    five-gas mixture specifications, 93
    generation of, 169
Nitrogen dioxide, 89, 95
    adsorption losses in static systems, 83
    dimerization, 190
    dynamic systems, 194, 195
    generation of, 72, 169, 173
    permeation device limitations, 157
    porous plug meters, 59
    pulse diluters, 126
    stability, cylinder material and, 95
    static systems, 190–193
    storage, 90, 91
    syringes, motor-driven, 122
Nitrous oxide, 95
Nitro olefins, 159
Nonideal gas laws, 8, 9
Nonrigid chambers, 85–88
Nylon permeation devices, 148

O

Operating pressure, characteristics of methods for producing gases and vapors, 2, 3
Orifice meters, 54–56, 111
Orifices, critical, 56–59
Organic compounds
    plastics resistant to, 88
    vapor removal, 19
Organo isocyanates, 148
Oxygen, generation of, 169
Ozone, 171, 173, 174
    adsorption losses in static systems, 83
    analysis, methods of, 188, 189
    generation methods, 185–187

P

Partially evacuated system, 101–104
Partial pressure methods, 94–100
  of calibration, 151
  gas mixture production, 94–100
Particulate removal, 16, 19, 20, 22
Pascal (Pa) unit, 10
Perchloroethylene, 159
Perfect gas law, deviation from, 239–241
Performance audits, 94
Peristaltic pumps, 122–124
Permeation methods
  calculations, 152–157
  calibration, 150–152
  device construction, 148–150
Permeation sources, commercial, 269
Phosgene, 162
  and mercury-sealed piston, 33
  porous plug meters, 59
Phosphine, 173
Phosphoric acid, 18
Phosphorus halides, 159
Phosphorus pentoxide, 12, 14
Piston pumps, 122–124
Pistons
  frictionless, 31–33
  liquid and gas, 126, 127
Pitot tubes, 29–31, 111
Plastic films
  chamber materials, 86–88
  permeation devices, 148, 150
Plugs, porous, 59–62
Polarography, 162
Polyamide permeation devices, 148, 150
Polyester
  chamber material, 88
  permeation devices, 148, 150
Polyethylene
  chamber material, 86–88
  permeation devices, 148, 150
Polyethylene terphthalate, 150
Polymers
  chamber materials, 86, 88, 90, 91
  permeation devices, 148, 150
Polypropylene, 148
Polythene, 150
Polyvinyl acetate, 150
Polyvinyl alcohol, 88
Polyvinyl chloride
  chamber materials, 86, 87
  mercury and, 33
Polyvinyl fluoride permeation devices, 148
Polyvinylidene chloride, 88, 150
Porous plugs, 59–62

Ports, injection, 126–129
Potassium hydroxide, 13, 14, 18
Preconditioning, 89
Pressure, see also Atmospheric pressure
  characteristics of methods for producing
    gases and vapors, 2, 3
  compressibility vs., for selected gases, 9
  and dessicant capacity, 12
  gas laws
    ideal gases, 4
    nonideal gases, 8, 9
  particle filtration, 19, 20
  static systems
    gravimetric methods, 92–94
    partially evacuated, 101–104
    partial pressure methods, 94–100
    volumetric methods, 100, 101
Pressurized systems, static, see Static systems
Primary standards, 25–34, see also Flow rate
    and volume measurements
Propyl bromide, 146
psi units, 10
Pulse diluters, 126, 127
Pumps, liquid, 122–124
Purification of air, see Air purification

R

Radioactive gases, porous plug meters, 59
Relative humidity
  equilibrium, of selected saturated salt
    solutions, 267, 268
  from wet- and dry-bulb thermometer
    readings, 255–258
Reservoirs, liquid, 116, 117
Rigid chambers, static systems
  in series, 83–85
  single, 67–83
Roller mixing, 92
Rotameters, 40–50, 111
Rotating stopcock, 111

S

Safety considerations, 9, 10
Salt solutions, equilibrium relative humidity of,
  267, 268
Sample decay, 87, 88
Sample introduction, see also Injection
    methods
  nonrigid chambers, 87
  single rigid chambers, 68–71
Saran Wrap, 86, 87, 90
Scientific notation, 10
Scotchpak, 86, 87, 90

Index

Series chambers, rigid, 67–83
Silica gel, 11, 13–15
Silicon rubber permeation devices, 148
Single dilution gas stream mixing, 110–114
SI units, 10
Soap bubble meters, 31–33
Soda lime, 20
Sodium hydroxide, 13, 14, 18, 20
Solid dessicants, 11–17
Spectophotometry, 162
Spirometers
    calibration of wet test meter against, 37, 38
    features of, 25–29
Squirrel cage rotor, 125
Standards, see Flow rate and volume measurements
Static systems, 67–104
    at atmospheric pressure, nonrigid chambers, 85–88
    at atmospheric pressure, series rigid chambers, 83–85
    at atmospheric pressure, single rigid chambers, 67–83
        concentration calculations, 73–78
        dispensers, miscellaneous, 72
        gas blenders, 72
        mixing devices, 72, 73
        sample introduction, 68–71
        volume dilution calculations, 78–82
    characteristics of, 2, 3
    humidity in, 11–212
    hydrogen cyanide, 198–200
    nitrogen dioxide, 190–103
    partially evacuated systems, 101–104
    pressurized, 88–101
        gravimetric methods, 92–94
        partial pressure methods, 94–100
        volumetric methods, 100, 101
Stopcock, rotating, 111
Strapping technique, 27–29
Stratification of gases, 93
Styrene, 76, 138, 146
    adsorption losses in static systems, 83
    chamber materials, 88
Sulfur dioxide, 148, 172
    and bubble meters, 32
    chemical reactions, 92
    porous plug meters, 59
    pulse diluters, 126
    removal of, 20
    stability, cylinder material and, 95
Sulfuric acid, 18
Syringe drive systems, 33–34, 118–122
Syringe injectors, 68–70
    commercial, characteristics of, 119–121

    dead volume determination, 104
    range of, 111

T

Tedlar, 86, 87
Teflon
    chamber material, 86–88
    permeation devices, 148, 150
Temperature
    and dessicant capacity, 12
    and evaporation methods, 158
    gas laws, ideal gases, 4
    gas mixing and stratification, 92, 93
    permeation device limitations, 148
    permeation rate vs., 151
    and vapor pressure of water, 251
    water vapor removal, 16
Test meters, wet, 34–38
Tetrachloroethylene, 146, 148
Tetrafluoroethylene, 150
Tetramethyl lead, 137
Thermal mixing, 92
Thermometers, wet- and dry-bulb, 255–258
Toluene, 148
Toluene diisocyanate, 137, 146
Traversing technique, 31
Tributyl phosphate, 159
Tributyl phosphene, 159
Trichloroethane, 146
Trichloroethylene, 76, 138, 159
Triethylene glycol, 18
Trimethyl borane, 162
Trioxane, 205
Tubing, pump, 122

U

Units of measurement, 10

V

Vacuum, partially evacuated static systems, 101–104
Vaporization for mixing, 72
Vaporization techniques
    dynamic system injection, 128–131
    for mixing, 72
Vapor pressure
    calculation of, constants for, 247–250
    concentration calculations, 163–165
    and evaporation methods, 158
    water, temperature and, 251
Vapors
    characteristics of methods for producing, 2, 3

organic, removal of, 19
viscosity of, 283, 284
Variable-area flow meter, 40–50, 111
Velocity, and dessicant capacity, 12
Vinyl chloride, 19, 88, 148
Viscosity
 and evaporation methods, 158
 of gases and vapors, 283, 284
Volatilization, see Vaporization
Volume
 atomic, for diffusion coefficient calculation, 263, 264
 characteristics of methods for producing gases and vapors, 2, 3
 gas laws, ideal gases, 4
Volume dilution calculations, static systems at atmospheric pressure, 78–82
Volume measurements, see Flow rate and volume measurements
Volumetric methods
 features of, 100, 101
 permeation device calibration, 151
Volumetric pipets, 69

W

Wall loss, 2, 3

Water
 and dessicant capacity, 12
 syringes, motor-driven, 122
 vapor pressure at various temperatures, 251
Water vapor, see also Humidity; Relative humidity
 mass of in saturated air at 1 atm, 253
 removal of, 11–18
  cooling, 16
  laboratory compressed-air system, 22
  liquid dessicants, 16–18
  solid dessicants, 11–17
Wet-bulb thermometer, relative humidity from, 255–258
Wet test meters, 34–38
 range of, 111
 rotameter calibration against, 45
Woesthoff gas-dosing apparatus, 126, 127

X

Xylene, 76, 138

Z

Zeolite, 20
Zinc chloride, 13